AI-Based Solutions for Engineering

Melda Yücel
İstanbul Aydin University, Turkey

Hasan Volkan Oral
Istanbul Aydin University, Turkey

IGI Global
Scientific Publishing
Publishing Tomorrow's Research Today

Vice President of Editorial Melissa Wagner
Director of Acquisitions Mikaela Felty
Director of Book Development Jocelynn Hessler
Production Manager Mike Brehm
Cover Design Jose Rosado

Published in the United States of America by
IGI Global Scientific Publishing
701 East Chocolate Avenue
Hershey, PA, 17033, USA
Tel: 717-533-8845
Fax: 717-533-7115
Website: https://www.igi-global.com E-mail: cust@igi-global.com

Library of Congress Cataloging-in-Publication Data

Names: Yucel, Melda, 1995- editor | Oral, Hasan, 1977- editor
Title: AI-based solutions for engineering / edited by Melda Yücel, Hasan Oral.
Description: Hershey, PA : IGI Global Scientific Publishing, [2026] | Includes bibliographical references and index. | Summary: "This book is based on the applications and novel solutions of engineering problems by using artificial intelligence and its methodologies such as machine learning via artificial neural networks, ensemble-learning models, decision trees etc., pattern and speech recognition, deep learning etc. For this reason, the main topic of this publication is to realize the solutions for different engineering problems with the contribution of artificial intelligence technology"-- Provided by publisher.
Identifiers: LCCN 2025004063 (print) | LCCN 2025004064 (ebook) | ISBN 9798337307466 h/c | ISBN 9798337307473 s/c | ISBN 9798337307480 ebook
Subjects: LCSH: Engineering--Data processing | Artificial intelligence--Industrial applications | Machine learning--Industrial applications | Internet of things--Industrial applications | Decision making--Data processing
Classification: LCC TA347.A78 A3876 2025 (print) | LCC TA347.A78 (ebook)
LC record available at https://lccn.loc.gov/2025004063
LC ebook record available at https://lccn.loc.gov/2025004064

British Cataloguing in Publication Data
A Cataloguing in Publication record for this book is available from the British Library.

Table of Contents

Preface xii

Chapter 1
Deep Learning From Foundations to Frontiers 1
S. Jansirani, Jansons Institute of Technology, India
M. Saravanan, KPR Institute of Engineering and Technology, India
Vishnu Kumar Kaliappan, KPR Institute of Engineering and Technology, India

Chapter 2
Simplifying Data With Robust Variable Selection for Enhanced Interpretability and Energy Efficiency 39
Jan Kalina, Institute of Computer Science, The Czech Academy of Sciences, Czech Republic

Chapter 3
TinyML Empowering Intelligent Edge Devices 71
Helen K. Joy, Christ University, India
Electa Alice Jayarani, Kammavari Sangha's Institute of Technology, India
R. Sridevi, Christ University, India

Chapter 4
Support Vector Machine in Predictive Maintenance of Belt-Drive System 103
Saahil Jamad, SRM Institute of Science and Technology, India
Anitha Kumari S., SRM Institute of Science and Technology, India
Josika Meenal Alamelu Muthu, SRM Institute of Science and Technology, India

Chapter 5
Exploring the Magnitude-Shape Plot Framework for Anomaly Detection in Crowded Video Scenes 137
Zuzheng Wang, Computer, Electrical and Mathematical Sciences and Engineering (CEMSE) Division, King Abdullah University of Science and Technology, Thuwal, Saudi Arabia
Fouzi Harrou, Computer, Electrical and Mathematical Sciences and Engineering (CEMSE) Division, King Abdullah University of Science and Technology, Thuwal, Saudi Arabia
Ying Sun, Computer, Electrical and Mathematical Sciences and Engineering (CEMSE) Division, King Abdullah University of Science and Technology, Thuwal, Saudi Arabia
Marc G. Genton, Computer, Electrical and Mathematical Sciences and Engineering (CEMSE) Division, King Abdullah University of Science and Technology, Thuwal, Saudi Arabia

Chapter 6
Artificial Intelligence-Based Automated System for Building Information Modelling Data Extraction and Analysis 175
Oluseye Olugboyega, Obafemi Awolowo University, Nigeria

Chapter 7
Accurate Estimation of Load-Carrying Capacity (Pu) in Steel Tube Structures 205
Abhishek Kumar, Lovely Professional University, India
Harpreet Singh Bedi, Lovely Professional University, India

Chapter 8
Artificial Intelligence in Wastewater Treatment: Advancing Processes With Nature-Based Solutions 221
Chaitanya Surya Chandra Nadimpalli, Raghu Engineering College, India

Chapter 9
Artificial Intelligence and Machine Learning in Additive Manufacturing 255
Koray Özsoy, Isparta University of Applied Sciences, Turkey
Bekir Aksoy, Isparta University of Applied Sciences, Turkey

Chapter 10
A Digital Twin and AI-Driven Framework for Optimizing Firefighter Escape Routes in Wildfire Scenarios: FAST 285
Walisa Romsaiyud, Sukhothai Thammathirat Open University, Thailand
Pirom Konglerd, Sukhothai Thammathirat Open University, Thailand
Sabsathit Takaew, Sukhothai Thammathirat Open University, Thailand

Compilation of References 323

About the Contributors 371

Index 377

Detailed Table of Contents

Preface xii

Chapter 1
Deep Learning From Foundations to Frontiers 1
S. Jansirani, Jansons Institute of Technology, India
M. Saravanan, KPR Institute of Engineering and Technology, India
Vishnu Kumar Kaliappan, KPR Institute of Engineering and Technology, India

In today's data-driven world, the ability to automatically recognize complicated patterns from enormous amounts of data is essential for many AI applications, such as recommendation systems, robotics, intelligent transportation systems, and healthcare. Therefore, developing an end-to-end framework using deep learning is crucial for advancing artificial intelligence and generating accurate predictions. The powerful features, such as handling big data, learning hierarchical data representations, and transfer learning, enable it to efficiently manage complex tasks, generalize to new data, and adapt to various domains, making it indispensable across numerous industries. This chapter explores the foundation and classification of deep learning including discriminative, generative, and hybrid learning approaches and also reviews the significance of regularization in deep learning, exploring its different types. Additionally, it highlights the specific deep learning techniques and its application.

Chapter 2
Simplifying Data With Robust Variable Selection for Enhanced Interpretability and Energy Efficiency 39
Jan Kalina, Institute of Computer Science, The Czech Academy of Sciences, Czech Republic

Machine learning, a cornerstone of artificial intelligence, equips engineers and scientists with powerful tools for uncovering valuable insights from complex, large-scale datasets across diverse applications. Engineering applications increasingly demand transparency and interpretability, particularly to understand the reasoning behind specific outcomes and decisions. This chapter addresses this need by exploring dimensionality reduction techniques that enhance model interpretability by focusing on the most significant variables. By employing robust, outlier-resistant methods, we enable a closer analysis of how individual data points impact results, making these approaches ideal for real-world data with inherent noise or contamination. The utility of these techniques is demonstrated on two datasets, one from gene expression analysis and the other from credit scoring. Furthermore, we discuss the relationship between dimensionality reduction and energy-efficient approximate computation, highlighting its relevance for engineering applications in AI.

Chapter 3
TinyML Empowering Intelligent Edge Devices 71
Helen K. Joy, Christ University, India
Electa Alice Jayarani, Kammavari Sangha's Institute of Technology, India
R. Sridevi, Christ University, India

Tiny Machine Learning (TinyML) bridges the gap between artificial intelligence and low-power embedded systems, enabling devices like microcontrollers to process data locally and operate autonomously. This chapter explores the foundational principles of TinyML, its applications across industries such as IoT, healthcare, industrial automation, and environmental monitoring, and the tools enabling its deployment. It also addresses challenges, including energy efficiency and model optimization, while providing insights into future advancements such as federated learning and neuromorphic computing. This chapter offers a comprehensive understanding of TinyML's transformative potential and its pivotal role in AI-based engineering solutions.

Chapter 4
Support Vector Machine in Predictive Maintenance of Belt-Drive System 103
Saahil Jamad, SRM Institute of Science and Technology, India
Anitha Kumari S., SRM Institute of Science and Technology, India
Josika Meenal Alamelu Muthu, SRM Institute of Science and Technology, India

Belt-drives are widely used as power transmission elements in light as well as heavy-duty industrial applications. Maintenance activities are imperative to ensure the performance and reliability of a belt-drive system. The vibrational signal patterns from the system can reveal failures such as belt misalignment and tension irregularities. These anomalies can be predicted and mitigated before they lead to costly downtime. The work presents Support Vector Machine (SVM) based predictive maintenance of the belt-drive system. The reason for choosing SVM for predictive maintenance of the belt-drive system is that it addresses the challenges posed by the sensor data, including the noise, high dimensionality and complex temporal dependencies. Also, SVM being a discriminant machine learning technique, returns the same optimal hyperplane parameter. This contrasts generative approaches such as Naive Bayes or Gaussian process classification. The SVM model developed has shown high accuracy in classifying healthy and faulty belt-drive systems.

Chapter 5
Exploring the Magnitude-Shape Plot Framework for Anomaly Detection in Crowded Video Scenes 137
Zuzheng Wang, Computer, Electrical and Mathematical Sciences and Engineering (CEMSE) Division, King Abdullah University of Science and Technology, Thuwal, Saudi Arabia
Fouzi Harrou, Computer, Electrical and Mathematical Sciences and Engineering (CEMSE) Division, King Abdullah University of Science and Technology, Thuwal, Saudi Arabia
Ying Sun, Computer, Electrical and Mathematical Sciences and Engineering (CEMSE) Division, King Abdullah University of Science and Technology, Thuwal, Saudi Arabia
Marc G. Genton, Computer, Electrical and Mathematical Sciences and Engineering (CEMSE) Division, King Abdullah University of Science and Technology, Thuwal, Saudi Arabia

This study explores video anomaly detection within a Functional Data Analysis framework, focusing on the application of the Magnitude-Shape (MS) Plot. Autoencoders are used to learn and reconstruct normal behavioral patterns from anomaly-free training data, resulting in low reconstruction errors for normal frames and higher errors for frames with potential anomalies. The reconstruction error matrix for each frame is treated as multivariate functional data, with the MS-Plot applied to analyze both magnitude and shape deviations, enhancing the accuracy of anomaly detection. Using its capacity to evaluate the magnitude and shape of deviations, the MS-Plot offers a statistically principled and interpretable framework for anomaly detection. The proposed methodology is evaluated on two widely used benchmark datasets, UCSD Ped2 and CUHK Avenue, demonstrating promising performance. It performs better than traditional univariate functional detectors (e.g., FBPlot, TVDMSS, Extremal Depth, and Outliergram) and several state-of-the-art methods.

Chapter 6
Artificial Intelligence-Based Automated System for Building Information Modelling Data Extraction and Analysis 175
Oluseye Olugboyega, Obafemi Awolowo University, Nigeria

The emergence of AI enables the use of AI-based automated systems for information extraction from building information models using machine learning methods, natural language processing, computer vision, and neural networks. This chapter analyzes the utilization of AI for the extraction and analysis of BIM data. The chapter examines the difficulties associated with manual data extraction from BIM and the advantages of employing AI for BIM data analysis and extraction. This chapter addressed the several strategies and algorithms employed in AI-based automated systems for extracting information from building information models. The chapter presented a case study on the effective application of AI in a BIM-based project. The chapter concludes with observations on upcoming trends and advancements in the use of artificial intelligence to extract meaningful insights from the building information models.

Chapter 7
Accurate Estimation of Load-Carrying Capacity (Pu) in Steel Tube Structures 205
Abhishek Kumar, Lovely Professional University, India
Harpreet Singh Bedi, Lovely Professional University, India

The accurate estimation of load-carrying capacity (Pu) in steel tube structures ensures structural integrity and safety. This chapter explores a machine learning-based approach for predicting Pu using key material and geometric parameters, including concrete strength, steel tube areas, yield strengths, and eccentric loading ratio. Traditional empirical models often fail to capture the complex nonlinear relationships among these variables. In this chapter, we experimented with (a) an advanced machine learning model (XBGBoost) that handles nonlinear relationships in structural data, (b) deep neural network model interactions between geometric and material properties, and (c) bayesian optimization to fine-tune ML models to improve predictive performance. With the lowest Mean Absolute Error (MAE), Mean Squared Error (MSE), and Root Mean Squared Error (RMSE) as well as the greatest R2 score (0.9811), the findings show that Bayesian Optimization applied to XGBoost performed the best.

Chapter 8
Artificial Intelligence in Wastewater Treatment: Advancing Processes With Nature-Based Solutions.. 221
Chaitanya Surya Chandra Nadimpalli, Raghu Engineering College, India

Wastewater treatment systems face rising challenges from urbanization, population growth, and stricter environmental norms. Traditional methods often struggle with energy use, process efficiency, and adaptability. In developing countries like India, limited resources and outdated infrastructure worsen the problem. Rapid urban and industrial growth further strains existing systems. Artificial intelligence (AI) offers a promising way forward by enabling real-time monitoring, predictive analytics, and optimized operations (Lee et al., 2022). In India's context, AI can support better decision-making and improve treatment performance despite resource constraints. By analyzing data continuously, AI systems can enhance efficiency, reduce waste, and promote sustainability (Corominas et al., 2018). The proposed AI-Nature-Based Solutions (NBS) framework combines modern technology with eco-friendly approaches, offering a sustainable path to address water management challenges in developing nations.

Chapter 9
Artificial Intelligence and Machine Learning in Additive Manufacturing 255
Koray Özsoy, Isparta University of Applied Sciences, Turkey
Bekir Aksoy, Isparta University of Applied Sciences, Turkey

This study explores the integration of Additive Manufacturing (AM) with Artificial Intelligence (AI) and Machine Learning (ML). While AM enables complex geometries, material efficiency, and design flexibility, it faces challenges such as poor surface quality, printing defects, limited material capabilities, and slow production speeds. Addressing these issues is crucial for AM to reach its full potential. AI and ML offer promising solutions through process optimization, monitoring, defect detection, material property prediction, and generative design. Advanced sensor systems combined with ML algorithms enable early detection of defects, allowing timely interventions to reduce waste and improve quality. This study provides a comprehensive of AI and ML applications in AM, highlights key advancements, and outlines future research. The findings reveal that AI-driven technologies significantly enhance efficiency and quality control in AM. Interdisciplinary collaboration, data infrastructure, and improved AI models are essential for maximizing the impact of AI-integrated AM in Industry 4.0.

Chapter 10
A Digital Twin and AI-Driven Framework for Optimizing Firefighter Escape Routes in Wildfire Scenarios: FAST 285
Walisa Romsaiyud, Sukhothai Thammathirat Open University, Thailand
Pirom Konglerd, Sukhothai Thammathirat Open University, Thailand
Sabsathit Takaew, Sukhothai Thammathirat Open University, Thailand

Wildfires pose critical threats to firefighter safety, requiring rapid decisions under hazardous, dynamic conditions. Existing AI-based evacuation systems often lack real-time prediction and integration with physical environments. This chapter introduces FAST (Firefighter AI Safety Digital Twin), an AI-driven system bridging that gap by combining deep learning and digital twin technology for time-sensitive wildfire response. FAST integrates four components: (1) a simulation environment modeling evolving wildfire scenarios; (2) an AI engine using multivariate LSTM and DQN to predict fire spread and optimize escape routes; (3) a bidirectional sync layer for real-time data exchange between virtual and physical systems; and (4) a performance assessment module. Experimental results show a prediction loss of 0.014, an 84% escape success rate, and a 260-second average evacuation time, confirming

FAST's effectiveness in enhancing situational awareness, reducing response time, and improving firefighter safety.

Compilation of References 323

About the Contributors 371

Index 377

Preface

INTRODUCTION

The convergence of artificial intelligence (AI) and engineering has transformed the landscape of problem-solving across a diverse range of technical domains. This edited volume, *AI-Based Solutions for Engineering*, was conceived to explore and showcase the capabilities of AI methodologies—such as machine learning, deep learning, artificial neural networks, ensemble-learning models, decision trees, and pattern and speech recognition—in addressing complex engineering challenges. With a strong interdisciplinary orientation, this book brings together novel applications and theoretical insights from civil and structural engineering, industrial and mechanical systems, electrical and electronic engineering, computer science, and even environmental sciences through the lens of nature-based and circular economy solutions.

The increasing ubiquity of AI and, more specifically, machine learning (ML), across scientific and practical domains has made these tools indispensable for researchers, engineers, and practitioners. Their ability to model nonlinear systems, adapt to dynamic environments, and provide reliable, repeatable outcomes makes them particularly well-suited for tackling real-world engineering problems. In this regard, the book is designed not only to illustrate AI's operational mechanisms and foundational principles but also to present concrete case studies and modeling frameworks grounded in real engineering contexts.

We have curated chapters that exemplify how AI methodologies can be employed to deliver efficient, innovative, and analytically sound solutions to problems previously considered too complex or time-consuming to solve using traditional approaches. Each contribution aims to balance theory with practice, incorporating computational models, simulations, and empirical validations that together make this volume a practical and scholarly resource. The versatility of AI applications—from optimizing mechanical systems to managing electrical circuits, or improving environmental sustainability through predictive analytics—underscores the depth and breadth of contributions found within this volume.

Importantly, this book also embraces the growing integration of *nature-based solutions* (NbS) and *circularity concepts* into engineering, particularly within the environmental sciences. As defined by the Food and Agriculture Organization (FAO, 2024), NbS refer to actions that sustainably manage and restore ecosystems while addressing pressing societal challenges such as climate change, water security, disaster mitigation, and biodiversity loss. In recent years, both NbS and circularity have gained traction not only in academic literature but also in public policy—evidenced by their inclusion in national strategies following the COP27 summit in Egypt. These themes mark a significant shift toward interdisciplinary and environmentally conscious engineering, and we are pleased to include this perspective in the book.

The relevance of this work is further magnified by the widespread institutionalization of AI and ML education. Universities across the globe now offer undergraduate and postgraduate programs that integrate AI into engineering curricula. Meanwhile, industry sectors, research centers, and technology firms increasingly depend on AI-driven insights and automation. Thus, this book is intended to serve as both an academic reference and a practical guide for students, faculty, researchers, and professionals in industry who seek to apply AI in engineering and beyond.

We believe that *AI-Based Solutions for Engineering* will contribute meaningfully to the growing body of literature that bridges artificial intelligence with applied sciences. Whether for those beginning their journey into AI or seasoned professionals looking to integrate new tools into their workflows, this volume offers comprehensive, multidisciplinary insights grounded in both theory and practice.

ORGANIZATION OF THE BOOK

Chapter 1: Deep Learning From Foundations to Frontiers

This chapter establishes the foundational landscape of deep learning and its transformative capabilities across numerous AI-driven applications, from intelligent transportation to healthcare. By dissecting core learning approaches—discriminative, generative, and hybrid—the chapter presents a cohesive view of how deep learning can be implemented to build end-to-end solutions. Additionally, it emphasizes the importance of regularization in model training and showcases specific techniques that enhance prediction accuracy and adaptability. As deep learning continues to evolve, this chapter sets the stage for understanding its core mechanics and far-reaching industrial impact.

Chapter 2: Simplifying Data with Robust Variable Selection for Enhanced Interpretability and Energy Efficiency

Focusing on interpretability and energy-efficient computation, this chapter addresses a key concern in engineering: understanding the influence of individual variables in complex machine learning models. The authors present robust dimensionality reduction techniques that retain critical data characteristics while enhancing model transparency, especially in the presence of noisy or contaminated data. By applying these methods to datasets from gene expression and credit scoring, the chapter not only demonstrates practical applications but also underlines their relevance to real-world AI systems that prioritize both interpretability and computational efficiency.

Chapter 3: TinyML Empowering Intelligent Edge Devices

This chapter dives into TinyML—an emerging field that empowers microcontrollers and other low-power devices to execute AI locally at the edge. It explores the unique intersection of AI, embedded systems, and real-time decision-making for applications in IoT, healthcare, and environmental monitoring. The chapter also evaluates challenges in energy consumption and model optimization, while shedding light on advanced trends like federated learning and neuromorphic computing. Readers gain a comprehensive overview of TinyML's growing significance in the development of autonomous and resource-efficient AI applications.

Chapter 4: Support Vector Machine in Predictive Maintenance of Belt-Drive System

Predictive maintenance is a cornerstone of modern industrial reliability, and this chapter applies Support Vector Machine (SVM) techniques to monitor belt-drive systems. By analyzing vibrational signal patterns to detect anomalies like belt misalignment and tension irregularities, the authors show how SVM's discriminative capabilities offer superior performance in complex, noisy environments. The chapter provides practical insight into fault classification and system health monitoring, contributing to cost reduction and operational continuity in industrial applications.

Chapter 5: Exploring the Magnitude-Shape Plot Framework for Anomaly Detection in Crowded Video Scenes

In this innovative chapter, the authors present the Magnitude-Shape (MS) Plot framework as an effective tool for video anomaly detection. Using autoencoders and treating reconstruction error matrices as multivariate functional data, the approach captures both the magnitude and shape deviations of anomalies in crowded scenes. Benchmark tests on UCSD Ped2 and CUHK Avenue datasets highlight its competitive edge over conventional detectors. This chapter stands out for its statistical rigor and interpretability, offering a meaningful advance in functional data analysis applied to computer vision.

Chapter 6: Artificial Intelligence based Automated System for Building Information Modelling Data Extraction and Analysis

This chapter presents a comprehensive overview of AI-powered systems designed for the extraction and analysis of Building Information Modeling (BIM) data. Moving beyond the limitations of manual data processing, the authors demonstrate how machine learning, natural language processing, and computer vision can automate and enhance information retrieval in architectural and construction contexts. Through a detailed case study, the chapter outlines implementation strategies and anticipates future innovations in AI-integrated BIM systems, making it an essential read for professionals in smart construction.

Chapter 7: Accurate Estimation of Load-Carrying Capacity (Pu) in Steel Tube Structures

Structural safety depends heavily on accurate estimations of load capacity, and this chapter introduces a machine learning-based approach for predicting Pu in steel tube structures. The authors compare XGBoost, deep neural networks, and Bayesian optimization techniques, revealing how advanced ML models outperform traditional empirical methods. Bayesian-tuned XGBoost achieved exceptional predictive performance, underscoring the value of AI in structural engineering design. This chapter serves as a model for how AI can enhance precision and safety in engineering analysis.

Chapter 8: Artificial Intelligence in Wastewater Treatment Advancing Processes with Nature-based Solutions

Addressing the mounting challenges in wastewater treatment, particularly in resource-constrained regions like India, this chapter merges artificial intelligence with nature-based solutions (NBS) to enhance sustainability. The authors discuss real-time monitoring, predictive analytics, and optimized operations powered by AI, offering a resilient and eco-friendly framework for water management. With the proposed AI-NBS model, the chapter presents a compelling case for technological and ecological integration in addressing pressing environmental and infrastructure concerns.

Chapter 9: Artificial Intelligence and Machine Learning in Additive Manufacturing

This chapter investigates the synergy between Additive Manufacturing (AM) and AI/ML technologies to solve challenges such as defects, limited materials, and production inefficiencies. With an emphasis on process monitoring, predictive modeling, and quality control, the chapter showcases how AI-driven solutions can elevate AM toward scalable and efficient production in Industry 4.0. The findings advocate for interdisciplinary collaboration and robust data infrastructure to further propel AI-enhanced manufacturing systems.

Chapter 10: A Digital Twin and AI-Driven Framework for Optimizing Firefighter Escape Routes in Wildfire Scenarios: FAST

In a powerful blend of AI and digital twin technology, this chapter introduces the FAST system for improving firefighter safety during wildfire emergencies. The system employs deep learning models like multivariate LSTM and DQN to predict fire spread and generate optimized evacuation routes in real time. Experimental results—such as an 84% escape success rate and 260-second average evacuation time—affirm its practical utility. This chapter demonstrates how intelligent simulation and real-time AI analytics can transform emergency response strategies.

CONCLUSION

As we conclude this volume, *AI-Based Solutions for Engineering*, we reflect on the breadth and depth of contributions that demonstrate the transformative role artificial intelligence plays in modern engineering. The chapters in this book traverse a diverse landscape—from foundational deep learning frameworks to highly specialized applications like predictive maintenance, anomaly detection, additive manufacturing, and sustainable wastewater treatment. Each contribution highlights how AI, through its many forms—machine learning, deep neural networks, support vector machines, TinyML, digital twins, and beyond—offers a dynamic toolkit for solving the complex and evolving challenges faced across engineering disciplines.

One of the unifying themes throughout this volume is the growing necessity for intelligent, adaptive, and interpretable systems. The ability of AI to uncover patterns in vast, noisy datasets, make real-time predictions, and support energy-efficient computation underlines its value not just as a theoretical construct, but as a practical enabler of innovation. Whether applied at the edge of embedded devices or integrated into large-scale civil and structural systems, the methodologies explored herein reveal the power of AI to deliver robust, accurate, and sustainable solutions.

Additionally, we are proud to have included perspectives that recognize the importance of aligning AI advancements with environmental responsibility, particularly through the inclusion of nature-based solutions and circularity principles. These approaches reflect a broader shift in engineering—toward systems that are not only intelligent and efficient but also regenerative and ecologically informed.

We hope this volume serves as a valuable resource for a wide audience, including students, researchers, practitioners, and policymakers. By presenting both theoretical underpinnings and practical case studies, we aim to inspire continued exploration and responsible application of AI in engineering. As the field continues to evolve, so too will the possibilities for intelligent, data-driven, and sustainable engineering solutions. It is our sincere belief that the insights and innovations presented in this book will contribute meaningfully to that future.

Chapter 1
Deep Learning From Foundations to Frontiers

S. Jansirani
https://orcid.org/0000-0002-7454-5340
Jansons Institute of Technology, India

M. Saravanan
https://orcid.org/0000-0002-2200-9159
KPR Institute of Engineering and Technology, India

Vishnu Kumar Kaliappan
https://orcid.org/0000-0003-2517-3074
KPR Institute of Engineering and Technology, India

ABSTRACT

In today's data-driven world, the ability to automatically recognize complicated patterns from enormous amounts of data is essential for many AI applications, such as recommendation systems, robotics, intelligent transportation systems, and healthcare. Therefore, developing an end-to-end framework using deep learning is crucial for advancing artificial intelligence and generating accurate predictions. The powerful features, such as handling big data, learning hierarchical data representations, and transfer learning, enable it to efficiently manage complex tasks, generalize to new data, and adapt to various domains, making it indispensable across numerous industries. This chapter explores the foundation and classification of deep learning including discriminative, generative, and hybrid learning approaches and also reviews the significance of regularization in deep learning, exploring its different types. Additionally, it highlights the specific deep learning techniques and its application.

DOI: 10.4018/979-8-3373-0746-6.ch001

INTRODUCTION

Today's digital era requires automating tasks and enhancing decision-making with intelligent systems. From historical advancements to the current computer revolution, the focus is on creating machines that mimic human abilities (Nisha.C.M & N. Thangarasu, 2023a). The emerging technology like artificial intelligence provides a huge impact in the society by automating tasks, fostering innovation, and addressing complex issues. Artificial intelligence comprises machine learning, deep learning, and generative AI, which enable computers to replicate or surpass human decision-making and autonomously solve complex problems. (Janiesch et al., 2021; Minsky, 1961). It is done by training a machine with numerous examples and crucial data in the form of pictures, numbers, text and videos to perform things that people do instinctively (Sundaram et al., 2021).

Machine learning is a subset of AI, primarily concerned with finding patterns in data and forecasting outcomes using prior knowledge gained without the need for explicit programming (Dabiri et al., 2018; Pothuganti, 2018; Walz, 2023). There are two stages involved in implementing ML algorithms which are training and testing (Sundaram et al., 2021). ML is divided into supervised, unsupervised, semi-supervised, and reinforcement learning based on the data used (Pothuganti, 2018; Walz, 2023), (Taye, 2023). In supervised learning, models can learn from labelled data and provide a result for a task like classification or regression (Jabir & Falih, 2021; Pothuganti, 2018; Walz, 2023). When working with unstructured data like images and videos, it's really very difficult or impossible to define labels or intended outputs. Hence unsupervised machine learning can be used to discover hidden patterns and gain a deeper understanding of video data like Object Detection, and Traffic Flow Analysis from unlabelled data (Walz, 2023). Semi-supervised learning combines elements of both supervised and unsupervised learning. For tasks involving sequential control, optimization, and decision-making, reinforcement learning is employed. To gain information, it enables agents to interact with their surroundings in order to get feedback in the form of rewards and penalties (Latif et al., 2023). The major demerits of ML are manual feature engineering where data scientists need to select the features manually. It's really difficult to capture features manually in high dimensional and unstructured complex data like video. Another major demerit is Machine learning models are typically task-specific and not easily transferable to other tasks. Building new models for different tasks requires significant effort and expertise. Lack of Interpretability, Underfitting, Overfitting, Data Quality, Computational Resources and Generalization are also the major challenges of machine learning (Pothuganti, 2018).

After the Industry 4.0 revolution (Jabir & Falih, 2021), there is a transformative shift across various sectors, including transportation, manufacturing, healthcare and energy (Hatcher & Yu, 2018; Sarker, 2021; Veres & Moussa, 2020) reshaping entire industries by leveraging the power of digital technologies, automation, and data-driven processes. Deep Learning (DL) plays a crucial role in this transformation by enabling the development of intelligent data-driven systems capable of handling unstructured, high-dimensional, and non-linear input representations. These capabilities are essential for creating automated solutions that meet the complex and dynamic needs of today's industries.

DL is a branch of ML, model and extracts high-level characteristics with the help of data and multiple-layered ANN (Mungoli, 2023a; Pothuganti, 2018). The basic components of deep learning involve input, hidden, and output layers, links, neurons, weight and bias. Neurons in different layers are connected by links. Each link is associated with weight. Weights influence strength and direction, as it travels from one neuron to another. Weights are changed throughout the training to reduce the error rate. Each layer has a specific number of neurons. The input layer, neuron count is determined by the number of features in the dataset. Input layer receives raw data, which is then sent to hidden layers. In the hidden layer, summation and activation functions are performed and the result is given to the output layer. Output layer, processes that input by doing a summation and activation function and produces a result related to regression or classification task (Alzubaidi et al., 2021; Hatcher & Yu, 2018; Sarker, 2021). Figure 1 shows a Deep learning model to classify the object and its features automatically based on the given input data. Due to its valuable performance features such as Universal Learning Approach, Robustness, Generalization and Scalability most of them have an interest to do research in Deep Learning (Alzubaidi et al., 2021).

Figure 1. Deep learning model for classification

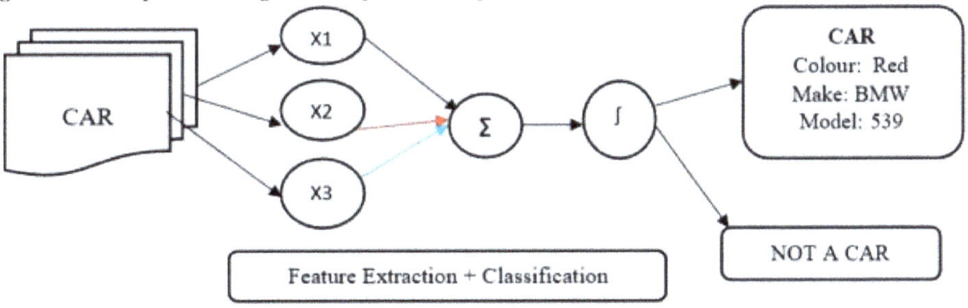

About Data

In deep learning, data's are essential, because it enables models to learn patterns and generalize well, in a way to reduce overfitting and increase accuracy. Large amounts of data are needed to supplement the difficulty of decision making when machine learning or deep learning algorithms are integrated into industrial operations (Sundaram et al., 2021). Data refers to an unprocessed, unorganized raw fact containing numerical and categorical values. Information is a processed, aggregated and organized data that has context, relevance and meaning. Insights involves the analysis and interpretation of data to extract meaningful patterns, trends, or relationships. The example for data, information and insights are shown in Figure 2.

Figure 2. Data vs information vs insights

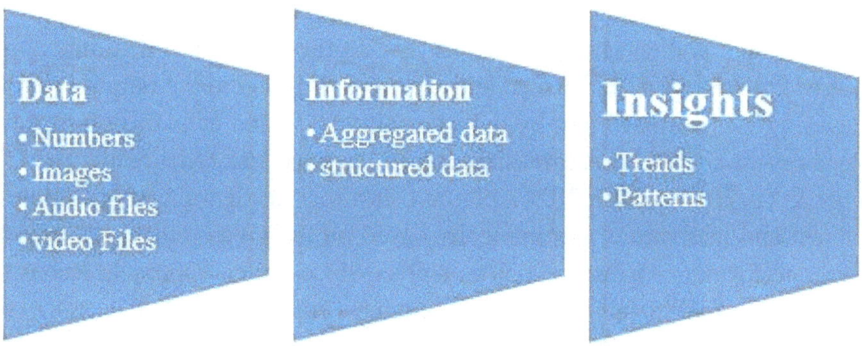

Data is a raw material to build an intelligent data driven system, in which data is collected using various sources like sensors, cookies, surveys, GPS and mobile devices, etc. Data gathered from the environment may have some types of noise due to various factors like radiation, less visibility conditions and heat. Therefore, identification of original data from the noisy data becomes a tedious task (Sundaram et al., 2021). Deep learning filters called pre-processing, play a major role for optimizing the data by data cleaning, scaling and transforming into a suitable form to enhance and create a generalized model (Koresh, 2024) are shown in Figure 3.

Figure 3. Pre-processing

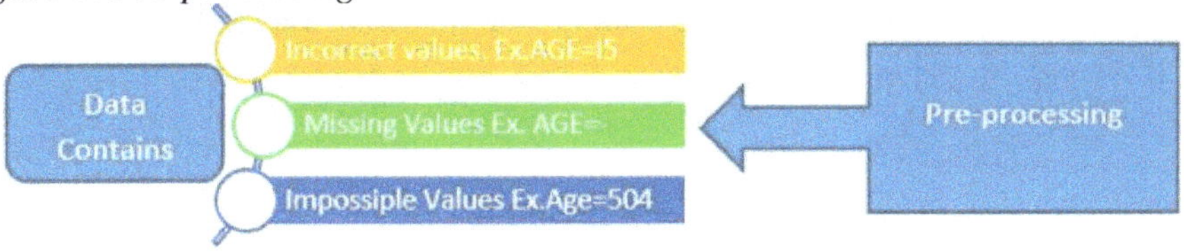

Features

Features are the characteristics, attributes, or variables in data used by deep learning models to make predictions or classifications. These can be numeric, categorical, or textual. Pre-processed data typically includes rows called as samples and columns called as features). With advances in data collection, datasets have become high-dimensional, containing more features than samples, leading to the "curse of dimensionality" (Chetana et al., 2020; Jia et al., 2022). This can cause data overload and hinder storage and processing. Dimensionality reduction, via feature selection or extraction, helps retain relevant information while improving model performance (Chetana et al., 2020; Jia et al., 2022; Velliangiri et al., 2019).

Feature Selection

It involves identifying the most relevant features from a dataset through four main steps. First, various feature subsets are generated, with a dataset of n features potentially producing up to 2^N subsets. Next, these subsets are evaluated based on criteria such as relevance, redundancy, and performance. The selection process stops when a full search is completed, a threshold is reached, accuracy starts to drop, or further feature changes yield no improvement. Finally, the chosen subset is validated using unseen data to ensure generalizability. Common methods used include filter, wrapper, and embedded techniques.

Feature Extraction

Feature extraction reduces dimensionality by transforming the original input space into a smaller set of derived features. It applies functions to generate new features from existing ones and is categorized into linear and non-linear methods. Linear methods, like Principal Component Analysis and Independent Component Analysis, use linear combinations of original features, while non-linear methods such as t-SNE, Autoencoders, Isomap, and Kernel PCA, capture more complex relationships. Though nonlinear techniques often yield higher accuracy, they are typically more complex to implement (Chetana et al., 2020; Jia et al., 2022; Nanga et al., 2021; Velliangiri et al., 2019).

CLASSIFICATION OF DEEP LEARNING

Deep learning effectively handles complex, unstructured data such as audio, text, images, and video by learning hierarchical features directly from raw inputs (Taye, 2023). A key advantage is end-to-end learning, where models map raw data to outputs without manual feature engineering. DL scales efficiently with large datasets using GPUs/TPUs and supports transfer learning, enabling fine-tuning of pre-trained models on smaller datasets for specific tasks (Alzubaidi et al., 2021; Kamilaris, 2018; Liu & Lang, 2019). DL approaches include Discriminative, Generative, and Hybrid learning shown in Figure 4. Discriminative models such as MLP, CNN, and RNN focus on classification using labeled data. Generative models produce synthetic data and optimize parameters to align it with original data. Hybrid models combine both approaches (Alzubaidi et al., 2021; Sarker, 2021).

Figure 4. Taxonomy of deep learning

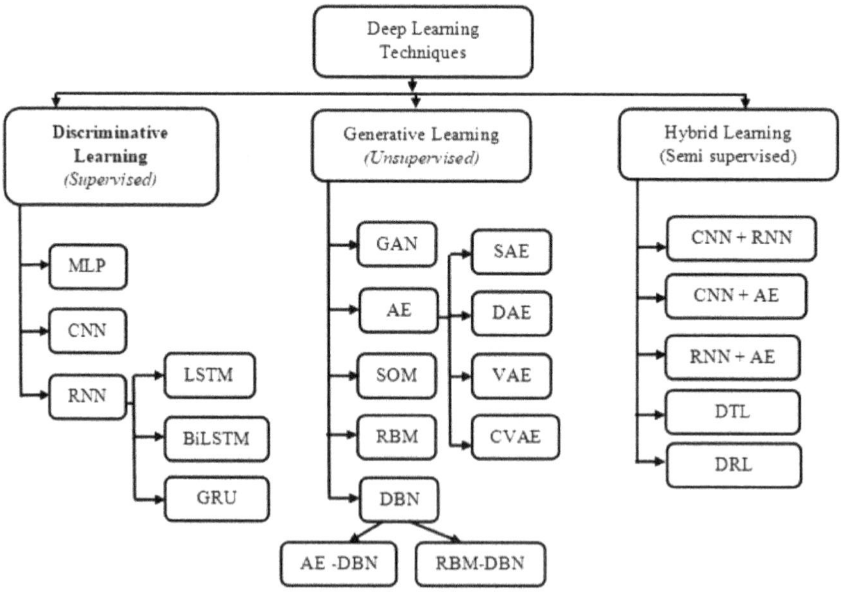

Multilayer Perceptron

It's a neural network with input, hidden, and output layers, commonly known as a multi-layer perceptron (Bikku, 2020; H. Liu & Lang, 2019). While designing a neural network architecture, consider elements such as the number of neurons in each layer, number of hidden layers, as well as the types of activation function. Input layer receives a vector of features as input and that will be processed in the hidden layer with the help of summation and various activation functions. Output layer processes the data received from the hidden layer and produces the result in terms of predicted value (Hazra & Anjaria, 2022). Error function or cost function is calculated by subtracting the target value with predicted value. The error rate can be reduced by updating the weight from output layer to input layer. This process is repeated until to get an optimal result.

Convolution Neural Network

It is a sort of DNN created primarily for reducing the need for manual feature engineering in the field of image and video processing. It can process hierarchical representations of input data, such as pictures and videos (Yamashita et al., 2018). It is typically composed of many layers namely input layer, convolution, pooling, Flatten and fully connected or Global Average Pooling Layers (Bikku & Satya Sree, 2020; Gholamalinezhad & Khosravi, 2020; Hazra & Anjaria, 2022; Purwono et al., 2022; Yamashita et al., 2018).

Input Layer

A machine interprets a picture as a matrix, or collection of integers. The light intensity at each number corresponds to a specific spot known as a pixel (Purwono et al., 2022). Input layer accepts pixels in the form of buffers and can store values up to three bits. Since 2^3 is equal to 8, it displays 8 colors. A color index is used to store colors in an image that has more than 8 colors. The color index retrieves color values depending on pixel value. As shown in Table 1, the RGB image frame buffer has a size of 3 (Bikku & Satya Sree, 2020). If a RGB image has a pixel value of (255,255,255), it means that the pixel is white which has full intensity in all three color channels, If the pixel value is (255, 0, 0), it symbolizes pure red, with full intensity in the red channel but no intensity in the channels of green and blue are shown in Figure 5.

Table 1. RGB image frame buffer of 3

S.No	Frame Buffer Bits			Color
	R	G	B	
1	0	0	0	Black
2	0	0	1	Blue
3	0	1	0	Green
4	0	1	1	Cyan
5	1	0	0	Red
6	1	0	1	Magenta
7	1	1	0	Yellow
8	1	1	1	White

Figure 5. Indexed color image

23	77	18	14	77	255
254	152	76	54	28	9
17	54	43	36	24	5
247	233	26	22	29	35
32	33	204	243	159	172
186	142	130	30	34	52

R	G	B
0	0	0
0	0	255
0	23	253
0	110	252
. . .	. . .	. . .
255	255	255

Convolution Layer

A key part of CNN design is the convolutional layer, which handles feature extraction. Typically, this involves combining linear and nonlinear processes, such as the activation function with the convolution operation (Haghighat et al., 2020; Yamashita et al., 2018). To recognize features, it processes the input data by applying filters known as kernels. A filter is a small array of numbers (Yamashita et al., 2018) or matrix that slides over the input data to perform element-wise multiplications and produce feature maps. These filters act as feature detectors, capturing patterns like edges, textures, and more complex structures in deeper layers. Hyper-parameters namely depth, stride and zero padding are used to achieve Optimization (O'Shea

& Nash, 2015). The distance between two consecutive kernel positions is called a stride (Yamashita et al., 2018). If stride equals one, the receptive field is extensively overlapped, resulting in exceptionally massive activations. Alternatively, increasing the stride will limit the amount of overlapping and provide an output with smaller spatial dimensions (O'Shea & Nash, 2015).

Pooling Layer

Pooling layers minimize the spatial dimensions of feature maps while keeping the most essential data (Gholamalinezhad & Khosravi, 2020; Mungoli, 2023a; Talaei Khoei et al., 2023; Teuwen & Moriakov, 2019; Zafar et al., 2022). It is used to control overfitting and minimize the number of parameters or weights, thereby decreasing the computational cost (Dabiri et al., 2018; Gholamalinezhad & Khosravi, 2020). The two most common pooling methods used in pooling layers to minimize spatial dimensions are average pooling and max pooling. (Yu et al., 2014; Zafar et al., 2022).

Max Pooling. The spatial size of a feature map is optimized with the help of max pooling. Kernel filter partitions the input into local regions and then maximum value is calculated from each of these regions are shown in Figure 6 (Qiu et al., 2021; Yu et al., 2014). This approach is well-liked as it lowers computing complexity and doesn't require any parameter tuning. The disadvantage of this method is, it only considers the largest element in each pooling zone, neglecting the rest. This might result in information loss and perhaps it will produce undesirable outcomes (Yu et al., 2014).

Figure 6. Max pooling

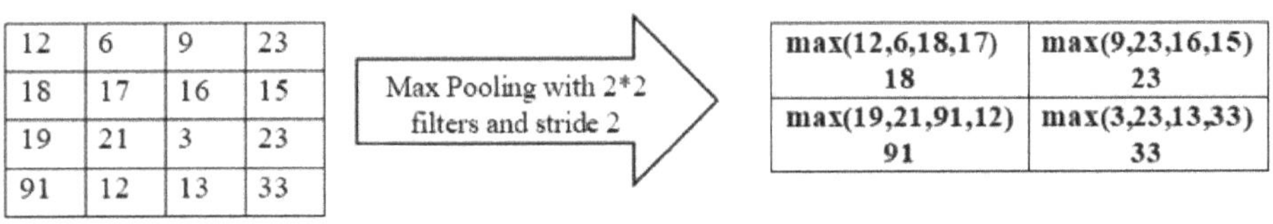

Average Pooling. It downsample given input by calculating average for specific regions within it. Kernel filter partitions the input into local regions, and average value is called for each local region are shown in Figure 7. It aids capturing the effect of every feature in the input data. In certain situations, average pooling is seen to be better than max pooling, as it may gather a wider range of information.

Figure 7. Average pooling

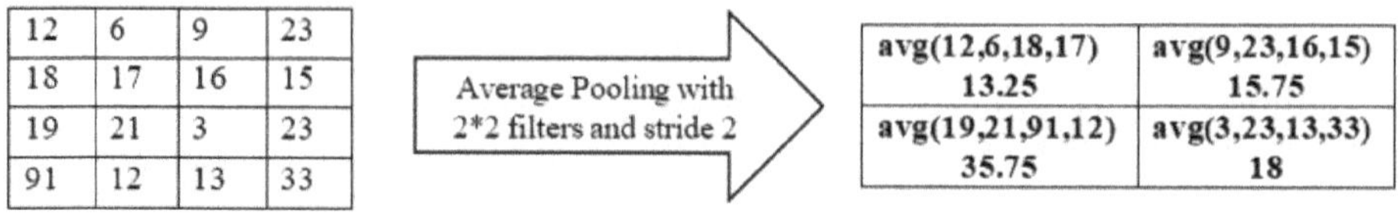

Mixed Max Average Pooling. It integrates average and maximum pooling by using weight factor. A mixing ratio, λ, which can be either 0 for average pooling or 1 for max pooling, determines which option to choose between max and average pooling are shown in Figure 8. Hybrid method, which is very useful for image classification applications. But the major drawback of this pooling system is fixed mixing ratio which can make it insensitive to important features.

Figure 8. Mixed max average pooling

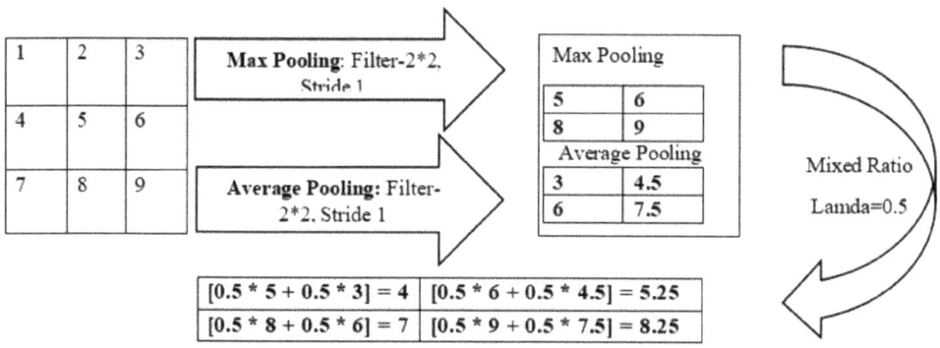

Tree Pooling. Tree pooling arranges the pooling process in a hierarchy of leaf, parent, and root nodes as shown in Figure 9. Leaf nodes use grouping filters to process input data, while parent nodes combine their children's outputs using a gate function, similar to how average pooling aggregates values. The root node then gathers all the information, capturing complex feature interactions and producing the final pooled output.

Figure 9. Tree pooling

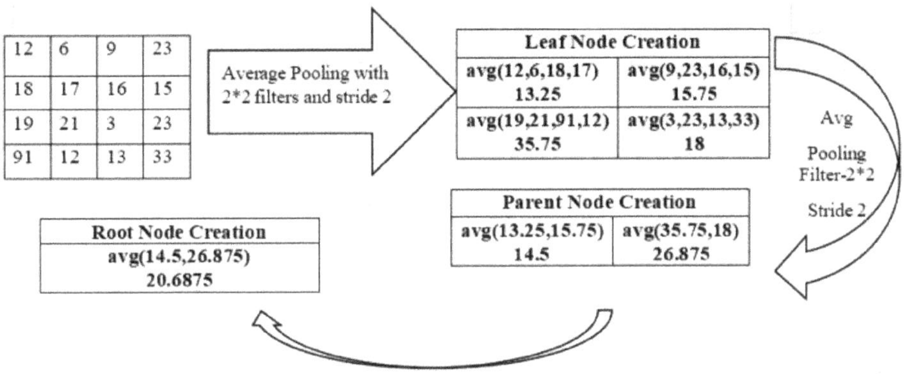

Tree Max Average Pooling. It combines tree pooling, max pooling, and average pooling. Tree pooling involves learning the organization of a group of filters, while max pooling selects the highest value, and average pooling computes average value in the particular region. In tree max average pooling, these three pooling methods are merged with weights to consider all of them simultaneously. This approach overcomes the limitations of using just max or average pooling individually by combining their strengths. The weights assigned to each pooling method are determined by a mixing ratio parameter, lambda (λ), which can be randomly chosen as either 0 (average pooling behavior) or 1 (max pooling behavior). This hybrid approach provides better results compared to max and average pooling.

Fractional Max Pooling. It is a stochastic approach used to modify the max pooling score by choosing the pooling area using randomness or pseudo-randomness. This method has been observed to be more efficient when the pooling regions overlap rather than being disjointed. Randomly choosing the pooling area can have a major impact on model performance, since fragmented fractional max pooling can lead to deterioration. It is recommended to design overlapping regions for more efficient results in Fractional Max Pooling

Rank-Based Pooling. It is a method that aims to address the limitations of traditional pooling techniques like max pooling and average pooling. It involves assigning weights to activations based on their ranks in a feature map. The nearest k activations are weighted with one, whereas the rest of the activations receive a weight of zero. Unlike the max and average pooling techniques, this method introduces extra free parameters that must be tuned to enhance performance. This method excels at reducing information loss and balancing max and average pooling, enhancing discriminative power for object recognition tasks.

Flatten Layer

It is used to convert feature maps from N-Dimensional representation into 1-Dimensional column vector representation, which can be used for image classification (Gholamalinezhad & Khosravi, 2020; Suriya et al., 2022).

Fully Connected layer

Once the downsampling can be done by the pooling layers the resultant N-Dimensional data are converted into 1-Dimensional column vector by flatten layer and that 1-Dimensional column vector is mapped with a subset of fully connected layers. Every input is connected to every output along with some specified weights and in the final fully connected layer the number of classes and output nodes are equal (Talaei Khoei et al., 2023; Yamashita et al., 2018). Fully connected layer helps for creating high-level abstractions by receiving mid- and low-level features and generating classification scores. However, it demands many parameters, resulting in increased memory usage and computational cost. This is due to the extensive connections between neurons in adjacent layers, which requires learning a multitude of weights. As a result, training time and costs are significantly increased (Alzubaidi et al., 2021).

Global Average Pooling (GAP). This layer is used to resolve issues typically found in fully connected layers. It reduces parameter count and connections by averaging all values within each feature map, yielding a single value per map shown in Figure 10 (Al-Sabaawi et al., 2020). Hence the dimensionality of the feature maps can be reduced while preserving essential features, thereby the network's representational power has been enhanced. This approach reduces memory usage, prevents overfitting, and speeds up training convergence. Therefore, the shift from fully connected layers to Global Average Pooling layers is a strategy to reduce computational costs and improve model efficiency. This method is commonly used in CNN architectures for tasks like image recognition and segmentation (Alzubaidi et al., 2021; Hatcher & Yu, 2018).

Figure 10. Global average pooling

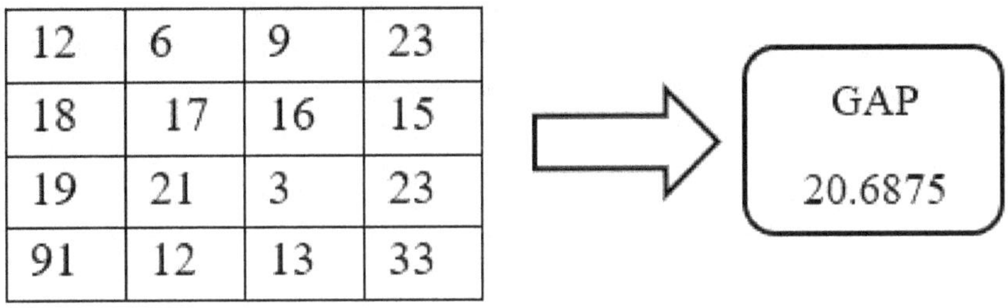

Activation Function

Neural network prediction accuracy relies on layer count and the specific activation function chosen (Sharma et al., 2020). Activation function introduces, non-linearity within neural networks, facilitating the learning of intricate patterns If activation functions are not present in the neural network, then it would function more like a linear model, which have limited ability to capture complex hierarchical patterns (S. Dong et al., 2021a; Sharma et al., 2020). In CNNs, the activation function is applied in both the convolutional and fully connected layers. The activation function makes decisions based on an assigned threshold value (Alzubaidi et al., 2021; Hatcher & Yu, 2018; Kamalov et al., 2021).

Table 2. Activation function and its mathematical form

S. No	Activation Function	Mathematical Form
1	Sigmoid	$\frac{1}{1+e^{-x}}$
2	Tanh	$\frac{e^{x}-e^{-x}}{e^{x}+e^{-x}}$
3	ReLU	$f(x) = \{0,\ x \leq 0\, x,\ x > 0$
4	Leaky ReLU	$f(x) = \{0.01x,\ x \leq 0\, x,\ x > 0$
5	Parametric Linear Units	$f(x) = \{\alpha x,\ x < 0\, x,\ x \geq 0$

The most commonly used CNN activation functions are Sigmoid, Tanh, ReLU, Leaky ReLU, Noisy ReLU, and Parametric Linear Units (Table 2) (Dong et al., 2021a; Hatcher & Yu, 2018; Purwono et al., 2022). Sigmoid function confines its output between zero and one, processing real numbers as input. Tanh function confines its output within the range of -1 to 1, akin to the Sigmoid function's behaviour. Both Sigmoid and tanh functions are often avoided, due to their tendency to cause the vanishing gradient problem, where gradients approach zero. A combination of sigmoid functions yields superior outcomes for classification problems. ReLU transforms input values to positives, with less computational burden than other functions. It's typically used in hidden layers rather than the output layer. Leaky ReLU can be used to address dead neurons within the network (Sharma et al., 2020).

Types of CNN Architecture

From 1989 to the present, numerous modifications have been made to the CNN architecture, including regularization, parameter optimization, and structural reformulation. LeNet-5 was designed for hand-written, machine-printed character recognition. AlexNet, introduced the use of Dropout and ReLU, significantly advancing CNN architectures. VGGNet increased depth and used small filter sizes, and it was known for its homogenous topology and simplicity. ResNet introduced skip links to make the network more robust against overfitting. DenseNet introduced dense connectivity, linking each layer to every other layer. MobileNet used an inverted residual structure to achieve high performance with fewer parameters (Alzubaidi et al., 2021). An overview of CNN architectures is shown in Table 3(Alzubaidi et al., 2021; Purwono et al., 2022; Teuwen & Moriakov, 2019).

Table 3. Overview of CNN architecture

S.No	Name	Architecture Type	Features	Depth	Convolution Layer	FC Layer	GAP Layer	Feature Extraction Stages	GPUs
1	LeNet-5	Classic CNN tailored for image classification tasks.	Handwritten digit recognition	7	3	2	Nil	5	1
2	AlexNet	Traditional CNN architecture featuring convolutional and fully connected layers	Introduction of ReLU activation, local response normalization, and dropout. Utilized large convolutional filters in the initial layers.	8	5	3	Nil	7	2
3	VGGNet	Classic CNN architecture with small 3x3 convolutional filters and max-pooling layers.	Uniform architecture with small filters, making it easy to understand. Deep but computationally expensive	16 or 19	13 or 16	3	Nil	5	1

continued on following page

Table 3. Continued

S.No	Name	Architecture Type	Features	Depth	Convolution Layer	FC Layer	GAP Layer	Feature Extraction Stages	GPUs
4	GoogLeNet	Inception architecture with the introduction of inception modules (1x1, 3x3, 5x5 convolutions).	Efficient use of parameters, reduced computational cost, and improved information flow with inception modules.	22	Not Fixed	Nil	Not Fixed	9	Not Fixed
5	ResNet	Residual Network with skip connections	Introduction of residual learning, which helps address the vanishing gradient problem and enables training of very deep networks.	Not Fixed	Not fixed	Nil	Not Fixed	Not Fixed	Not Fixed
6	DenseNet	Densely Connected Convolutional Networks	Feature reuse through dense connectivity, reduces the number of parameters, and enhances feature propagation.	201	Not fixed	Nil	Not fixed	Not fixed	Not fixed
7	MobileNet	It employs depth-wise separable convolutions to reduce computational complexity.	Designed for mobile and edge devices, efficient use of parameters, and reduced computational requirements	53	Not fixed	Nil	Not fixed	Not fixed	Not fixed

Recurrent Neural Network

A Recurrent Neural Network (RNN) is an ANN designed to process sequential or time-series data using feedback loops, as shown in Figure 11 (Haghighat et al., 2020; Mungoli, 2023a; Talaei Khoei et al., 2023; Walz, 2023). Each neuron in an RNN retains information from previous inputs, enabling temporal memory (Bikku & Satya Sree, 2020; Sarker, 2021). Feedback connections loop outputs back as inputs, supporting sequence modeling (Merz & Actuarial, 2023). Unlike feedforward networks, RNNs share parameters across time steps and use the same weights within each layer (Merz & Actuarial, 2023; Walz, 2023). Bidirectional RNNs process data in both forward and backward directions to capture complete context (Walz, 2023). To capture long-term dependencies, gradient flow is crucial. Random weight initialization may work for simple tasks but fails on complex ones due to the need for more precise parameters (Hochreiter et al., 2001). Small gradients hinder learning by preventing effective parameter updates, especially over longer sequences, leading to the vanishing gradient problem (Y. Wang et al., 2020). LSTM and Bidirectional LSTM Neural Networks effectively mitigate this issue (Alzubaidi et al., 2021; C. Liu et al., 2021; Sarker, 2021). RNNs are commonly used in tasks such as word prediction, music generation, image captioning, and speech recognition (Bikku & Satya Sree, 2020).

Figure 11. Max pooling the basic structure of recurrent neural network

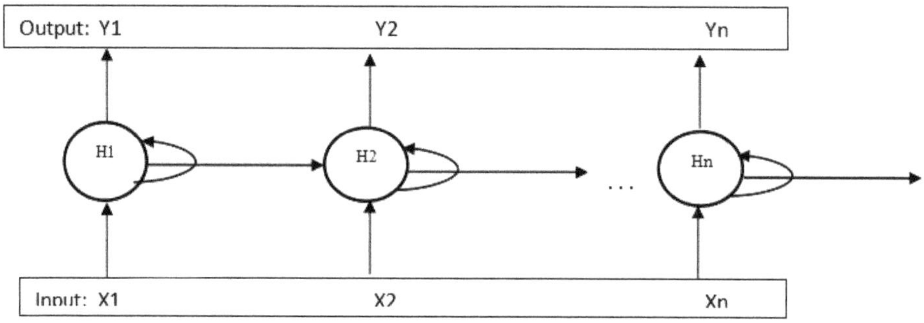

Long Short-Term Memory

It includes memory cells, input/output gates, and a forget gate, enabling it to learn long-term dependencies (Ashraf et al., 2021; Nguyen et al., 2018; Y. Wang et al., 2019). The memory cells are capable of storing data for long durations, allowing the system to maintain its state over extended periods. The flow of information through this system is regulated by non-linear gating units(Van Houdt et al., 2020). The forget gate decides which information from the previous cell state is retained and what is discarded. The input gate determines what new data should be added to the memory cell, while the output gate decides which data from the memory cell is sent to the next state are shown in Figure 12. (Sarker, 2021).

Figure 12. Block diagram of LSTM block

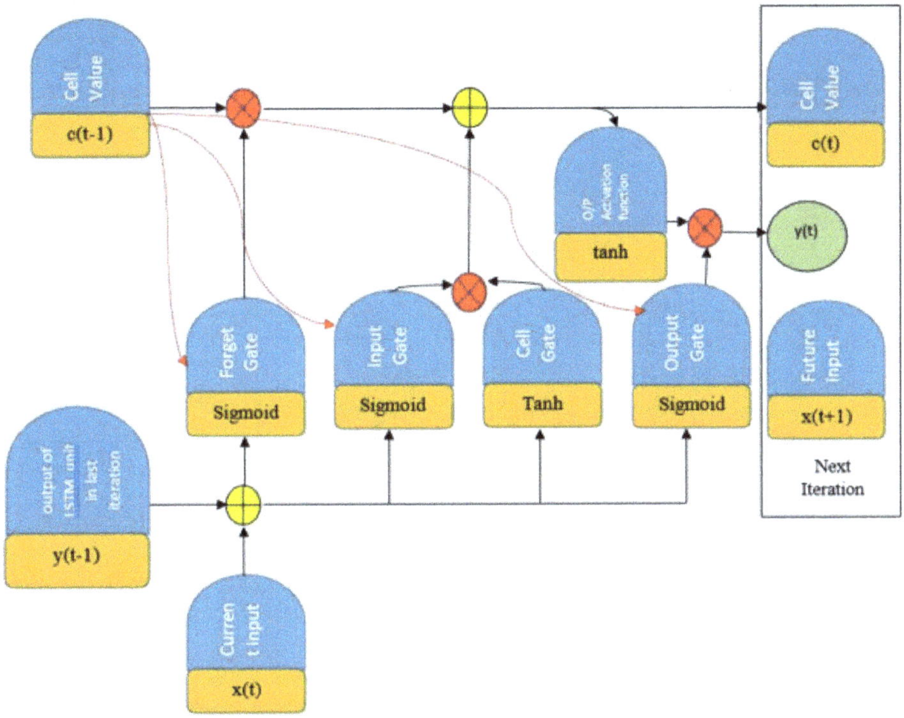

LSTM networks excel at processing, predicting and categorizing time series data. The major drawback of LSTM is that it's very challenging to extract dynamic and nonlinear features hiding in the processed data (Changrui Xie, Runjie Yao, Zhengbang Liu, Lingyu Zhu, 2022). Because these models process data in a single forward direction, relying on past data to make predictions. It cannot use future information from the current state, meaning that the output layer only utilizes previous contexts for generating predictions (Jung & Yeon, 2021).

Bidirectional LSTM

It is an extension of standard LSTM, used to extract dynamic, nonlinear, spatial and temporal features from the data effectively (Changrui Xie, Runjie Yao, Zhengbang Liu, Lingyu Zhu, 2022; Sarker, 2021). BiLSTM uses two LSTM networks trained simultaneously: one processes sequences from the start, and other from the end. These sequences are then concatenated to form a single output vector at each time step, which integrates past and future information (Y. H. Li et al., 2020). This combined output vector can then be passed to a subsequent layer, such as a fully connected layer, for further processing as shown in Figure 13.(Kai Zhang, Wei Song, Lizhen Liu, Xinlei Zhao, 2019).

Figure 13. Block diagram of BiLSTM

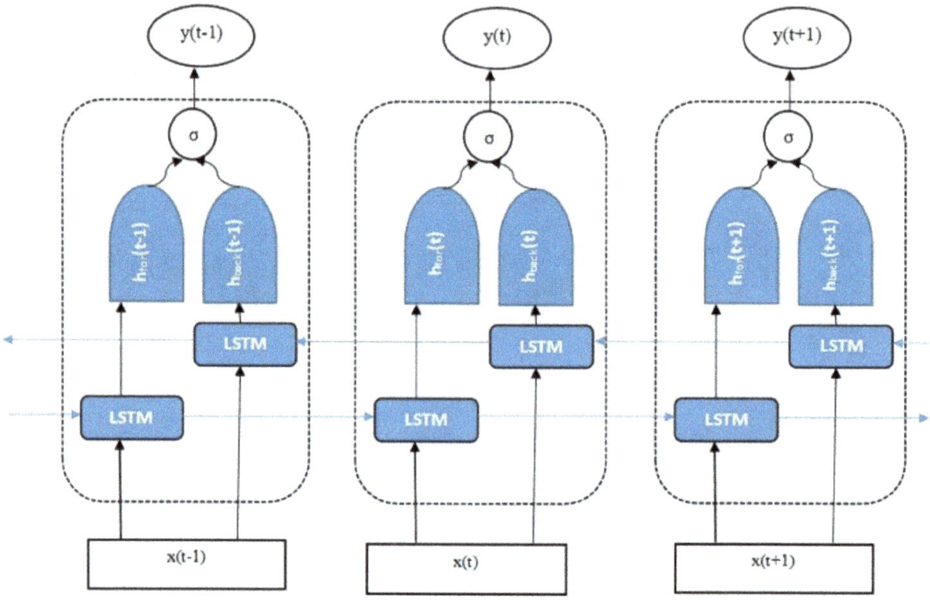

Gated Recurrent Units (GRU)

GRU controls information flow and mitigates vanishing/exploding gradients in traditional RNNs. Its input layer size depends on feature space, and output neurons match output space. The hidden layer, consisting of memory cells, performs core operations shown in Figure 14. GRU uses a single hidden state, combining the input and forget gates into an update gate to reduce training time and improve performance (ArunKumar et al., 2022; Y. Wang et al., 2019). It merges hidden and cell states into one and, with only update and reset gates (half the gates of LSTM), is a more efficient LSTM variant. The update gate regulates how much past information is retained, while the reset gate decides what to forget (ArunKumar et al., 2022; Obaidur Rahman et al., 2019; Sarker, 2021).

Figure 14. Structure of GRU

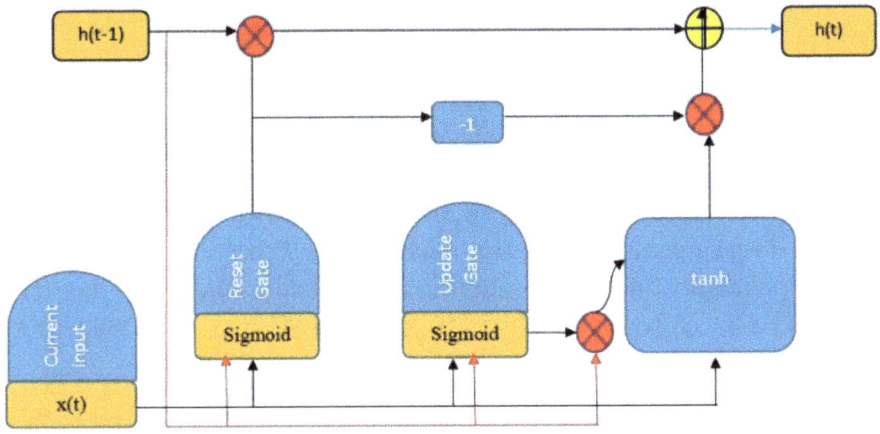

Table 4. Comparison of LSTM, BiLSTM, and GRU

S.No	Model	LSTM	BiLSTM	GRU
1	Description	Can handle long-term dependencies, effective in capturing context from the beginning of the input.	Captures past and future context by processing inputs in both forward and backward directions.	Designed to handle vanishing gradient problems with a simpler structure.
2	**Demerits**	More complex and computationally expensive than simpler RNN models.	Increased complexity due to processing in two direction	Less effective for tasks requiring long-term memory and complex context.
3	Structure	Contains memory cell, input and output gates	Two LSTM networks connected in opposite directions	Contains reset and update gates
4	Bidirectional	No	Yes	No
5	Long Term Memory	Yes	Yes	No
6	Complexity	More Complex	More Complex	Less Complex
7	Training Time	Longer	Longer	Faster

A comparison of various Recurrent Neural Network algorithms, including LSTM, BiLSTM, and GRU are shown in Table 4 (M. Tarwani & Edem, 2017), is presented, highlighting their use in analyzing time series data.

Generative Adversarial Network

GANs are neural networks for generative modeling, creating new samples based on the input data. They use adversarial learning which has both generator and discriminator. Generator creates synthetic samples based on random input, and a discriminator distinguishes between generated and real data (Aggarwal et al., 2021). Fine-tuning the generator hyperparameters aims for optimal results. Convergence occurs when the discriminator can't differentiate as shown in Figure 15. GANs apply across image processing, language modeling, and other domains. Challenges include interpretability, training instability, mode collapse, and overfitting (Esan, Dorcas Oladayo, Pius Adewale Owolawi, 2023; Hazra & Anjaria, 2022; Mikołajczyk, Agnieszka, 2018; M.Tarwani & Edem, 2017; K. Wang et al., 2017).

Figure 15. Structure of generative adversarial network model

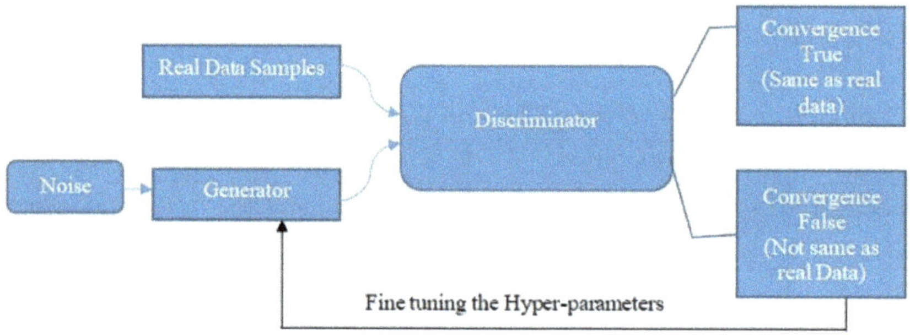

Autoencoder

Autoencoders are neural networks that compress input data into compact representations and then reconstruct them to resemble the original (Bank et al., 2020; S. Zhang et al., 2018). They map input vectors to code vectors using recognition weights and reconstruct inputs via generative weights (Chen & Guo, 2023). Common in unsupervised tasks like dimensionality reduction, feature learning, and anomaly detection, they learn representations automatically (Ashraf et al., 2021; Berahmand et al., 2024; Hazra & Anjaria, 2022; Pengzhi Li & Pei, 2023; Sarker, 2021). Autoencoders consist of an encoder and decoder as shown in Figure 16, and include types such as Sparse, Denoising, Contractive, Variational, and Conditional VAEs. Sparse Autoencoders enforce sparsity in latent space using regularization such as L1 or KL-divergence. Denoising Autoencoders learn to reconstruct clean data from noisy inputs. Contractive Autoencoders ensure robustness by penalizing the Jacobian (Bank et al., 2020). VAEs model latent variables probabilistically, while CVAEs add conditioning variables like labels (G. Dong et al., 2018). Table 5 summarizes these types.

Figure 16. Flow diagram of auto encoder

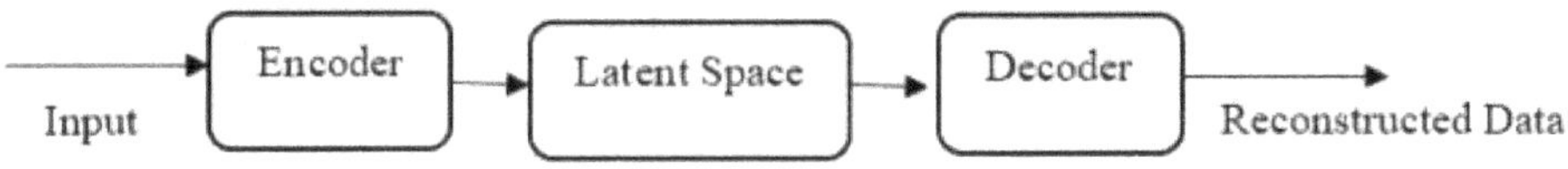

Table 5. Overview of autoencoders

S.No	Feature	Sparse Autoencoder	Denoising Autoencoder	Contractive Autoencoder	Variational Autoencoder	Conditional Variational Autoencode
1	Purpose	Encourages sparse representation in latent space	Removes noise from input data during encoding	Less sensitive to small perturbations in input	Learn data distribution in latent space	Introduce conditions into the encoder and decoder in VAE
2	Reconstruction Error	Penalizes high reconstruction error	Focuses on minimizing reconstruction error	Imposes penalty on Jacobian of the network	Minimizes reconstruction error	Minimizes a conditional lower bound on data distribution
3	Dimensionality Reduction	Can be used for dimensionality reduction	Can be used for dimensionality reduction	Emphasizes learning features less sensitive to small changes	Learns low-dimensional manifold	Learns low-dimensional latent space with specific conditions
4	Regularization	Encourages sparse representations	Adds noise to input data	Imposes penalty on Jacobian of network	Introduces latent space regularization	Introduces conditional distribution in the model
5	Latent Space	Sparse representation with few active units	Encodes clean version without noise	Encodes input to be less sensitive to perturbations.	Learns distribution of data in latent space	Learns distribution of data with specific conditions

Restricted Boltzmann Machine (RBM)

The Boltzmann Machine consists of two layers namely visible and hidden layers. All nodes in these layers are fully interconnected to learn arbitrary probability distributions across binary vectors. Boltzmann Machines encounter difficulties in training and scalability due to the intricate interactions among visible and hidden units. Restricted Boltzmann Machines (RBMs) simplify the Boltzmann Machine by limiting connections between visible and hidden layers as shown in Figure 17. RBMs are preferred for deep learning due to their simpler architecture, making them easier to train and scale. They feature a bipartite graph structure without connections within layers, facilitating straightforward modelling and serving as foundational components in deep probabilistic models (Goodfellow et al., 2016; Sarker, 2021).

Figure 17. Delineates the distinctions between Boltzmann machines and restricted Boltzmann machines

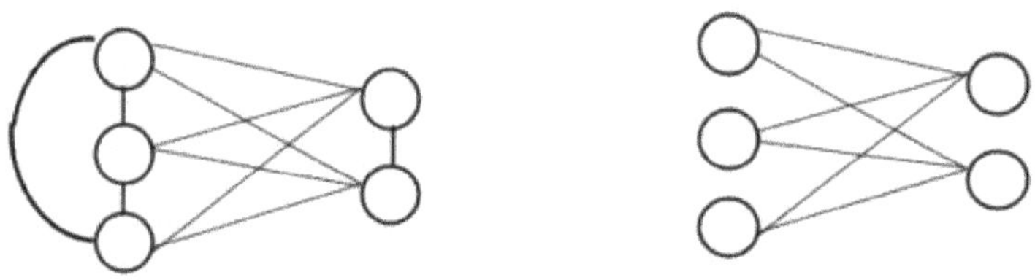

Deep Boltzmann Machine

DBM is a deep, generative model, containing multiple layers of latent variables in an undirected graphical structure. Unlike RBM's, DBMs contain multiple layers of hidden variables, and are capable of capturing complex dependencies between variables. They are typically used for tasks such as document modeling and are trained using the joint probability of variables, parameterized by an energy function. Additionally, DBM's can be trained using the multi-prediction approach, allowing for better classification performance (Goodfellow et al., 2016).

Deep Belief Network

DBN is a generative model structured with multiple layers of latent variables interconnected between adjacent layers (Hazra & Anjaria, 2022). The term "deep belief network" refers to models where the deepest layer has undirected connections, and each consecutive pair of layers has downward directed connections. Latent variables are usually binary, while visible units can be binary or real-valued. Deep Belief Networks use Annealed Importance Sampling to estimate log-likelihood, which helps assess the quality of their generative model. However, calculating log-likelihood is challenging due to inter-layer interactions and the intractable partition function in the top two undirected layers. Elementary Structure of Deep Boltzmann Machine and Deep Belief Network are shown in Figure 18.

Figure 18. Structure of deep Boltzmann machine and deep belief network

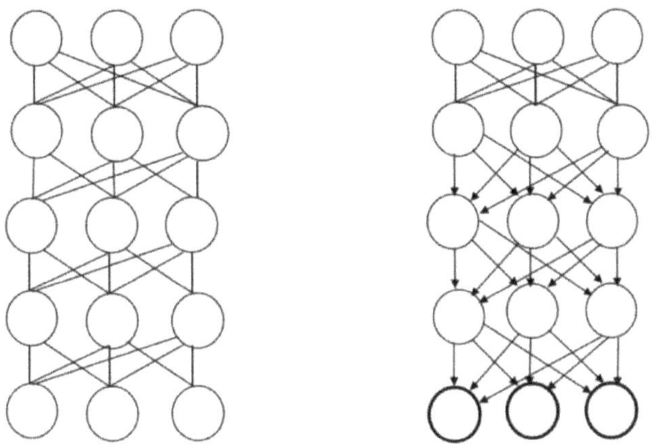

Autoencoder-Deep Belief Network (AE-DBN)

AE-DBN combines deep belief networks with autoencoders. AE are unsupervised learning models, in which it utilizes neural networks for representation learning. Stacked autoencoders are built by chaining each autoencoder's hidden layer to the next, thereby constructing a multi-layer generative model. AE-DBN aims to expedite unsupervised training via contrastive divergence, enabling hierarchical representation learning of complex data structures (Sarker, 2021).

Restricted Boltzmann Machines - Deep Belief Network (RBM – DBN)

RBM-DBN, a Deep Belief Network composed of restricted Boltzmann machines, aims to hierarchically represent input data. Its main goal is to accelerate unsupervised training using contrastive divergence for each sub-network. RBM-DBN excels at detecting deep patterns and improving reasoning abilities, particularly in distinguishing normal and erroneous data (Sarker, 2021) (Claude Sammut, n.d.).

Deep Transfer Learning (DTL)

Deep transfer learning uses knowledge from other tasks to reduce learning costs, enabling high-accuracy results with limited data and efficient ML on edge devices (Iman et al., 2023). DTL efficiently applies previously learned model knowledge to new tasks with minimal training or fine-tuning. It enables training deep neural networks with limited data and categorizes into instance-based, mapping-based, network-based, and adversarial-based approaches as shown in Figure 19. DTL techniques find utility across various domains like natural language processing, sentiment analysis, visual recognition, speech recognition, and spam filtering (Iman et al., 2023; Sarker, 2021).

Figure 19. General structure of deep transfer learning

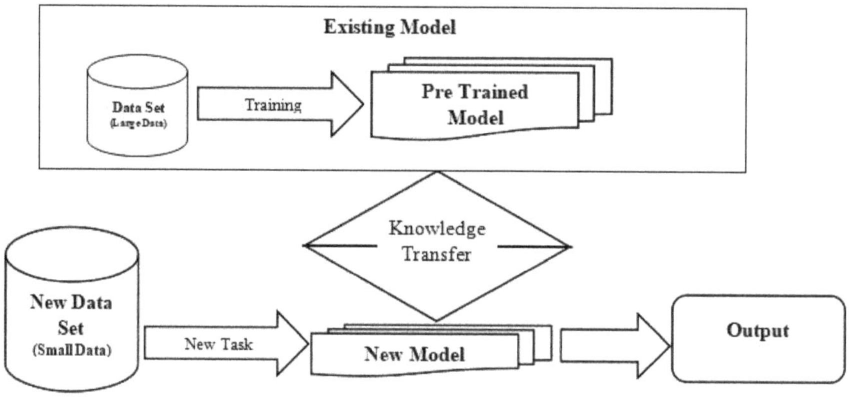

Deep Reinforcement Learning (DRL)

DRL combines reinforcement learning and deep learning to train agents through interactions with environments and feedback in the form of rewards or penalties (Haydari & Yilmaz, 2022). Deep learning is employed in DRL to handle high-dimensional data, such as images or sensor inputs, to facilitate decision-making in virtual environments as shown in Figure 20. (Arulkumaran et al., 2017; Sarker, 2021).

Figure 20. A basic structure of deep reinforcement learning

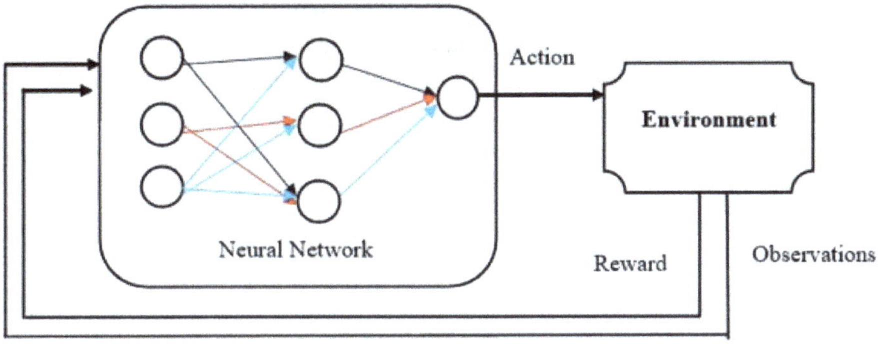

RECENT ADVANCES AND EMERGING DIRECTIONS IN DEEP LEARNING

Transformers and Attention Mechanisms

Transformers use a self-attention mechanism for prediction, which enables the model to weigh the importance of different elements in the input sequence when generating outputs. Unlike RNNs, which process data sequentially, Transformers handle the entire sequence simultaneously, allowing for parallel computation and better modeling of long-range dependencies. In the self-attention mechanism, each input token is transformed into three vectors namely Query (Q), Key (K), and Value (V). The attention scores are calculated by comparing the Query with all Keys, and these scores are used to compute a weighted sum of the Value vectors. This process allows the model to dynamically focus on the most relevant parts of the sequence during learning. Transformers have shown strong performance in engineering applications such as time-series forecasting in smart grids, machine condition monitoring, and event sequence modeling in industrial processes (Hertel et al., 2023; Wen et al., 2022).

Graph Neural Networks (GNNs)

GNNs are specialized DL models designed to operate on data structured as graphs, where nodes represent entities such as sensors, machines, or locations, and edges denote the relationships or interactions between them. Each node carries a feature vector such as temperature, voltage, or traffic density, that contributes to modeling the system's overall structure and dynamics. The core operation in GNNs is called message passing, where each node updates its feature representation by combining information from its neighbors. This enables the network to capture both local dependencies from immediate neighbors and global patterns from distant nodes via multiple layers. Different GNN architectures define this aggregation process in unique ways. Graph Convolutional Networks typically use a weighted average of neighboring node features. Graph Attention Networks enhance this process by learning dynamic attention weights, allowing the model to prioritize more relevant neighbors during aggregation. As layers are stacked, nodes gain access to increasingly wider graph contexts, allowing for the modeling of long-range dependencies and complex topological interactions. GNNs have been successfully applied in various domains, including fault detection in electrical power grids, traffic congestion prediction, and predictive maintenance of industrial equipment (Liao et al., 2021; Jiang & Luo, 2021).

Diffusion Models

Diffusion models are a class of generative models that synthesize data through a two-stage process involving a forward noising phase and a reverse denoising phase. In the forward stage, Gaussian noise is incrementally added to clean data across multiple time steps until it resembles pure noise. During training, the model learns to reverse this process by iteratively removing the noise and reconstructing the original data distribution. This reverse process is typically modeled using deep neural architectures such as U-Net or Transformer-based networks. U-Net's encoder–decoder structure with skip connections enables it to capture both local details and global context, while Transformer architectures excel at modeling long-range dependencies in structured or sequential data. Due to their iterative and probabilistic nature, diffusion models often produce more diverse and high-quality outputs than traditional generative models like GANs and VAEs. In engineering applications, they have been used for generating synthetic sensor data to address class imbalance, detecting anomalies in noisy time-series data, and modeling probabilistic environmental signals (Liu et al., 2025; Wyatt et al., 2022).

Federated Learning (FL)

Federated Learning is a decentralized machine learning paradigm that enables multiple edge devices or clients to collaboratively train a shared model without transferring their raw data to a central server. Instead, each client independently trains a local model on its private dataset and periodically transmits model updates such as gradients or parameters to a central aggregator. The server aggregates these updates, typically using a method like Federated Averaging, to refine the global model. This cycle of local training and global aggregation is repeated iteratively until the model converges. FL is particularly advantageous in scenarios where data privacy, bandwidth limitations, or regulatory constraints prevent centralized data collection. Real-world applications include privacy-preserving fault diagnostics across distributed industrial environments, decentralized learning in sensor networks, and secure predictive maintenance in sectors such as aerospace and automotive (Pruckovskaja et al., 2023; Becker et al., 2022).

Hybrid Architectures (Wavelet-CNNs, WD–LSTM)

Hybrid deep learning models that integrate wavelet decomposition with neural network architectures have demonstrated significant effectiveness in processing complex, noisy, and non-stationary signals commonly encountered in engineering systems. Wavelet decomposition transforms input signals into multiple time-frequency components, separating them into low-frequency components called approximations, that capture overall trends or long-term behaviors, and high-frequency components called details, that represent sharp changes, noise, or transient events. This separation is achieved using techniques like the Discrete Wavelet Transform, which applies a pair of filters—a low-pass filter for extracting approximations and a high-pass filter for extracting details—followed by downsampling. This enables the model to analyze both slow-changing and rapidly fluctuating information across various scales. In Wavelet-CNN models, the decomposed sub-signals are fed into convolutional layers to extract localized spatial patterns within each frequency band. Alternatively, WD–LSTM architectures apply wavelet transforms first and then model temporal dependencies across the decomposed signals using Long Short-Term Memory networks. This multi-resolution fusion not only improves the model's robustness to noise but also enhances its ability to learn hierarchical and temporal features. These hybrid approaches have been effectively applied in domains such as vibration-based fault detection, energy load forecasting, and seismic signal interpretation (Frizzo-Stefenon et al., 2023).

Self-Supervised and Foundation Models

Self-supervised learning is a paradigm that enables models to learn rich and informative representations from unlabeled data by solving pretext tasks designed to uncover underlying structure in the input. These tasks might involve predicting missing parts of a signal, reconstructing masked data, or solving spatial and temporal relationships, thereby encouraging the model to understand the intrinsic patterns in the data without relying on manual annotations. Once pre-trained in this manner, the model can be fine-tuned with a relatively small amount of labeled data for specific applications. Foundation models, on the other hand, are large-scale architectures trained on diverse and massive datasets to develop broadly transferable knowledge. These models, once trained, can be adapted to a wide range of downstream tasks, including those in engineering domains, with minimal additional training. This makes them particularly effective in low-resource or cross-domain scenarios. In engineering contexts, SSL and foundation models have been used for extracting meaningful insights from unlabeled sensor logs, facilitating transfer learning where annotated data is limited, and enabling cross-modal learning across modalities such as text, sensor streams, and visual data (Zhang et al., 2023; Gui et al., 2023). Key features of emerging deep learning techniques are presented in Table 6.

Table 6. Key features of emerging deep learning techniques

S.No	DL Technique	Features	Applications	Key References
1	Transformers	Self-attention, parallel processing, long-range dependency modeling	Time series forecasting, fault diagnosis	Wen et al., 2022; Hertel et al., 2023
2	Graph Neural Networks	Message passing, neighborhood aggregation, attention on graph edges	Traffic prediction, power grid monitoring	Liao et al., 2021; Jiang & Luo, 2021
3	Diffusion Models	Noise injection and removal, probabilistic generation, high-quality synthesis	Sensor data simulation, anomaly detection	Liu et al., 2025; Wyatt et al., 2022
4	Federated Learning	Decentralized training, privacy-preserving, model aggregation	Predictive maintenance, industrial IoT	Becker et al., 2022; Pruckovskaja et al., 2023
5	Hybrid Architectures	Wavelet decomposition, multi-resolution feature fusion	Vibration fault detection, load forecasting	Frizzo-Stefenon et al., 2023
6	Self-Supervised/ Foundation Models	Pretext tasks, transferability, minimal labeled data required	Sensor log analysis, cross-modal learning	Gui et al., 2023; Zhang et al., 2023

CONCLUSION

Deep learning technology is playing an increasingly important role in perception, prediction, simulation, and control of various field's by providing powerful tools and techniques to enhance the efficiency of the model. This survey offers a systematic and in-depth examination of deep learning technology, which is recognized as a pivotal element of artificial intelligence and data science. The discussion starts with machine learning fundamentals, progresses to deep learning, covers data handling, preprocessing, and feature management, followed by an exploration of supervised, unsupervised, and hybrid deep learning techniques.

In the supervised learning section, we begin with a basic multilayer perceptron, detailing its input, hidden, and output layers. We then move on to convolutional neural networks for efficient feature extraction from grid-like data such as images and videos. The discussion covers CNN components, starting with the input layer, progressing to convolutional layers, various pooling methods, flattening layers, fully connected layers, and global average pooling layers. Convolutional layers utilize kernels to effectively extract features from the input data. To extract the most relevant information and reduce computational complexity and spatial dimensions, pooling layers are employed. We then address the fully connected layer, noting that its major challenges of time and cost are alleviated by the Global Average Pooling layer. Activation functions are emphasized for their crucial role in capturing complex hierarchical patterns in deep learning; without them, neural networks would behave like linear models. Next, we delve into Recurrent Neural Networks, which process sequential or time-series data through feedback loops, allowing each neuron to retain previous information via memory cells and various types of gates. To address long-term dependency issues, we explore LSTM, Bi-LSTM, and GRU. Generative Adversarial Networks are discussed for generating synthetic data, and various types of autoencoders are examined for tasks such as dimensionality reduction, data denoising, anomaly detection, feature learning, and data generation. Following that, investigation is done on various models, starting with the Restricted Boltzmann Machine for unsupervised learning, which effectively extracts features and reduces dimensionality. Next, we explored the Deep Boltzmann Machine, which learns complex data distributions through energy-based methods and hierarchical latent variables. The Deep Belief Network was discussed, showcasing its use of stacked RBMs for robust unsupervised learning and feature extraction. We also delved into Deep Transfer Learning, which enhances performance by adapting pre-trained models to new tasks through knowledge transfer between related domains. Additionally, Deep Reinforcement Learning was covered, illustrating how it enables agents to learn optimal decision-making strategies through environmental interactions. Our discussion also extended to recent deep learning trends such as Transformers, GNNs, and diffusion models, showcasing their applicability across various engineering domains.

REFERENCES

Aggarwal, A., Mittal, M., & Battineni, G. (2021). Generative adversarial network: An overview of theory and applications. *International Journal of Information Management Data Insights*, *1*(1), 100004.DOI: 10.1016/j.jjimei.2020.100004

Al-Sabaawi, A., Muayad Ibrahim, H., & Mohsin, Z., & sabah Alzubaidi, L. (2020). Amended Convolutional Neural Network with Global Average Pooling for Image Classification. *20th International Conference Intelligent Systems Design and Applications.*

Alzubaidi, L., Zhang, J., Humaidi, A. J., Al-Dujaili, A., Duan, Y., Al-Shamma, O., Santamaría, J., Fadhel, M. A., Al-Amidie, M., & Farhan, L. (2021). Review of deep learning: Concepts, CNN architectures, challenges, applications, future directions. *Journal of Big Data*, *8*(1), 53. DOI: 10.1186/s40537-021-00444-8 PMID: 33816053

Andreas Kamilaris, F. X. P.-B. (2018). Deep learning in agriculture: A survey. *Computers and Electronics in Agriculture*, *147*, 70–90. DOI: 10.1016/j.compag.2018.02.016

Arulkumaran, K., Deisenroth, M. P., Brundage, M., & Bharath, A. A. (2017). A Brief Survey of Deep Reinforcement Learning. *IEEE Signal Processing Magazine*, *34*(6), 26–38. DOI: 10.1109/MSP.2017.2743240

ArunKumar, K. E., Kalaga, D. V., Mohan Sai Kumar, C., Kawaji, M., & Brenza, T. M.ArunKumar. (2022). Comparative analysis of Gated Recurrent Units (GRU), long Short-Term memory (LSTM) cells, autoregressive Integrated moving average (ARIMA), seasonal autoregressive Integrated moving average (SARIMA) for forecasting COVID-19 trends. *Alexandria Engineering Journal*, *61*(10), 7585–7603. DOI: 10.1016/j.aej.2022.01.011

Ashraf, J., Bakhshi, A. D., Moustafa, N., Khurshid, H., Javed, A., & Beheshti, A. (2021). Novel Deep Learning-Enabled LSTM Autoencoder Architecture for Discovering Anomalous Events from Intelligent Transportation Systems. *IEEE Transactions on Intelligent Transportation Systems*, *22*(7), 4507–4518. DOI: 10.1109/TITS.2020.3017882

Bank, D., Koenigstein, N., & Giryes, R. (2020). *Autoencoders. Machine Learning for Data Science Handbook: Data Mining and Knowledge Discovery Handbook.* Springer International Publishing.

Becker, S., Styp-Rekowski, K., Stoll, O. V. L., & Kao, O. (2022). *Federated learning for autoencoder-based condition monitoring in the industrial internet of things.* arXiv. https://arxiv.org/abs/2206.08187 DOI: 10.1109/BigData55660.2022.10020836

Berahmand, K., Daneshfar, F., Salehi, E. S., Li, Y., & Xu, Y. (2024). Autoencoders and their applications in machine learning: A survey. *Artificial Intelligence Review*, *57*(2), 28. DOI: 10.1007/s10462-023-10662-6

Bikku, T. (2020). Multi-layered deep learning perceptron approach for health risk prediction. *Journal of Big Data*, *7*(1), 50. DOI: 10.1186/s40537-020-00316-7

Bikku, T., & Satya Sree, K. P. N. V. (2020). DEEP LEARNING APPROACHES FOR CLASSIFYING DATA: A REVIEW. *Journal of Engineering Science and Technology*, *15*(4), 2580–2594.

Chen, S., & Guo, W. (2023). Auto-Encoders in Deep Learning—A Review with New Perspectives. In *Mathematics* (Vol. 11, Issue 8). MDPI. DOI: 10.3390/math11081777

Chetana, V. L., Kolisetty, S. S., & Amogh, K. (2020). A Short Survey of Dimensionality Reduction Techniques. In *Recent Advances in Computer Based Systems, Processes and Applications* (pp. 3–14). CRC Press. DOI: 10.1201/9781003043980-2

Claude Sammut, G. I. W. (n.d.). *Encyclopedia of Machine Learning*. Springer-Verlag New York Inc.

Dabiri, S., Heaslip, K. P., Abbas, M. M., Gerdes, R. M., & Ramakrishnan, N. (2018). *Application of Deep Learning in Intelligent Transportation Systems.*

Dong, G., Liao, G., Liu, H., & Kuang, G. (2018). A Review of the Autoencoder and Its Variants: A Comparative Perspective from Target Recognition in Synthetic-Aperture Radar Images. *IEEE Geoscience and Remote Sensing Magazine*, *6*(3), 44–68. DOI: 10.1109/MGRS.2018.2853555

Dong, S., Wang, P., & Abbas, K. (2021a). A survey on deep learning and its applications. In *Computer Science Review* (Vol. 40). Elsevier Ireland Ltd., DOI: 10.1016/j.cosrev.2021.100379

Esan, D. O., & Owolawi, P. A. C. T. (2023). Generative Adversarial Networks: Applications, Challenges, and Open Issues. *IntechOpen*. https://doi.org/http://dx.doi.org/10.5772/intechopen.113098

Frizzo-Stefenon, S., Seman, L., & Mariani, V. (2023). Machine fault detection using a hybrid CNN-LSTM attention-based model. *Sensors (Basel)*, *23*(9), 4512. DOI: 10.3390/s23094512 PMID: 37177716

Gholamalinezhad, H., & Khosravi, H. (2020). Pooling Methods in Deep Neural Networks, a Review. *ArXiv Preprint ArXiv:2009.07485*.

Goodfellow, I., Bengio, Y., & Courville, A. (2016). *Deep Learning*. MIT Pres.

Gui, J., Chen, T., Zhang, J., Cao, Q., Sun, Z., Luo, H., & Tao, D. (2023). A survey on self-supervised learning: Algorithms, applications, and future trends. *arXiv*. https://arxiv.org/abs/2302.00487

Haghighat, A. K., Ravichandra-Mouli, V., Chakraborty, P., Esfandiari, Y., Arabi, S., & Sharma, A. (2020). Applications of Deep Learning in Intelligent Transportation Systems. *Journal of Big Data Analytics in Transportation*, *2*(2), 115–145. DOI: 10.1007/s42421-020-00020-1

Hatcher, W. G., & Yu, W. (2018). A Survey of Deep Learning: Platforms, Applications and Emerging Research Trends. *IEEE Access : Practical Innovations, Open Solutions*, *6*, 24411–24432. DOI: 10.1109/ACCESS.2018.2830661

Haydari, A., & Yilmaz, Y. (2022). Deep Reinforcement Learning for Intelligent Transportation Systems: A Survey. *IEEE Transactions on Intelligent Transportation Systems*, *23*(1), 11–32. DOI: 10.1109/TITS.2020.3008612

Hazra, T., & Anjaria, K. (2022). Applications of game theory in deep learning: A survey. *Multimedia Tools and Applications*, *81*(6), 8963–8994. DOI: 10.1007/s11042-022-12153-2 PMID: 35496996

Hertel, L., de Prado, M. T., & Ferrara, E. (2023). Transformer training strategies for forecasting multiple load time series. *IEEE Transactions on Smart Grid.*

Hochreiter, S., Bengio, Y., Frasconi, P., & Schmidhuber, J. (2001). Gradient Flow in Recurrent Nets: the Difficulty of Learning Long-Term Dependencies. *A Field Guide to Dynamical Recurrent Neural Networks. IEEE Press, 1.*

Iman, M., Arabnia, H. R., & Rasheed, K. (2023). A Review of Deep Transfer Learning and Recent Advancements. In *Technologies* (Vol. 11, Issue 2). MDPI. DOI: 10.3390/technologies11020040

Janiesch, C., Zschech, P., & Heinrich, K. (2021). Machine learning and deep learning. *Electronic Markets*, *31*(3), 685–695. DOI: 10.1007/s12525-021-00475-2

Jia, W., Sun, M., Lian, J., & Hou, S. (2022). Feature dimensionality reduction : A review. *Complex & Intelligent Systems*, *8*(3), 2663–2693. DOI: 10.1007/s40747-021-00637-x

Jiang, Z., & Luo, F. (2021). Graph neural networks for traffic forecasting: A survey. *IEEE Transactions on Intelligent Transportation Systems*, *22*(11), 7028–7046.

Jung, S., & Yeon, M. (2021). Comparison of Deep Learning Techniques for River Streamflow Forecasting. *IEEE Access : Practical Innovations, Open Solutions*, *9*, 71805–71820. DOI: 10.1109/ACCESS.2021.3077703

Kamalov, F., Nazir, A., Safaraliev, M., Cherukuri, A. K., & Zgheib, R. (2021). Comparative analysis of activation functions in neural networks. *2021 28th IEEE International Conference on Electronics, Circuits, and Systems, ICECS 2021 - Proceedings*. DOI: 10.1109/ICECS53924.2021.9665646

Koresh, H. J. D. (2024). Impact of the preprocessing steps in deep learning-based image classifications. *National Academy Science Letters*, *47*(6), 1–3. DOI: 10.1007/s40009-023-01372-2

Latif, S., Cuayáhuitl, H., Pervez, F., Shamshad, F., Ali, H. S., & Cambria, E. (2023). A survey on deep reinforcement learning for audio-based applications. *Artificial Intelligence Review*, *56*(3), 2193–2240. DOI: 10.1007/s10462-022-10224-2

Le, X. H., Ho, H. V., Lee, G., & Jung, S. (2019). Application of Long Short-Term Memory (LSTM) neural network for flood forecasting. *Water (Basel)*, *11*(7), 1387. DOI: 10.3390/w11071387

Li, H., Li, J., Guan, X., Liang, B., Lai, Y., & Luo, X. (2019). Research on Overfitting of Deep Learning. *Proceedings - 2019 15th International Conference on Computational Intelligence and Security, CIS 2019*, 78–81. DOI: 10.1109/CIS.2019.00025

Li, P., & Pei, Y. J. L. (2023). A comprehensive survey on design and application of autoencoder in deep learning. *Applied Soft Computing, Elsevier*, 110176.

Li, Y. H., Harfiya, L. N., Purwandari, K., & Der Lin, Y. (2020). Real-time cuffless continuous blood pressure estimation using deep learning model. *Sensors (Basel)*, *20*(19), 1–19. DOI: 10.3390/s20195606 PMID: 33007891

Liao, Q.. (2021). A review of graph neural networks in power systems. *IEEE Transactions on Power Systems*, *36*(6), 5024–5039.

Liu, C., Tang, D., Zhu, H., & Nie, Q. (2021). A novel predictive maintenance method based on deep adversarial learning in the intelligent manufacturing system. *IEEE Access : Practical Innovations, Open Solutions*, *9*, 49557–49575. DOI: 10.1109/ACCESS.2021.3069256

Liu, H., & Lang, B. (2019). Machine learning and deep learning methods for intrusion detection systems: A survey. *Applied Sciences (Basel, Switzerland)*, *9*(20), 4396. DOI: 10.3390/app9204396

Liu, X., & Zhang, Z.. (2025). A survey on diffusion models for anomaly detection. *IEEE Transactions on Neural Networks and Learning Systems*.

Merz, M., & Actuarial, S. (2023). Recurrent Neural Networks. *StatisticalFoundationsofActuarialLearninganditsApplications,Springer, Actuarial*, 381–406.

Mikołajczyk. Agnieszka, M. G. (2018). Data augmentation for improving deep learning in image classification problem. *IEEE*, 117–122.

Minsky, M. (1961). Steps Toward Artificial Intelligence*. *PROCEEDINGS OF THE IRE*, 8–29.

Mungoli, N. (2023a). Advancements in Deep Learning: A Comprehensive Study of the Latest Trends and Techniques in Machine Learning. *International Journal of Advanced Engineering Technologies and Innovations*, *04*, 43–64.

Nanga, S., Bawah, A. T., Acquaye, B. A., Billa, M., Baeta, F. D., Odai, N. A., Obeng, S. K., & Nsiah, A. D. (2021). Review of Dimension Reduction Methods. *Journal of Data Analysis and Information Processing. Journal of Data Analysis and Information Processing*, *9*(3), 189–231. DOI: 10.4236/jdaip.2021.93013

Nguyen, H., Kieu, L. M., Wen, T., & Cai, C. (2018). Deep learning methods in transportation domain: A review. *IET Intelligent Transport Systems*, *12*(9), 998–1004. DOI: 10.1049/iet-its.2018.0064

Nisha, C. M., & Thangarasu, N. (2023a). Deep learning algorithms and their relevance: A review. *International Journal of Data Informatics and Intelligent Computing*, *2*(4), 1–10. DOI: 10.59461/ijdiic.v2i4.78

Nisha, C. M., & Thangarasu, N. (2023b). Deep learning algorithms and their relevance: A review. *International Journal of Data Informatics and Intelligent Computing*, *2*(4), 1–10. DOI: 10.59461/ijdiic.v2i4.78

O'Shea, K., & Nash, R. (2015). An Introduction to Convolutional Neural Networks. *ArXiv Preprint ArXiv:1511.08458.*

Obaidur Rahman, M., Sabir Hossain, M., Shafiul Alam Forhad, M., Junaid, T.-S., & Kamal Hossen, M. (2019). Predicting Prices of Stock Market using Gated Recurrent Units (GRUs) Neural Networks. *IJCSNS International Journal of Computer Science and Network Security, 19*(1).

Pothuganti, S. (2018). Review on over-fitting and under-fitting problems in Machine Learning and solutions. *Article in International Journal of Advanced Research in Electrical Electronics and Instrumentation Engineering*, *7*(9). Advance online publication. DOI: 10.15662/IJAREEIE.2018.0709015

Pruckovskaja, V., Weißenfeld, A., & Heistracher, C.. (2023). *Federated learning for predictive maintenance and quality inspection in industrial applications. Proceedings of PHM Europe 2023.* https://arxiv.org/abs/2304.11101 DOI: 10.1109/PHM58589.2023.00064

Purwono, Ma'arif, A., Rahmaniar, W., Fathurrahman, H. I. K., Frisky, A. Z. K., & Haq, Q. M. U. (2022). Understanding of Convolutional Neural Network (CNN): A Review. *International Journal of Robotics and Control Systems, 2*(4), 739–748. DOI: 10.31763/ijrcs.v2i4.888

Qiu, L., Zhang, D., Tian, Y., & Al-Nabhan, N. (2021). Deep learning-based algorithm for vehicle detection in intelligent transportation systems. *The Journal of Supercomputing*, *77*(10), 11083–11098. DOI: 10.1007/s11227-021-03712-9

Sarker, I. H. (2021). Deep Learning: A Comprehensive Overview on Techniques, Taxonomy, Applications and Research Directions. In *SN Computer Science* (Vol. 2, Issue 6). Springer. DOI: 10.1007/s42979-021-00815-1

Sharma, S., Sharma, S., & Athaiya, A. (2020). Activation Functions in Neural Networks. *International Journal of Engineering Applied Sciences and Technology*, *4*(12), 310–316. DOI: 10.33564/IJEAST.2020.v04i12.054

Shorten, C., & Khoshgoftaar, T. M. (2019). A survey on Image Data Augmentation for Deep Learning. *Journal of Big Data*, *6*(1), 60. DOI: 10.1186/s40537-019-0197-0

Sundaram, K. M., Hussain, A., Sanjeevikumar, P., Holm-Nielsen, J. B., Kaliappan, V. K., & Santhoshi, B. K. (2021). Deep Learning for Fault Diagnostics in Bearings, Insulators, PV Panels, Power Lines, and Electric Vehicle Applications - The State-of-the-Art Approaches. In *IEEE Access* (Vol. 9, pp. 41246–41260). Institute of Electrical and Electronics Engineers Inc., DOI: 10.1109/ACCESS.2021.3064360

Suriya, M., Chandran, V., & Sumithra, M. G. (2022). Enhanced deep convolutional neural network for malarial parasite classification. *International Journal of Computers and Applications*, *44*(12), 1113–1122. DOI: 10.1080/1206212X.2019.1672277

Talaei Khoei, T., Ould Slimane, H., & Kaabouch, N. (2023). Deep learning: systematic review, models, challenges, and research directions. In *Neural Computing and Applications* (Vol. 35, Issue 31, pp. 23103–23124). Springer Science and Business Media Deutschland GmbH. DOI: 10.1007/s00521-023-08957-4

Tarwani, M., K., & Edem, S. (. (2017). Survey on Recurrent Neural Network in Natural Language Processing. *International Journal of Engineering Trends and Technology*, *48*(6), 301–304. DOI: 10.14445/22315381/IJETT-V48P253

Taye, M. M. (2023). Understanding of machine learning with deep learning: Architectures, workflow, applications and future directions. *Computers*, *12*(5), 91. DOI: 10.3390/computers12050091

Teuwen, J., & Moriakov, N. (2019). Convolutional neural networks. In *Handbook of Medical Image Computing and Computer Assisted Intervention* (pp. 481–501). Elsevier., DOI: 10.1016/B978-0-12-816176-0.00025-9

Van Houdt, G., Mosquera, C., & Nápoles, G. (2020). A review on the long short-term memory model. *Artificial Intelligence Review, 53*(8), 5929–5955. DOI: 10.1007/s10462-020-09838-1

Velliangiri, S., Alagumuthukrishnan, S., & Thankumar Joseph, S. I. (2019). A Review of Dimensionality Reduction Techniques for Efficient Computation. *Procedia Computer Science, 165*, 104–111. DOI: 10.1016/j.procs.2020.01.079

Veres, M., & Moussa, M. (2020). Deep Learning for Intelligent Transportation Systems: A Survey of Emerging Trends. In *IEEE Transactions on Intelligent Transportation Systems* (Vol. 21, Issue 8, pp. 3152–3168). Institute of Electrical and Electronics Engineers Inc. DOI: 10.1109/TITS.2019.2929020

Walz, W. (2023). Machine Learning for Brain Disorders Series Editor. *Humana Press*.

Wang, K., Gou, C., Duan, Y., Lin, Y., Zheng, X., & Wang, F. (2017). Generative Adversarial Networks : Introduction and Outlook. *IEEE/CAA Journal of Automatica Sinica, 4*(4), 588–598.

Wang, Y., Addepalli, S., & Zhao, Y. (2020). Recurrent neural networks and its variants in remaining useful life prediction. *IFAC-PapersOnLine, 53*(3), 137–142. DOI: 10.1016/j.ifacol.2020.11.022

Wang, Y., Zhang, D., Liu, Y., Dai, B., & Lee, L. H. (2019). Enhancing transportation systems via deep learning: A survey. In *Transportation Research Part C: Emerging Technologies* (Vol. 99, pp. 144–163). Elsevier Ltd. DOI: 10.1016/j.trc.2018.12.004

Wen, Y., Li, Y., Ye, Y., Dai, Z., & Wang, H. (2022). Transformers in time series: A survey. *Journal of Systems Science and Complexity*. https://arxiv.org/abs/2202.07125

Wyatt, M., Marino, J., & Kingma, D.. (2022). *AnoDDPM: Anomaly detection with denoising diffusion probabilistic models*. NeurIPS 2022 Workshop on Anomaly Detection. https://arxiv.org/abs/2206.08545

Xie, C., Yao, R., Liu, Z., & Zhu, L. X. C. (2022). Process performance prediction based on spatial and temporal feature extraction through bidirectional LSTM. *Computer Aided Chemical Engineering, 49*, 1615–1620. https://doi.org/https://doi.org/10.1016/B978-0-323-85159-6.50269-4

Yamashita, R., Nishio, M., Do, R. K. G., & Togashi, K. (2018). Convolutional neural networks: an overview and application in radiology. In *Insights into Imaging* (Vol. 9, Issue 4, pp. 611–629). Springer Verlag. DOI: 10.1007/s13244-018-0639-9

Yu, D., Wang, H., Chen, P., & Wei, Z. (2014). Mixed pooling for convolutional neural networks. *The 9th International Conference on Rough Sets and Knowledge Technology (RSKT'14), Springer*, 364–375. DOI: 10.1007/978-3-319-11740-9_34

Zafar, A., Aamir, M., Mohd Nawi, N., Arshad, A., Riaz, S., Alruban, A., Dutta, A. K., & Almotairi, S. (2022). A Comparison of Pooling Methods for Convolutional Neural Networks. *Applied Sciences (Basel, Switzerland)*, *12*(17), 8643. DOI: 10.3390/app12178643

Zhang, H., Zhang, L., & Jiang, Y. (2019). Overfitting and Underfitting Analysis for Deep Learning Based End-to-end Communication Systems. *11th International Conference on Wireless Communications and Signal Processing (WCSP)*, 1–6. DOI: 10.1109/WCSP.2019.8927876

Zhang, K., Song, W., Liu, L., Zhao, X., & Du, C. (2019). *Bidirectional Long Short-Term Memory for Sentiment Analysis of Chinese Product Reviews*. IEEE. DOI: 10.1109/ICEIEC.2019.8784560

Zhang, K., Wen, Q., Zhang, C., Cai, R., Jin, M., Liu, Y., Zhang, J., Liang, Y., Pang, G., Song, D., & Pan, S. (2023). Self-supervised learning for time series analysis: Taxonomy, progress, and prospects. *IEEE Transactions on Pattern Analysis and Machine Intelligence*. DOI: 10.1109/TPAMI.2023.3264798 PMID: 38598381

Zhang, S., Zhai, J., Chen, J., & He, Q. (2018). Autoencoder and its various variants. *2018 IEEE International Conference on Systems, Man, and Cybernetics (SMC)*, 415–419. DOI: 10.1109/SMC.2018.00080

Chapter 2
Simplifying Data With Robust Variable Selection for Enhanced Interpretability and Energy Efficiency

Jan Kalina
https://orcid.org/0000-0002-8491-0364
Institute of Computer Science, The Czech Academy of Sciences, Czech Republic

ABSTRACT

Machine learning, a cornerstone of artificial intelligence, equips engineers and scientists with powerful tools for uncovering valuable insights from complex, large-scale datasets across diverse applications. Engineering applications increasingly demand transparency and interpretability, particularly to understand the reasoning behind specific outcomes and decisions. This chapter addresses this need by exploring dimensionality reduction techniques that enhance model interpretability by focusing on the most significant variables. By employing robust, outlier-resistant methods, we enable a closer analysis of how individual data points impact results, making these approaches ideal for real-world data with inherent noise or contamination. The utility of these techniques is demonstrated on two datasets, one from gene expression analysis and the other from credit scoring. Furthermore, we discuss the relationship between dimensionality reduction and energy-efficient approximate computation, highlighting its relevance for engineering applications in AI.

DOI: 10.4018/979-8-3373-0746-6.ch002

INTRODUCTION

While grand, futuristic concepts of artificial intelligence shaping humanity's transition to a knowledge-driven society (Tegmark, 2017) may still belong more to the realm of science fiction for the time being, it is undeniable that AI holds significant potential to exert a profound impact on global society in the near future. When AI is tasked with extracting practical and actionable knowledge from available data while considering uncertainty, machine learning with its statistical models becomes essential. Today, numerous innovative machine learning techniques are accessible for uncovering valuable insights from data, including large datasets, across various applications. The integration of high-performance computing (HPC) could further improve the ability to implement complex algorithms more efficiently. For instance, several AI technologies have been actively employed in combating the COVID-19 pandemic (Lalmuanawma et al., 2020). Additionally, the rapidly advancing field of scientific computing, which utilizes cutting-edge technology to solve scientific challenges, is often referred to as computational science (Holder & Eichholz, 2019). The growing prominence of computational science has also garnered increased attention from professionals in the field (Quarteroni, 2018).

Although commonly used machine learning techniques in scientific computing are often considered "black boxes," many practical applications require a deeper understanding of how and why specific conclusions (such as decisions) are reached. This drives the growing need for explainable artificial intelligence and highlights the increasing focus on explainable machine learning, which aims to interpret data in a more understandable manner. Understanding the limitations of AI is also a crucial ethical concern, as the inability to rigorously explain why certain algorithms produce specific outcomes presents an important ethical dilemma (Jacobson et al., 2020). Two key strategies (or objectives) for enhancing the transparency of machine learning models, which are often used together, are dimensionality reduction and improving robustness to outliers. The latter approach helps minimize the impact of individual outliers and enables the evaluation of the influence of specific data points. Data-driven approaches to dimensionality reduction methods can contribute to explainability if they show how the most important features of the data are chosen and how they help simplify the information (Tian et al., 2025).

The diverse types and formats of data available today necessitate a wide range of advanced machine learning techniques for effective analysis. A key challenge is properly analyzing data with a large number of variables (features), which has become an increasingly important issue. As the complexity of data continues to rise, data analysts are recognizing the importance of using complexity reduction methods to ensure reliable and efficient analysis (Fordellone, 2019). The volume of data in various domains is growing rapidly, and it is widely acknowledged that complex data holds significant potential to impact research and everyday activities (López-Robles et al., 2019). Complexity reduction refers to any strategy that simplifies data analysis, which may involve extracting meaningful features from images (such as brain scans, as seen in Bučková et al. (2020)), voice recordings, narrative text (e.g., health reports), and more. Dimensionality reduction, a specific form of complexity reduction, is particularly relevant for numerical data, i.e., data represented by numbers. Naturally, dimensionality reduction methods are also highly applicable to Big Data, which is considered a valuable resource with significant untapped potential for decision-making and societal relevance (Bradlow et al., 2017).

In engineering, the need for advanced techniques that efficiently handle complex, high-dimensional data is more pressing than ever. As industries increasingly rely on data-driven decision-making, the integration of machine learning with engineering applications has become a vital tool for optimizing systems, improving design processes, and enhancing operational efficiency. Engineers are tasked with not only extracting actionable insights from large datasets but also ensuring that the results are interpretable, transparent, and resilient to noise or outliers. This is where methods like dimensionality reduction come into play, providing a means to simplify complex data without losing critical information, ultimately facilitating better decision-making and fostering innovation in engineering solutions.

This chapter aims to outline the principles of robust dimensionality reduction, highlight some promising methods, and provide examples using real-world data that demonstrate the reliability of these techniques. Throughout the chapter, robust dimensionality reduction approaches are presented as a distinctive methodology for creating explainable (understandable) data analyses using machine learning. Specifically, reducing dimensionality enhances the interpretation of the insights gained from data by emphasizing the most important variables. While outliers can disproportionately influence and potentially distort standard data analysis, robust statistical methods can mitigate their impact. Additionally, the relationship between dimensionality reduction and low-energy approximate computation is explored. The chapter will showcase robust methods designed for contaminated data through classification and regression tasks, illustrated by three real-world datasets: gene expression measurements, neighborhood crime rates, and tourism infrastructure data.

TYPES OF DIMENSIONALITY REDUCTION METHODS

Dimensionality reduction is commonly recommended in various applications to simplify the analysis of complex multivariate data by removing the impact of redundant variables (Wilson, 2018). Throughout this chapter, we will assume the availability of n observations (samples, measurements). Dimensionality reduction is widely used in machine learning and is particularly essential for Big Data (Blazques & Domenech, 2018) or high-dimensional datasets, where the number of variables often exceeds n significantly. Without dimensionality reduction, one would be forced to rely on computationally intensive methods. However, it's important to note that dimensionality reduction can also be advantageous for datasets with relatively fewer variables (Martinez et al., 2017).

In general, dimensionality reduction offers several benefits, such as simplifying subsequent computations. Parsimonious models, which focus on a small set of relevant variables, help enhance the interpretability of the results, for instance, by clustering variables into homogeneous groups. Some methods even reduce or eliminate correlations between variables. Additionally, dimensionality reduction may improve the performance of subsequent analyses (e.g., decision-making) when compared to using the full dataset. However, it's important to note that dimensionality reduction can lead to the loss of relevant information if the set of variables is reduced too much, potentially resulting in biased outcomes.

We distinguish between two main types of dimensionality reduction methods: supervised and unsupervised. Supervised methods are designed for datasets that consist of two or more groups, where information about group membership is utilized. In contrast, unsupervised methods are applied when the data are from a single group. When the data belong to multiple groups, but the group labels are unknown (even in the training dataset), unsupervised methods must be used. Using an unsupervised approach for data from known groups is suboptimal.

Furthermore, we differentiate between feature extraction, which creates new combinations of variables, and variable selection (or feature selection), which identifies a smaller subset of important variables while discarding the rest.

Feature extraction methods aim to identify the most relevant (linear or nonlinear) combinations of measurements, replacing the original dataset with a smaller set of these combinations. Feature extraction is always performed as a preprocessing step, before the actual data analysis begins. Common methods include principal component analysis (PCA), factor analysis, correspondence analysis, independent component analysis, partial least squares regression, multivariate scaling (Greene, 2017), and techniques based on information theory. Unlike variable selection, feature extraction generally ensures that variables become decorrelated, provides reliable predictions (especially when the data is not affected by outliers), and yields results with strong local robustness (stability). It's important to note that PCA, as one of the most widely used dimensionality reduction methods, is an unsupervised technique and, therefore, is not appropriate for datasets containing two or more distinct groups (i.e., mixtures of populations). While PCA is often praised for its ability to identify decorrelated components (Kalina & Rensová, 2015), many machine learning algorithms do not require data to be uncorrelated.

Variable selection methods can be classified into three categories:

- Prior (a priori, preliminary), i.e. before the data analysis of interest (regression, classification, clustering, instrumental variables estimation, etc.),
- Intrinsic (sparse), using a suitable regularization within the data analysis of interest,
- Posterior (post hoc analysis), i.e. after the data analysis of interest.

Some methods can be used both prior to and after the main analysis. For instance, hypothesis testing and various forms of backward elimination can be performed before or after the analysis. However, it's important to recognize that hypothesis testing for dimensionality reduction cannot be applied directly, as repeated use of the same procedure increases the likelihood of a Type I error.

Prior variable selection is a preparatory or supportive step performed before the main analysis task (e.g., regression modeling or learning a classification rule). It enables the analysis of a smaller set of relevant variables, facilitating the interpretation of individual parameters and identifying which variables are most important for predicting the response. By selecting relevant variables before the analysis, future measurements under similar conditions can focus on fewer variables, potentially reducing the financial or organizational costs of experiments (or measurements). Various variable selection methods that can ignore redundant variables include wrappers, filters, embedded methods, the minimum redundancy maximum relevance approach, and techniques based on information theory. A key group of tools for this purpose includes hypothesis testing, which may require distributional assumptions (as in specific regression models) or can be carried out without such assumptions (e.g., using nonparametric bootstrap methods). A recent discussion of Staerk et al. (2024) claims that prior variable selection becomes predominant in some fields including biomedical applications.

Intrinsic variable selection (also known as sparse variable selection) is performed within regression or classification tasks using regularization techniques (Fan et al., 2020). Recently developed methods of this type tend to produce sparse solutions, utilizing only a subset of the variables while disregarding the others (Su & Wu, 2024). Although regularization can provide local robustness to minor fluctuations in the data, it's important to note that regularized methods can be sensitive to the presence of outliers.

Posterior variable selection is typically carried out using hypothesis tests or information theory-based methods. In a supervised context, particularly when data come from two or more groups, hypothesis testing is used to rank the variables according to the strength of evidence against the null hypothesis, rather than assessing the significance of individual variables through p-values. The issue of multiple testing is commonly addressed using the false discovery rate (FDR), which ensures that the proportion of false-positive results among all significant tests remains below 5%. The widely used method to control the FDR, as defined by Benjamini & Hochberg (1995) for independent tests, helps manage this problem. An extension proposed by Benjamini & Yekutieli (2001) can be applied even for dependent statistics, even if the dependence structure is unknown. However, posterior variable selection in model selection tasks has faced criticism due to its lack of stability (Heinze et al., 2017).

Bayesian dimension reduction represents an important type of method that leverages probabilistic modeling to reduce the dimensionality of complex datasets while incorporating uncertainty in the process. By framing the reduction as a latent variable model, it identifies the most significant variables that capture the underlying structure of the data. This is especially valuable in applications where the data may be noisy or contaminated, as it provides robustness against outliers and ensures that the essential features are preserved. The Bayesian framework allows for the integration of prior knowledge, offering a transparent and interpretable approach to dimensionality reduction. In fields like gene expression analysis and credit scoring, Bayesian dimension reduction not only improves model interpretability but also facilitates a deeper understanding of how individual data points influence outcomes, making it an essential tool for engineers seeking to enhance the explainability of AI models in real-world scenarios.

Generally, it is difficult to outline a universal method for identifying the best dimensionality reduction technique for a specific dataset. Heinze et al. (2017) made attempts to systematically compare different dimensionality reduction methods and provide practical recommendations for handling this issue. Selecting the appropriate method for Big Data remains a challenging problem, with many open questions. Each method comes with its own set of assumptions, and some methods are better suited for economic data analysis, while others are frequently used in biomedicine, and so on. Therefore, dimensionality reduction should be customized to the specific task or problem within a particular field (such as medicine, economics, or engineering) and the statistical analysis task at hand (e.g., regression or classification). When the analysis demands clear interpretability, prior or sparse variable selection should be preferred over feature extraction (Olson, 2017).

ROBUST METHODS FOR DIMENSIONALITY REDUCTION

Principles

Many widely used machine learning techniques are susceptible to outliers in the data, which has led to the development of robust statistical methods over recent decades (Huber, 2009). Standard dimensionality reduction methods are also sensitive to outliers, highlighting the need for robust alternatives that remain unaffected by such anomalies (Filzmoser & Nordhausen, 2021). This issue extends to common variable selection methods, making the development of their robust counterparts particularly important. Our primary focus is on methods that are resilient to outliers, meaning they are globally robust in terms of their breakdown point. Users typically favor methods that are both computationally stable and well-suited for high-dimensional datasets. However, relatively little attention has been given to local robustness, particularly in response to small changes in the data near the center of mass of the main data distribution (Jurečková et al., 2019). In fact, locally robust methods have been available primarily for some of the simplest statistical models. Growing attention is being paid to reducing dimensionality in nonlinear statistical models (Chen, 2024).

Robust Linear Regression

Let us revisit the standard linear model, where the goal is to estimate the regression parameters. The lasso estimator is an L1-regularized version of least squares, which introduces sparsity into the model. It performs intrinsic variable selection, completely disregarding redundant variables, making the lasso a popular choice for regression modeling in high-dimensional data settings. However, the lasso is not robust to outliers, and the same limitation applies to ridge regression estimators. To achieve robustness in these regularized estimates, principles from robust statistics must be applied.

The LTS-lasso estimator of Alfons et al. (2013) represents a robust regularized estimator for linear regression and can be described as a regularized (or penalized) version of the LTS estimator, which trims away a certain percentage of observations, i.e., those with the largest squared residuals. Using $h = 3n/4$ and denoting the estimated parameters by $b = \left(b_1, \ldots, b_p\right)^T$, LTS-lasso is defined by

$$min\left[\sum_{i=1}^{h} u_{(i)}^2 + \lambda \sum_{j=1}^{p} |b_j|\right].$$

It is natural to consider an L1-regularized version of the LWS estimator denoted here as LWS-lasso.

An adaptation of the algorithm for computing LTS may be used for computing both LTS-lasso and LWS-lasso. Because Alfons et al. (2013) suggested using cross-validation for finding a suitable value of the regularization parameter for LTS-lasso, we use 10-fold cross-validation here to find suitable regularization parameters of both LTS-lasso and LWS-lasso.

The loss of the LWS-lasso estimator is the loss of the LWS plus a penalization equal to the sum of absolute values of all parameters multiplied by a regularization (penalization) parameter, i.e.

$$min\left[\sum_{i=1}^{n} w_i u_{(i)}^2 + \lambda \sum_{j=1}^{p} |b_j|\right]$$

with given magnitudes of the weights. Extensions of robust estimation to generalized linear models, which include hypothesis testing for variable selection, include the work of Kalogridis et al. (2023).

Robust Training of Neural Networks

In nonlinear regression models, where the response is represented as a function of p given regressors (predictors), neural networks offer a nonlinear extension of traditional linear regression estimators. Commonly used methods for training regression neural networks using least squares tend to be biased in the presence of contaminated data and are also susceptible to adversarial examples. This lack of robustness becomes particularly pronounced when dealing with data that contains a large number of regressors. This issue is also evident in two prominent types of multilayer feedforward networks: multilayer perceptrons (MLPs) and radial basis function (RBF) networks (Kaveh, 2024).

Recently, there has been growing interest in developing robust (resistant) methods for training neural networks (Luo et al., 2020). However, only a limited number of robust methods have been proposed for training MLPs. Many of these methods become computationally impractical when p increases, and as a result, they have primarily been demonstrated on simple examples with a small p. Their training typically requires to compute

$$min \sum_{i=1}^{n} u_i^2$$

across all values of parameters, where $u_1, \ldots, u_n$ are residuals as above. The first robust alternatives were proposed by replacing the sum of squared residuals with the loss functions corresponding to the median (Aladag et al., 2014) or LTS estimator. Recent robust alternatives to training neural networks were proposed in Yilmaz et al. (2023) or Bas et al. (2024).

A robust variant of the multilayer perceptron (MLP), based on the LTS estimator and referred to as LTS-MLP, was introduced by Kalina & Vidnerová (2020). Given the favorable characteristics of the LWS estimator in linear regression, as discussed earlier, we now turn our attention to the robust loss function associated with the

LWS estimator within the framework of regression neural networks. We explore the LWS-MLP with trimmed linear weights. This version can be formally defined as the LWS estimator applied to minimize across all the parameters of the network. It can be stated that the LTS-MLP and LWS-MLP differ from the traditional MLP solely in their use of the LTS or LWS loss functions, respectively. Figure 1 demonstrates the non-robustness of a standard MLP in comparison to the more robust LWS-MLP using a contaminated dataset.

Regularized neural networks exploiting the L1-regularization are popular extensions of plain networks, which ensure some of the parameters to be equal exactly to 0. The L1-regularized MLP is defined by

$$min \sum_{i=1}^{n} u_i^2 + \lambda \sum_{j=1}^{d} |\theta_j|,$$

where $\theta = (\theta_1, \ldots, \theta_d)^T$ is the vector of all (estimated) parameters of the network; this will be denoted here as MLP-L1.

The regularized versions of LTS-MLP and LWS-MLP are referred to as LTS-MLP-L1 and LWS-MLP-L1, respectively. To the best of our knowledge, these variants have not been previously explored in the literature. Their computation involves a back-propagation algorithm utilizing a stochastic gradient descent method. The loss function for LWS-MLP-L1 consists of the LWS estimator loss, with an added penalty term that is the sum of the absolute values of all the MLP parameters, scaled by a regularization (penalization) parameter. To identify the optimal value for the regularization parameter, we employ 10-fold cross-validation.

Dimension reduction for neural networks can also be performed by applying hypothesis testing techniques, particularly for feature selection (Binson et al., 2024). By testing the statistical significance of individual features, one can identify and retain only the most relevant variables, effectively reducing the dimensionality before training the model. This approach helps eliminate redundant or irrelevant features, which can improve model efficiency and interpretability. Although hypothesis testing is not typically used for direct dimensionality reduction in neural networks, it can complement other methods by guiding the selection of important features and enhancing the overall performance of the model.

Figure 1. Illustration of the plain MLP (left) and a robust version denoted as LWS-MLP (right) evaluated for a contaminated artificial dataset. The architecture was selected to contain 2 hidden layers containing 16 and 2 neurons, respectively. A sigmoid activation function is used in the hidden layers, and a linear output layer with a single neuron is used

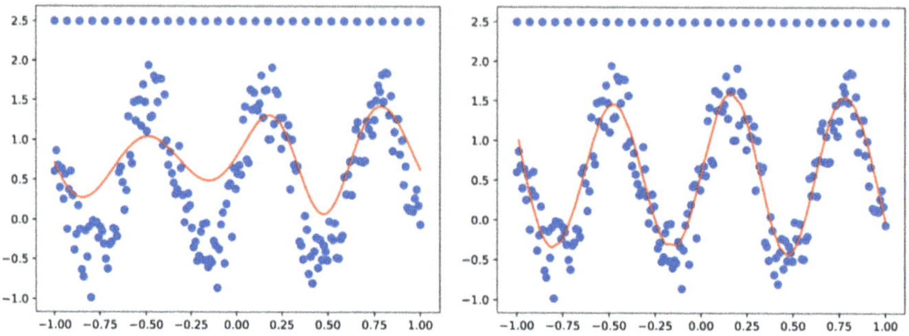

Robust Classification Exploiting Centroids

In many applications, classification based on centroids (also known as prototypes or templates) is both effective and straightforward, making it easy to understand. While centroids are widely used in image processing, they also perform well with other types of data, including high-dimensional datasets that are not image-based (Delaigle & Hall, 2012). In this context, let's consider two groups of p-dimensional continuous observations. Typically, when the goal is to classify data into two groups, the centroid of the first group is computed as the mean of its observations, and the Pearson correlation coefficient serves as the similarity measure between the centroid and a given observation. An observation is assigned to the first group if its correlation coefficient exceeds a specific threshold.

A method for optimization of centroids was recently proposed by Kalina & Matonoha (2020). This supervised learning method exploits the weighted correlation coefficient r_w, i.e. weighted extension of Pearson correlation coefficient. Let us consider weights $w_1,\ldots,w_n$ and denote the weighted mean of a vector x by $\bar{x}_w$. We may recall r_w between two vectors x and y to be defined as

$$r_w(x,y) = \frac{\sum_{i=1}^{n} w_i(x_i - \bar{x}_w)(y_i - \bar{y}_w)}{\sqrt{\sum_{i=1}^{n}\left[w_i(x_i - \bar{x}_w)^2\right] \sum_{i=1}^{n}\left[w_i(y_i - \bar{y}_w)^2\right]}}.$$

Optimization tasks typically involve minimizing a nonlinear loss function, which corresponds to a regularized margin-like distance (Herb, 2024). This distance is calculated for the worst-case pair across the training dataset. Kalina & Matonoha (2020) compared two optimization strategies: [I] a linear (approximate) method for solving the optimization task, and [II] a nonlinear approach using the interior point method. Their comparison demonstrated that both optimization approaches were computationally stable, with the linear approximation producing solutions that were nearly identical to those from the nonlinear method. Furthermore, optimizing the weights can also be performed, encouraging many pixels to be assigned zero weights, thus yielding a sparse solution. This weight optimization can be seen as an analogy to centroid optimization, where the linear (approximate) method once again proves to be reliable and stable. The optimal centroid can be used without weight adjustments, but fine-tuning the weights can enhance classification performance. The robustness of this method has not yet been explored and will be empirically assessed later in this chapter. Figure 2 illustrates the results of these optimization procedures applied to mouth localization in facial images.

Figure 2. Left: a mouth centroid suitable for mouth localization obtained by the optimization procedure proposed in Kalina & Matonoha (2020). Right: weights corresponding to this centroid

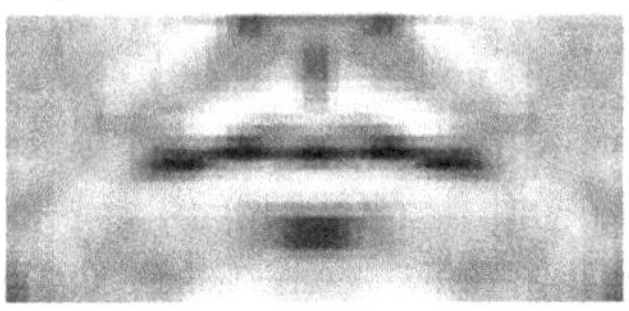
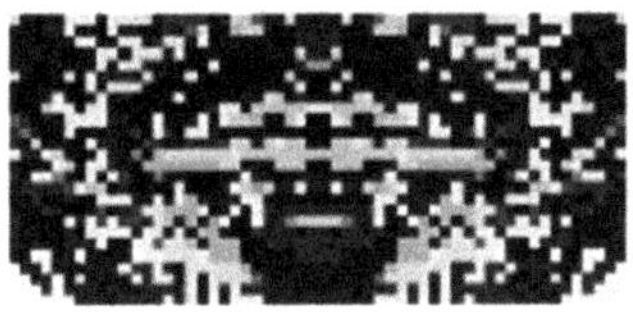

Regression and Correlation Resistant to Outliers

In the linear regression model, the classical least squares estimator is notoriously well known to be too vulnerable to the presence of outliers in the data (Jurečková et al., 2019). We now recall the least weighted squares (LWS) estimator (Kalina, 2015) of parameters of linear regression, which will be later exploited as an inspiration or basis for several robust approaches to dimensionality reduction. We consider the standard linear regression model

$$Y_i = \beta_1 X_{i1} + \ldots + \beta_p X_{ip} + e_i, i = 1, \ldots, n,$$

which may be expressed in the matrix notation as $Y = X\beta + e$. Here, β is the d-dimensional vector of parameters and the i-th row of X will be denoted as X_i for

$i = 1,\ldots,n$. If we consider a fixed estimate b of β, it will be useful to denote the residual corresponding to the i-th observation as

$$u_i(b) = Y_i - b_1 X_{i1} - \ldots - b_p X_{ip}, i = 1,\ldots,n.$$

Ordering squared residuals as

$$u^2_{(1)}(b) \leq u^2_{(2)}(b) \leq \ldots \leq u^2_{(n)}(b),$$

the LWS estimator is formally defined by

$$arg\min_{b \in R^p} \sum_{i=1}^{n} w_i u^2_{(i)}(b).$$

The LWS estimator is an attractive option for practical data analysis due to its strong global robustness (when appropriate weights are selected), high local robustness, and high efficiency when the data is not contaminated. As a highly robust method, the LWS estimator is characterized by a high breakdown point. The breakdown point refers to the minimal proportion of data that can cause an estimator to produce extreme values when set to arbitrary values (cf. Statti et al., 2017). The LWS estimator relies on implicit weights assigned to individual observations and does not require prior outlier detection. While it does not automatically detect outliers, there is an available adaptation for incorporating outlier detection (Kalina, 2018). Additionally, the LWS estimator demonstrates favorable properties in the context of nonlinear regression models. Figure 3 provides an illustration comparing the LWS estimator with the traditional least squares method.

The LWS estimator is defined for the standard linear regression model, where the response Y is explained by p independent variables (regressors, predictors). The formal definition of the LWS estimators of the parameters of the linear regression model requires the user to specify a sequence of magnitudes of non-increasing weights, which are assigned to individual observations only after the optimal permutation. Particularly, the most reliable observations (with the smallest absolute residuals) obtain the largest weight, while the outliers (with the largest absolute residuals) obtain the smallest weights (possibly zero). The LWS estimator remains consistent for any non-increasing weights (Kalina, 2015). The computation of the LWS estimator may exploit a reliable approximate algorithm of Rousseeuw & van Driessen (2006). All computations within this chapter consider LWS with weights with magnitudes equal to *1, 1-1/n, …, n/4,0,…0,* where 25% of observations obtain exactly zero weights. These weights will be called trimmed linear weights (Kalina et al., 2019). Formally, these are defined using the weight function

$$\psi(t) = \left(1 - \frac{t}{\tau}\right) 1[t < \tau], t \in [0,1], \tau = 3/4.$$

Figure 3. Illustration of robust regression on a contaminated dataset from Rousseeuw & Leroy (1987). The least squares fit is heavily influenced by the cluster of 6 outliers and the LWS fit estimates very well the non-outlying data points

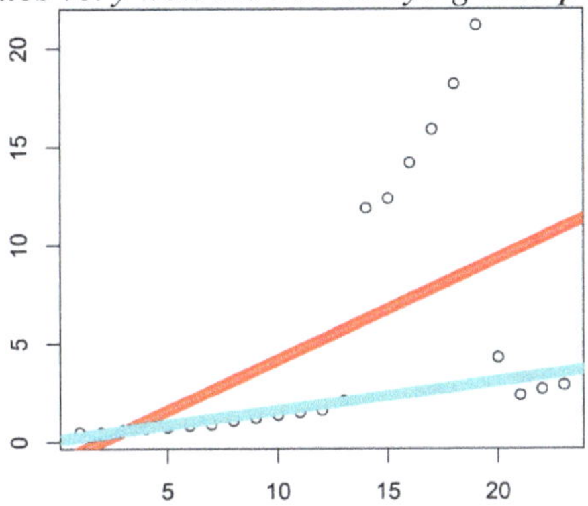

The least trimmed squares (LTS) estimator is a special case with weights only equal to 1 or 0; the user has to specify the trimming constant h so that e.g. specifying $h=0.75$ means that 75% of the observations obtain the weight 1 and the remaining 25% are trimmed away completely.

While Pearson correlation coefficient is well known to be highly vulnerable to the presence of outliers in the data, its various robust versions have been available (Shevlyakov & Oja, 2016). Let us recall the implicitly weighted robust correlation coefficient denoted here as r_{LWS}, which is based on the LWS estimator. Formally, r_{LWS} proposed and investigated in Kalina (2018) is obtained as the weighted Pearson correlation coefficient with the weights that are found as optimal by the LWS estimator in the linear regression model *YX*. This symbolic notation denotes the model, where *Y* plays the role of the response explained by *X*. Known properties of r_{LWS} include its high breakdown point and asymptotic normality.

APPLICATION 1: BOSTON HOUSING DATASET

The Boston Housing dataset, available in public repositories such as the UCI repository (Dheeru & Taniskidou, 2017), has been analyzed using standard machine learning techniques, including in Chapter 3 of Chollet (2018). This dataset contains no missing values, allowing us to work with all 506 cases, each representing a piece of real estate within the greater Boston urban area. The goal of this analysis is to model and predict the per capita crime rate for different neighborhoods using the available regressors.

Although the original dataset includes 14 variables, we focus on 11 continuous ones, excluding features 4, 7, and 9. Among the regressors used for each neighborhood are variables such as the average number of rooms per dwelling, the median value of owner-occupied homes, the percentage of the population with a lower socio-economic status, and the concentration of nitrogen oxides in the air. The methods for this analysis were implemented in Keras (Chollet, 2024).

The goal of this analysis is to compare both standard and robust neural networks, with a particular emphasis on contrasting regularized and non-regularized versions. We utilize a multilayer perceptron (MLP) with two hidden layers, consisting of 16 and 8 neurons, respectively. Each hidden layer uses a sigmoid activation function, while the output layer is linear. All other networks in the comparison share this same architecture.

For the regularized models, we determine the optimal regularization parameter using 10-fold cross-validation. In the case of LTS-MLP and related networks based on the LTS loss function, we set the parameter $h=0.75n$.

Additionally, we perform backward elimination for variable selection in the standard MLP model. Backward elimination is a well-known technique in linear regression model selection. Here, we use this method, referred to as Back-MLP, to identify the 5 most important variables. The process begins with the full model containing all regressors. Then, we systematically eliminate one regressor at a time, computing the mean squared error (MSE) or trimmed MSE (TMSE) at each step. We retain the regressor that results in the smallest prediction error, continuing this process until only the 5 most relevant variables remain. These measures are formally defined as

$$MSE = \frac{1}{n}\sum_{i=1}^{n} u_i^2$$

and

$$TMSE = \frac{1}{h}\sum_{i=1}^{h} u_{(i)}^2.$$

We continue by removing another regressor from the model, repeating the process until a model with 5 regressors is achieved. The prediction error for this resulting model is then recorded for the standard MLP. This same procedure is applied to LTS-MLP and LWS-MLP models, leading to the novel approaches denoted as

Back-LTS-MLP and Back-LWS-MLP, respectively. The results from these models are summarized in Table 1, which was computed using 10-fold cross-validation.

Table 1. Results of application 2 (Boston Housing dataset): comparison of MSE and TMSE for different versions of multilayer perceptrons

Method	MSE	TMSE
MLP	57.9	5.3
MLP-L1	59.2	5.4
Back-MLP	59.3	5.4
LTS-MLP	67.2	4.3
LTS-MLP-L1	68.2	4.3
Back-LTS-MLP	68.1	4.4
LWS-MLP	70.8	4.5
LWS-MLP-L1	71.9	4.5
Back-LWS-MLP	72.0	4.6

In the table, we present the values of the mean squared error (MSE) alongside the values of its robust counterpart, the trimmed mean square error (TMSE). The TMSE is defined by considering a given percentage of observations, which in this case is set to $h=0.75n$. This means that TMSE is computed as the sum of squared residuals only over the 75% of observations with the smallest squared residuals.

The standard versions of neural networks outperform others when evaluated using conventional MSE. However, this does not imply that the robust methods are unsuitable, as MSE is sensitive to the presence of outliers. When considering data contaminated by outliers, only robust versions of MSE remain meaningful. In this context, when TMSE is used, the methods LTS-MLP-L1 and LWS-MLP-L1 outperform the others. The superior performance of these robust methods can be attributed to the presence of outliers in the dataset. A robustness analysis, not included here, indicates that the dataset contains around 10% severe outliers.

The L1 regularization in LTS-MLP-L1 and LWS-MLP-L1 serves to prune the networks, though this results in only a slight reduction in prediction accuracy. The Back-LTS-MLP and Back-LWS-MLP methods enhance the explainability of the models with only a marginal decrease in predictive performance. It is important to note that while L1 regularization in these models trims some parameters, it does not eliminate any variables entirely from the model. Thus, L1 regularization does not perform true variable selection in this case; true variable selection would occur only in a linear regression model, which is explored in Application 3.

APPLICATION 2: GENE EXPRESSION DATA

The Dataset

The gene expression (GE) dataset from the cerebrovascular stroke study conducted at the Center for Biomedical Informatics in Prague was analyzed using hypothesis testing by Marozzi et al. (2020). This dataset includes 38,950 GE values, measured as continuous variables in a paired design involving 24 pairs of individuals. It consists of 24 individuals who had a cerebrovascular stroke and 24 control individuals without any cardiovascular disease. Measurements for stroke patients were taken as soon as possible after the event. Each GE measurement corresponds to a specific gene transcript. For computational efficiency, only the first $p=3000$ genes from the dataset were considered. The goal is to develop a classification method to distinguish between the two groups, enabling the assignment of a set of GE measurements from a new (potentially independent) individual to either the stroke patient group or the control group. The centroid and weight optimization were performed using C++, while R was used for all other computations.

Methodology

The centroid-based classifier was applied in various forms, with the initial centroid being calculated as the average of the measurements from the 24 stroke patients (excluding the control group). Initial weights were set to be equal for each variable, with each assigned a weight of 1/3000, ensuring their sum equals 1. The parameters for optimizing the centroids and weights were taken from the values used in Kalina & Matonoha (2020). Two enhancements to the standard centroid-based classification methods were introduced here, both aimed at improving the interpretability of the results. These innovations, referred to as thresholded optimal centroid and binarized optimal weights, have not previously been explored in the context of centroid-based classification. We now describe the construction of the thresholded optimal centroid and binarized optimal weights.

Starting with an initial centroid (denoted as c), a new centroid is computed. If the value of c for a particular gene is below a specified threshold, the value of the new centroid for that gene is set to 0. For all other genes, the new centroid matches the initial centroid. Specifically, when normalizing the centroid so that the sum of all values equals 1, a threshold of 0.00005 is used. This thresholded optimal centroid is applied in conjunction with equal weights.

After computing the optimal centroid with the corresponding optimal weights, the question arises as to how much information would be lost if the optimal weights were binarized. To address this, we modify the optimal weights (denoted as w) by applying a simple strategy: If the value of w for a particular gene is below a predefined threshold (in this case, 0.0001), the weight is set to 0. For all remaining genes, the new weights are assigned a constant value (the same for all genes). This constant is chosen so that the sum of all the weights equals 1.

For comparison purposes, we also perform Principal Component Analysis (PCA) on the entire dataset (i.e., combining data from both groups). Linear Discriminant Analysis (LDA) is then used to construct the classification rule based only on the first 10 principal components. Using PCA enables us to apply LDA in its standard form without the need for using one of the many regularized versions of LDA (Kalina & Duintjer Tebbens, 2015).

Robustness Evaluation

In addition to the raw data, we consider three contaminated versions of the dataset to assess the robustness of the methods. For the contamination process, we introduce severe outliers at three different contamination levels: 4%, 8%, and 12% of the values for each observation (individual). For each contaminated individual, the genes (variables) to be modified are selected randomly, and independent random variables from a normal distribution with a mean of 0 and a standard deviation of 1 are generated and added to the original values. When working with contaminated data, the centroids and weights are computed based solely on the contaminated data, meaning that the methods have no access to the raw (non-contaminated) data at any point.

Results

The results of the gene expression measurements analysis, as shown in Table 2, present classification accuracies, which are defined as the ratios of correctly classified samples to the total number of samples. The table also indicates the number of genes used in the classification rule for each method. Comparisons with standard machine learning techniques are not included, as these were already extensively studied by Kalina & Matonoha (2020).

Let us first discuss the results obtained using the raw data. The approach using the optimized centroid with optimized weights significantly outperforms the results obtained with the initial centroid. The centroid-based approach, which uses Pearson correlation as a measure, yields better results compared to the initial centroid, but it is still outperformed by the method that incorporates the optimized centroid.

Table 2. Results of application 1 (gene expression dataset) for raw and contaminated data.

Centroid	Weights	Classification accuracy	Number of used genes
Raw data (no contamination)			
Initial	Initial	0.73	3000
Optimal	Initial	0.81	3000
Thresholded optimal	Initial	0.77	986
Optimal	Optimal	0.85	247
Optimal	Binarized optimal	0.85	196
Initial centroid using r_{LWS}		0.77	3000
PCA⇒LDA		0.69	3000
Contamination level 0.04			
Initial	Initial	0.69	3000
Optimal	Initial	0.81	3000
Thresholded optimal	Initial	0.77	1034
Optimal	Optimal	0.85	261
Optimal	Binarized optimal	0.85	213
Initial centroid using r_{LWS}		0.77	3000
PCA⇒LDA		0.65	3000
Contamination level 0.08			
Initial	Initial	0.67	3000
Optimal	Initial	0.79	3000
Thresholded optimal	Initial	0.75	1095
Optimal	Optimal	0.83	259
Optimal	Binarized optimal	0.83	202
Initial centroid using r_{LWS}		0.73	3000
PCA⇒LDA		0.63	3000
Contamination level 0.12			
Initial	Initial	0.63	3000
Optimal	Initial	0.77	3000
Thresholded optimal	Initial	0.73	1215
Optimal	Optimal	0.81	283
Optimal	Binarized optimal	0.81	220
Initial centroid using r_{LWS}		0.69	3000
PCA⇒LDA		0.58	3000

The binarized optimal weights, introduced as a novel extension of the centroid-based approach, provide the same classification performance as the more complex optimal weights. This suggests that binary weights retain essentially the same information as the optimized weights. In other words, while the optimization of weights is a delicate process that captures subtle nuances, the binarization of the weights—despite being a simpler transformation—does not result in a significant loss of classification accuracy.

The thresholded optimal centroid, however, fails to maintain the same level of classification performance. It does not provide satisfactory results and cannot be recommended for use in this context.

Interpretation and Insights

Optimizing the centroid yields very promising results, and further simplifying the optimal centroid can improve its interpretability. This simplification can be achieved by optimizing the weights, which serves as a form of posterior variable selection. It turns out that a small subset of variables carries the majority of the relevant information for classification. As a result, it is possible to create a more understandable classification model with only a slight reduction in performance, and in some cases, even improve the interpretability of the subsequent data analysis.

When it comes to contaminated data, centroid-based approaches that rely on initial, non-optimal centroids are highly susceptible to the influence of outliers. However, centroid-based classification using an optimized centroid with optimal weights proves to be more robust, outperforming other methods under these conditions.

In contrast, Principal Component Analysis (PCA) does not perform well in this context. As an unsupervised dimensionality reduction technique, PCA is not designed for data from two distinct groups (such as stroke patients and control subjects). It focuses on explaining overall variability, rather than the separation between groups, which makes it less suitable for this type of analysis. Additionally, the vulnerability of PCA to outliers further limits its effectiveness, highlighting a significant disadvantage compared to centroid-based methods.

Outcome

In conclusion, the analysis demonstrates that centroid-based classifiers are well-suited to solve the classification task for the gene expression dataset. The centroid-based classification method (with or without weights) proves to be an explainable approach, significantly improving classification results compared to the initial (naïve) centroid. The optimization of weights allows for posterior variable selection, and the optimal centroid with optimal weights can be seen as a classifier with sparse (intrinsic) variable selection, relying on approximately 1000 genes. When binarized weights are used, the classification rule is based on around 200 relevant genes. In this case, simplifying the information significantly enhances explainability without sacrificing classification performance.

APPLICATION 3: A WORLD TOURISM DATASET

As the third application, the Travel and Tourism Competitiveness Index (TTCI) dataset is considered. This publicly available dataset reported by the World Economic Forum for the year 2021 is available on the website of Uppink (2022). The tourist service infrastructure measured across $n=141$ countries of the world is modeled as a response of $p=12$ characteristics (pillars) of TTCI. All the 12 regressors are continuous; the examples of individual regressors reported for individual countries include e.g. indicators of the business environment, international openness, air transport infrastructure, health, or hygiene. We used R software for the computations.

The methods used here for the analysis of the TTCI dataset include the least squares, lasso, LTS-lasso, and the novel LWS-lasso. For LTS-lasso, we trim away 25% of the data values (countries). For regularized methods, we use 10-fold cross-validation to find a suitable value of the regularization parameter. Table 3 presents values of MSE and TMSE (using $h=0.85n$) computed by various regression methods. The values were obtained in 10-fold cross-validation. Only TMSE as a robust version of MSE should be considered for these data because they contain severe outliers. A detailed analysis namely reveals that the dataset contains 5 outliers (France, the United States, Spain, China, and Italy), which have a high influence on the resulting estimates. Robust sparse estimators (LTS-lasso and the novel LWS-lasso) outperform other methods in terms of TMSE. The approach with PCA using 4 principal components does not reveal any interesting structure in the data. We can say that the contribution of all regressors to the first principal components is approximately equal and the other principal components are too difficult to interpret. The left part of Table 4 shows the ranking of all 12 variables based on their importance (or relevance) in the regression task, specifically for explaining the response variable. To clarify, the ordering method used here is similar to the usual variable selection available for the lasso estimator. The ordering is determined by a key property of the lasso estimator: as the regularization parameter increases, the impact of each individual regressor diminishes, causing variables to gradually be removed from the model. This approach is applied to all the regularized methods in this analysis. For the least squares method, however, we determined the variable ranking using backward selection, performed through t-tests.

The results reveal that the robust regularized methods outperform all other methods, including standard least squares and lasso. Interestingly, there is little difference between the two robust methods, LTS and LWS, in terms of performance. Both robust approaches provide superior results compared to the standard least squares or lasso methods, highlighting the advantages of using robust regularization techniques for this regression task.

Table 3. Results of application 3 (TTCI dataset): prediction error measures MSE and TMSE

Method	MSE	TMSE
Least squares	0.449	0.363
Lasso	0.461	0.366
LTS-lasso	0.453	0.327
LWS-lasso	0.451	0.334
PCA⇒ least squares	0.510	0.398

Table 4. Results of application 3 (TTCI dataset): analysis of the role of individual variables in the model. Left: ordering of all variables according to their relevance for explaining the variability of the response, starting from the most relevant to the least relevant. Right: ordering only for the variables that are found as relevant (by using the optimal regularization parameter or by t-tests)

Method	Ordering of all variables	Ordering only for relevant variables
Least squares	6, 5, 9, 1, 10, 3, 4, 2, 7, 8, 12, 11	6, 5, 9, 1, 10
Lasso	6, 5, 9, 1, 3, 10, 4, 7, 2, 11, 12, 8	6, 5, 9, 1, 3
LTS-lasso	5, 6, 3, 10, 9, 7, 1, 8, 11, 2, 12, 4	5, 6, 3, 10, 9, 7
LWS-lasso	5, 6, 3, 10, 9, 7, 8, 11, 1, 12, 4, 2	5, 6, 3, 10, 9, 7

Figure 4. Horizontal axis: regularization parameter with values between 0 and 1. Vertical axis: LWS-lasso estimates of regression parameters for all 12 regressors evaluated for individual values of the regularization parameters

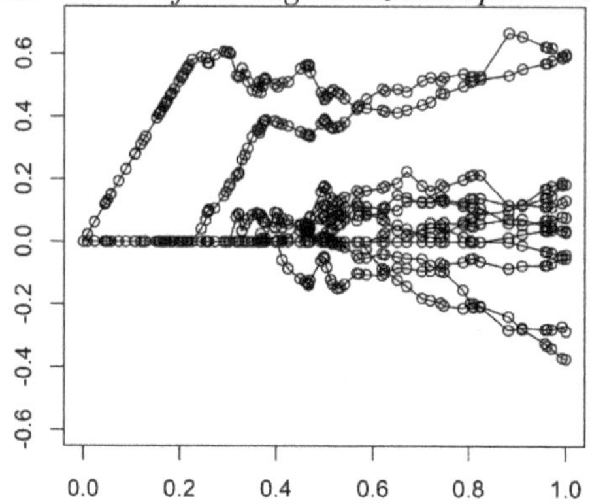

When using regularized methods with optimal regularization parameters obtained through 10-fold cross-validation, the methods do not rely on all 12 variables but instead select only a subset of them, performing sparse variable selection. These selected variables are shown in the right part of Table 4. For the least squares method, however, a different approach is employed; we conducted backward selection using t-tests instead.

The interpretation of these results reveals that the number of relevant variables identified varies across the different methods. Notably, the most relevant variables identified by the robust approaches differ from those selected by the non-robust methods. This discrepancy can likely be attributed to the presence of outliers in the dataset. As such, the sparse robust variable selection tools, such as LTS-lasso and LWS-lasso, are considered more reliable for this dataset. These methods are viewed as resilient and trustworthy, ensuring the explainability of the regression model by excluding redundant variables.

Figure 4 illustrates how the estimated parameters for each of the 12 regressors change in relation to the regularization parameter. In the case with no regularization, all regressors are included, but as the level of regularization increases, dimensionality reduction occurs, and only the most relevant variables are retained.

ENERGY COMPLEXITY

The methods discussed in this chapter can be viewed within the broader framework of low-energy approximate computation. In the experiments presented, it becomes clear that robustness, sparsity, and energy-efficient computation do not conflict with the pursuit of optimal performance. In fact, these principles complement each other, suggesting that it is possible to achieve highly efficient computation without sacrificing accuracy.

When we consider dimensionality reduction, it becomes apparent that reducing the number of features or parameters in a model leads to lower energy complexity. In the context of deep neural networks (DNNs), dimensionality reduction can be achieved through various techniques, such as sparsity induction (via regularization) and variable selection, both of which reduce the number of active neurons or connections in the network. This reduction in the number of computations directly impacts the energy required for training and inference, making the process more efficient.

Energy-efficient computation in DNNs often involves eliminating redundant computations or using approximations that retain essential information while minimizing resource usage. This ties into the concept of sparsity, where only a subset of the network's parameters are active at any given time. Sparsity not only reduces energy consumption but also can lead to faster processing times, as fewer operations are needed.

In summary, the interplay between dimensionality reduction, sparsity, and energy efficiency is crucial in the design of computationally efficient deep learning models. By carefully managing dimensionality and promoting sparsity, it is possible to build models that are not only robust and accurate but also energy-efficient, ensuring that the trade-off between computational complexity and performance is optimally balanced.

Shallow neural networks are valuable tools in machine learning, but they often struggle with datasets that have a very large number of observations, particularly when the number of observations exceeds the number of features. In such cases, deep neural networks (DNNs) are essential for handling the complexity and scale of the data. While shallow networks are ill-suited for data contaminated with outliers (Dey et al., 2019), deep neural networks typically incorporate mechanisms for more robust training, making them better equipped to handle noisy or outlier-laden data.

However, deep learning models, particularly DNNs, are often referred to as "black boxes" because their decision-making processes are not easily interpretable. The architecture of these networks is either chosen manually by the user or optimized through techniques like cross-validation to achieve the best performance on independent data. Some deep neural networks are essentially an extension of shallow multilayer perceptrons (MLPs) with additional hidden layers, which require more advanced algorithms for training but do not fundamentally differ from shallow networks in their conceptual approach. Convolutional neural networks (CNNs) are a popular subset of deep networks known for their applications in image processing, leveraging specialized layers to capture spatial hierarchies in data.

Dimensionality reduction is a common strategy for reducing computational complexity in machine learning, particularly when dealing with large datasets. This approach is particularly valuable in scenarios where resources are constrained, such as running intensive computations on mobile devices with limited memory and battery power. However, reducing the size of the data is not the only method for decreasing computational demand. Modifying the underlying computations or adopting smaller, more energy-efficient architectures can also be effective. A balance between model performance (e.g., approximation accuracy) and network size can be achieved by integrating other criteria that control the network's size and complexity. Regularization techniques such as weight decay, activity regularization, or dropout, along with tools like weight binarization or data quantization, can help reduce computational costs.

Sparsity is another approach that can lead to more efficient computation while also improving model interpretability. By selectively keeping only the most relevant connections, sparse networks minimize unnecessary computations and enhance the model's explainability. This is why regularization plays a crucial role in approximate computing: it reduces computational demands while ensuring that the model remains robust, accurate, and interpretable. In summary, a thoughtful combination of dimensionality reduction, regularization, sparsity, and energy-efficient architectures can optimize both performance and computational efficiency in machine learning, especially when working with deep neural networks.

With the advancements in supercomputing, energy efficiency has become a key focus in the development and training of deep learning models, especially when dealing with Big Data. Li et al. (2018) explored energy-efficient models for services demanding significant energy consumption, specifically in the context of internet-of-things (IoT) platforms. Mahmoud et al. (2018) also contributed by investigating energy-efficient architectures within the Cloud of Things (CoT). Additionally, Lakshmanaprabu et al. (2019) highlighted a clinical decision support system for diagnosing chronic kidney disease. This system utilizes deep neural networks trained on Big Data, which includes data collected from wearable devices and publicly available datasets such as those from the UCI repository. The paper emphasizes achieving energy efficiency by integrating energy models suitable for operation within cloud-based IoT platforms.

In the broader context of approximate neurocomputing, several open challenges remain, particularly in reducing the energetic complexity of neural networks, especially during the training phase of convolutional neural networks (CNNs). One promising direction is to combine neural architecture search (NAS) with strategies aimed at reducing the number of variables and parameters in the network. Multi-objective Bayesian optimization could play a crucial role in this, enabling the simultaneous optimization of performance and energy demands (Mrázek et al., 2016).

Additionally, when optimizing for energy efficiency, it is important to consider the energetic demands of computations, which could be incorporated as part of the optimization objectives.

Applications for mobile devices, in particular, hold great potential for reducing energy consumption, which could further contribute to the development of "green machine learning" (Xu et al., 2021). These methods could significantly enhance the sustainability of machine learning processes, especially in energy-constrained environments. Moreover, another important area of research involves developing robust neural networks with energy-efficient capabilities. By integrating robust loss functions with penalization mechanisms focused on energy consumption, we could create neural network architectures that are both resilient to noise and outliers, while also optimized for low energy usage.

CONCLUSION

Artificial intelligence holds immense potential to transform society into a knowledge-driven one. To realize this potential, AI relies on machine learning algorithms that are capable of extracting valuable insights from available data. In complex decision-making tasks, particularly those under uncertainty, humans often struggle to arrive at rational, optimal solutions (Steyvers & Kumar, 2024), whereas AI tools can surpass human capabilities. Decision support systems, which are among the most promising AI tools, have already been successfully applied to such tasks (Kalina et al., 2013; Bridge & Dodds, 2019). These systems require machine learning algorithms that are coupled with dimensionality reduction techniques (Kalina & Zvárová, 2013). This combination ensures that only the most relevant variables are used, transforming complex, multivariate data into actionable knowledge.

This chapter focuses on procedures for reducing data dimensionality that are robust to the presence of outliers. While it cannot cover all available variable selection methods, it highlights intrinsic and posterior variable selection methods, with particular emphasis on novel techniques such as LWS-lasso. Dimensionality reduction always involves balancing simplicity with the retention of relevant information.

The real-life examples presented in this chapter aim to demonstrate the potential of dimensionality reduction in multivariate data. In all three applications discussed—related to classification or regression tasks—robust variable selection methods consistently deliver the best results. When performed correctly and robustly, variable selection helps separate useful signal from noise, offering practitioners an interpretable approach that avoids being misled by outliers. Unlike feature extraction, which typically transforms data into abstract representations, variable selection provides clear, understandable insights into which variables are truly relevant to the task at hand.

Looking ahead, the authors plan to conduct further research on the performance and robustness of the methods presented in this chapter, with a focus on expanding them to more complex scientific computing tasks, particularly in the context of deep learning (Goodfellow et al., 2016). This research will contribute to the rise of robust statistical learning, or robust machine learning, by promoting the use of robust estimation techniques within machine learning models. Many machine learning tasks, such as deep fake detection using centroids, robust template matching with convolutional neural networks (CNNs) (Sun et al., 2021), and the application of filters in CNN layers, have not yet explored the potential of robust methods.

Another promising area of future research involves combining various data types and dimensionality reduction methods. While there have been some attempts to integrate different dimensionality reduction techniques (Ayesha et al., 2020), the robustness of such combined approaches has not yet been explored. Additionally, the limited availability of suitable big datasets has hindered the development of systematic comparisons of dimensionality reduction methods, especially for Big Data (Sadiq et al., 2021). Overcoming these challenges will be crucial for advancing the field of dimensionality reduction and robust machine learning.

ACKNOWLEDGMENT

The work was supported by the grant 25-15490S of the Czech Science Foundation. The author would like to thank Petra Vidnerová for Figure 1 and technical assistence; she was supported by the program Strategy AV21 „AI: Artificial Intelligence for Science and Society". We thank the Institute of Computer Science of the Czech Academy of Sciences for its support and congratulate it on 50 years of excellence.

REFERENCES

Alfons, A., Croux, C., & Gelper, S. (2013). Sparse least trimmed squares regression for analyzing high-dimensional large data sets. *The Annals of Applied Statistics*, *7*(1), 226–248. DOI: 10.1214/12-AOAS575

Ayesha, S., Hanif, M. K., & Talib, R. (2020). Overview and comparative study of dimensionality reduction techniques for high dimensional data. *Information Fusion*, *59*, 44–58. DOI: 10.1016/j.inffus.2020.01.005

Bas, E., Egrioglu, E., & Cansu, T. (2024). Robust training of median dendritic artificial neural networks for time series forecasting. *Expert Systems with Applications*, *238*, 122080. DOI: 10.1016/j.eswa.2023.122080

Benjamini, Y., & Hochberg, Y. (1995). Controlling the false discovery rate: A practical and powerful approach to multiple testing. *Journal of the Royal Statistical Society. Series B, Statistical Methodology*, *57*(1), 289–300. DOI: 10.1111/j.2517-6161.1995.tb02031.x

Benjamini, Y., & Yekutieli, D. (2001). The control of the false discovery rate in multiple testing under dependency. *Annals of Statistics*, *29*(4), 1165–1188. DOI: 10.1214/aos/1013699998

Binson, V. A., Thomas, S., Subramoniam, M., Arun, J., Naveen, S., & Madhu, S. (2024). A review of machine learning algorithms for biomedical applications. *Annals of Biomedical Engineering*, *52*(5), 1159–1183. DOI: 10.1007/s10439-024-03459-3 PMID: 38383870

Blazques, D., & Domenech, J. (2018). Big Data sources and methods for social and economic analyses. *Technological Forecasting and Social Change*, *130*, 99–113. DOI: 10.1016/j.techfore.2017.07.027

Bradlow, E. T., Gangwar, M., Kopalle, P., & Voleti, S. (2017). The role of Big Data and predictive analytics in retailing. *Journal of Retailing*, *93*(1), 79–95. DOI: 10.1016/j.jretai.2016.12.004

Bridge, J. & Dodds, J.C. (2019). Managerial decision making. Routledge, Abingdon-on-Thames.

Bučková, B., Brunovský, M., Bareš, M., & Hlinka, J. (2020). Predicting sex from EEG: Validity and generalizability of deep-learning-based interpretable classifier. *Frontiers in Neuroscience*, *14*, 589303. DOI: 10.3389/fnins.2020.589303 PMID: 33192274

Chen, L. P. (2024). Variable selection and estimation for misclassified binary responses and multivariate error-prone predictors. *Journal of Computational and Graphical Statistics*, *33*(2), 407–420. DOI: 10.1080/10618600.2023.2218428

Chollet, F. (2018). *Deep learning with Python*. Manning Publications Co.

Chollet, F. (2024). Keras, version 3.0.0. Github repository. [https://github.com/keras-team/keras]

Delaigle, A., & Hall, P. (2012). Achieving near perfect classification for functional data. *Journal of the Royal Statistical Society. Series A, (Statistics in Society)*, *74*, 267–286.

Dey, P., Gopal, M., Pradhan, P., & Pal, T. (2019). On robustness of radial basis function network with input perturbation. *Neural Computing & Applications*, *31*(2), 523–537. DOI: 10.1007/s00521-017-3086-5

Dheeru, D., & Taniskidou, E. K. (2017). *UCI Machine Learning Repository*. University of California. [http://archive.ics.uci.edu/ml]

Fan, J., Ke, Y., & Wang, K. (2020). Factor-adjusted regularized model selection. *Journal of Econometrics*, *216*(1), 71–85. DOI: 10.1016/j.jeconom.2020.01.006 PMID: 32269406

Filzmoser, P., & Nordhausen, K. (2021). Robust linear regression for high-dimensional data: An overview. *Wiley Interdisciplinary Reviews: Computational Statistics*, *13*(4), e1524. DOI: 10.1002/wics.1524

Fordellone, M. (2019). *Statistical analysis of complex data. Dimensionality reduction and classification methods*. LAP LAMBERT Academic Publishing.

Goodfellow, I., Bengio, Y., & Courville, A. (2016). *Deep learning*. MIT Press.

Greene, W. H. (2017). *Econometric analysis* (8th ed.). Pearson.

Heinze, G., Wallisch, C., & Dunkler, D. (2017). Variable selection–A review and recommendations for the practicing statistician. *Biometrical Journal. Biometrische Zeitschrift*, *60*(3), 431–449. DOI: 10.1002/bimj.201700067 PMID: 29292533

Herb, K. (2024). *Engineering mathematics and artificial intelligence: Foundations, methods, and applications*. Routledge.

Holder, A., & Eichholz, J. (2019). *An introduction to computational science*. Springer. DOI: 10.1007/978-3-030-15679-4

Huber, P. J. (2009). *Robust statistics* (2nd ed.). Wiley. DOI: 10.1002/9780470434697

Jacobson, N. J., Bentley, K. H., Walton, A., Wang, S. B., Fortgang, R. G., Millner, A. J., Coombs, G.III, Rodman, A. M., & Coppersmith, D. D. L. (2020). Ethical dilemmas posed by mobile health and machine learning in psychiatry research. *Bulletin of the World Health Organization*, *98*(4), 270–276. DOI: 10.2471/BLT.19.237107 PMID: 32284651

Jurečková, J., Picek, J., & Schindler, M. (2019). *Robust statistical methods with R* (2nd ed.). CRC Press. DOI: 10.1201/b21993

Kalina, J. (2015). Three contributions to robust regression diagnostics. Journal of Applied Mathematics. *Statistics and Informatics*, *11*(2), 69–78.

Kalina, J. (2018). A robust pre-processing of BeadChip microarray images. *Biocybernetics and Biomedical Engineering*, *38*(3), 556–563. DOI: 10.1016/j.bbe.2018.04.005

Kalina, J., & Duintjer Tebbens, J. (2015). Algorithms for regularized linear discriminant analysis. Proceedings of the 6th International Conference on Bioinformatics Models Methods, and Algorithms (BIOINFORMATICS '15), Scitepress, Lisbon, 128-133. DOI: 10.5220/0005234901280133

Kalina, J., & Matonoha, C. (2020). A sparse pair-preserving centroid-based supervised learning method for high-dimensional biomedical data or images. *Biocybernetics and Biomedical Engineering*, *40*(2), 774–786. DOI: 10.1016/j.bbe.2020.03.008

Kalina, J., & Rensová, D. (2015). How to reduce dimensionality of data: Robustness point of view. *Serbian Journal of Management*, *10*(1), 131–140. DOI: 10.5937/sjm10-6531

Kalina, J., Seidl, L., Zvára, K., Grünfeldová, H., Slovák, D., & Zvárová, J. (2013). System for selecting relevant information for decision support. *Studies in Health Technology and Informatics*, *186*, 83–87. DOI: 10.3233/978-1-61499-240-0-83 PMID: 23542973

Kalina, J., Vašaničová, P., & Litavcová, E. (2019). Regression quantiles under heterosceasticity and multicollinearity: Analysis of travel and tourism competitiveness. Ekonomický časopis. *Journal of Economics*, *67*, 69–85.

Kalina, J., & Vidnerová, P. (2020). On robust training of regression neural networks. In Aneiros, G., Horová, I., Hušková, M., & Vieu, P. (Eds.), *Functional and high-dimensional statistics and related fields. IWFOS 2020. Contributions to Statistics* (pp. 145–152). Springer. DOI: 10.1007/978-3-030-47756-1_20

Kalina, J., & Zvárová, J. (2013). Decision support systems in the process of improving patient safety. In Moumtzoglou, A., & Kastania, A. (Eds.), *E-health technologies and improving patient safety: Exploring organizational factors* (pp. 71–83). IGI Global. DOI: 10.4018/978-1-4666-2657-7.ch005

Kalogridis, I., Claeskens, G., & Van Aelst, S. (2023). Robust and efficient estimation of nonparametric generalized linear models. *Test*, *32*(3), 1055–1078. DOI: 10.1007/s11749-023-00866-x PMID: 37363066

Kaveh, A. (2024). *Applications of artificial neural networks and machine learning in civil engineering*. Springer. DOI: 10.1007/978-3-031-66051-1

Lakshmanaprabu, S. K., Mohanty, S. N., Krishnamoorthy, S., Uthayakumar, J., & Shankar, K. (2019). Online clinical decision support system using optimal deep neural networks. *Applied Soft Computing*, *81*, 105487. DOI: 10.1016/j.asoc.2019.105487

Lalmuanawma, S., Hussain, J., & Chhaakchhuak, L. (2020). Applications of machine learning and artificial intelligence for Covid-19 (SARS-CoV-2) pandemic: A review. *Chaos, Solitons, and Fractals*, *139*, 110059. DOI: 10.1016/j.chaos.2020.110059 PMID: 32834612

Li, Y., Orgerie, A.C., Rodero, I., Amersho, B.L., Parashar, M. et al. (2018). End-to-end models for edge cloud-based IoT platforms: Application to data stream analysis in IoT. [https://hal.archives-ouvertes.fr/hal-01673501]

López-Robles, J. R., Rodríguez-Salvador, M., Gamboa-Rosales, N. K., Ramirez-Rosales, S., & Cobo, M. J. (2019). The last five years of Big Data Research in economics, econometrics and finance: Identification and conceptual analysis. *Procedia Computer Science*, *162*, 729–736. DOI: 10.1016/j.procs.2019.12.044

Luo, X., Li, Y., Wang, W., Ban, X., Wang, J. H., & Zhao, W. (2020). A robust multilayer extreme learning machine using kernel risk-sensitive loss criterion. *International Journal of Machine Learning and Cybernetics*, *11*(1), 197–216. DOI: 10.1007/s13042-019-00967-w

Mahmoud, M. M. E., Rodrigues, J. J. P. C., Ahmed, S. H., Shah, S. C., Al-Muhtadi, J. F., Korotaev, V. V., & De Albuquerque, V. H. C. (2018). Enabling technologies on cloud of things for smart healthcare. *IEEE Access : Practical Innovations, Open Solutions*, *6*, 31950–31967. DOI: 10.1109/ACCESS.2018.2845399

Marozzi, M., Mukherjee, A., & Kalina, J. (2020). Interpoint distance tests for high-dimensional comparison studies. *Journal of Applied Statistics*, *47*(4), 653–665. DOI: 10.1080/02664763.2019.1649374 PMID: 35707487

Martinez, W. L., Martinez, A. R., & Solka, J. L. (2017). *Exploratory data analysis with MATLAB* (3rd ed.). Chapman & Hall/CRC.

Mrázek, V., Sarwar, S. S., Sekanina, L., Vašíček, Z., & Roy, K. (2016). Design of power-efficient approximate multipliers for approximate artificial neural networks. *2016 IEEE/ACM International Conference on Computer-Aided Design (ICCAD)*, 1-7. DOI: 10.1145/2966986.2967021

Olson, D. L. (2017). *Descriptive data mining*. Springer. DOI: 10.1007/978-981-10-3340-7

Quarteroni, A. (2018). The role of statistics in the era of big data: A computational scientist' perspective. *Statistics & Probability Letters*, *136*, 63–67. DOI: 10.1016/j.spl.2018.02.047

Rousseeuw, P. J., & Leroy, A. M. (1987). *Robust regression and outlier detection*. Wiley. DOI: 10.1002/0471725382

Rousseeuw, P. J., & Van Driessen, K. (2006). Computing LTS regression for large data sets. *Data Mining and Knowledge Discovery*, *12*(1), 29–45. DOI: 10.1007/s10618-005-0024-4

Rusiecki, A. (2013). Robust learning algorithm based on LTA estimator. *Neurocomputing*, *120*, 624–632. DOI: 10.1016/j.neucom.2013.04.008

Sadiq, M. T., Yu, X., & Yuan, Z. (2021). Exploiting dimensionality reduction and neural network techniques for the development of expert brain-computer interfaces. *Expert Systems with Applications*, *164*, 114031. DOI: 10.1016/j.eswa.2020.114031

Shevlyakov, G. L., & Oja, H. (2016). *Robust correlation. Theory and applications*. Wiley. DOI: 10.1002/9781119264507

Staerk, C., Byrd, A., & Mayr, A. (2024). Recent methodological trends in epidemiology: No need for data-driven variable selection? *American Journal of Epidemiology*, *193*(2), 370–376. DOI: 10.1093/aje/kwad193 PMID: 37771042

Statti, F., Sued, M., & Yohai, V. J. (2018). High breakdown point robust estimators with missing data. *Communications in Statistics. Theory and Methods*, *47*(21), 5145–5162. DOI: 10.1080/03610926.2017.1388396

Steyvers, M., & Kumar, A. (2024). Three challenges for AI-assisted decision-making. *Perspectives on Psychological Science*, *19*(5), 722–734. DOI: 10.1177/17456916231181102 PMID: 37439761

Su, E. C. Y., & Wu, H. M. (2024). Dimension reduction and visualization of multiple time series data: A symbolic data analysis approach. *Computational Statistics*, *39*(4), 1937–1969. DOI: 10.1007/s00180-023-01440-7

Sun, L., Sun, H., Wang, J., Wu, S., Zhao, Y., & Xu, Y. (2021). Breast mass detection in mammography based on image template matching and CNN. *Sensors (Basel)*, *2021*(8), 2855. DOI: 10.3390/s21082855 PMID: 33919623

Tegmark, M. (2017). *Life 3.0: Being human in the age of artificial intelligence.* Alfred A. Knopf.

Tian, Y., Jin, X., Zhu, G., Hu, Y., Wang, Y., & Huang, Z. (2025). Data-driven dimension reduction for high-dimensional random vibration systems with cubic nonlinearity. *Nonlinear Dynamics*, *113*(5), 4163–4177. DOI: 10.1007/s11071-024-10039-6

Uppink, L. (2022). Travel & Tourism Development Index 2021. World Economic Forum, Geneva. [https://www.weforum.org/publications/travel-and-tourism-development-index-2021]

Wilson, P. W. (2018). Dimension reduction in nonparametric models of production. *European Journal of Operational Research*, *267*(1), 349–367. DOI: 10.1016/j.ejor.2017.11.020

Xu, J., Zhou, W., Fu, Z., Zhou, H., & Li, L. (2021). A survey on green deep learning. ArXiv:2111.05193.

Yilmaz, A., & Yolcu, U. (2023). A robust training of dendritic neuron model neural network for time series prediction. *Neural Computing & Applications*, *35*(14), 10387–10406. DOI: 10.1007/s00521-023-08240-6

Chapter 3
TinyML Empowering Intelligent Edge Devices

Helen K. Joy
https://orcid.org/0000-0001-7776-3856
Christ University, India

Electa Alice Jayarani
Kammavari Sangha's Institute of Technology, India

R. Sridevi
https://orcid.org/0000-0001-6620-141X
Christ University, India

ABSTRACT

Tiny Machine Learning (TinyML) bridges the gap between artificial intelligence and low-power embedded systems, enabling devices like microcontrollers to process data locally and operate autonomously. This chapter explores the foundational principles of TinyML, its applications across industries such as IoT, healthcare, industrial automation, and environmental monitoring, and the tools enabling its deployment. It also addresses challenges, including energy efficiency and model optimization, while providing insights into future advancements such as federated learning and neuromorphic computing. This chapter offers a comprehensive understanding of TinyML's transformative potential and its pivotal role in AI-based engineering solutions.

DOI: 10.4018/979-8-3373-0746-6.ch003

INTRODUCTION

In a world driven by data and intelligent automation, machine learning (ML) has become a cornerstone of modern technology. However, traditional ML models often require vast computing resources, high-performance processors, and continuous internet connectivity, limiting their deployment to cloud servers or powerful edge devices. But what if we could bring the power of AI to even the smallest of devices—microcontrollers that fit within the palm of your hand and operate on mere milliwatts of power?

This is precisely where **Tiny Machine Learning (TinyML)** comes into play. TinyML is a groundbreaking technology that enables machine learning models to run efficiently on ultra-low-power embedded systems. It bridges the gap between AI capabilities and resource-constrained hardware, allowing devices to process data and make intelligent decisions in real-time without relying on cloud computing.

What is TinyML?

Imagine a world where your coffee maker knows exactly when you wake up and starts brewing your favorite blend, or where a tiny sensor in a remote forest can detect the first signs of a wildfire and alert authorities before it spreads. This is the promise of Tiny Machine Learning (TinyML), a cutting-edge technology that brings the power of artificial intelligence (AI) to the smallest of devices.

Figure 1. Empowering small devices with AI

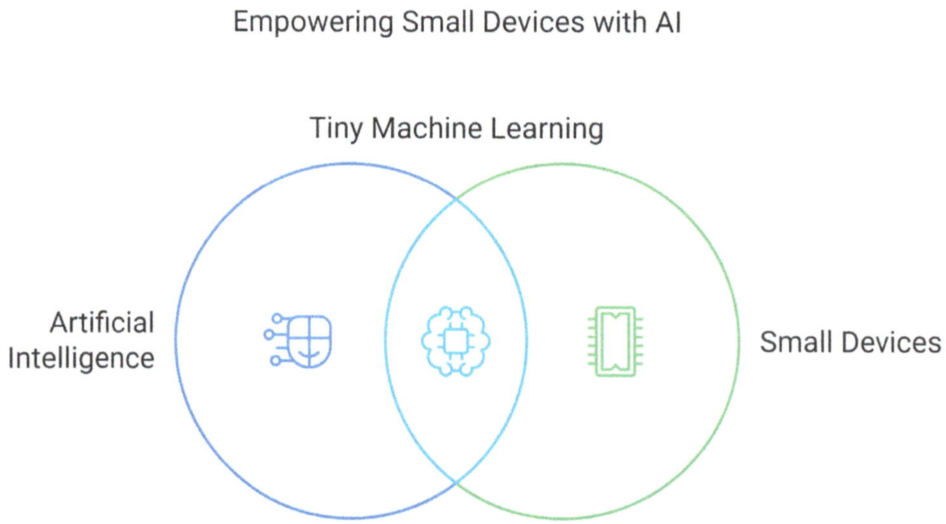

Figure 2. Tiny ML defined

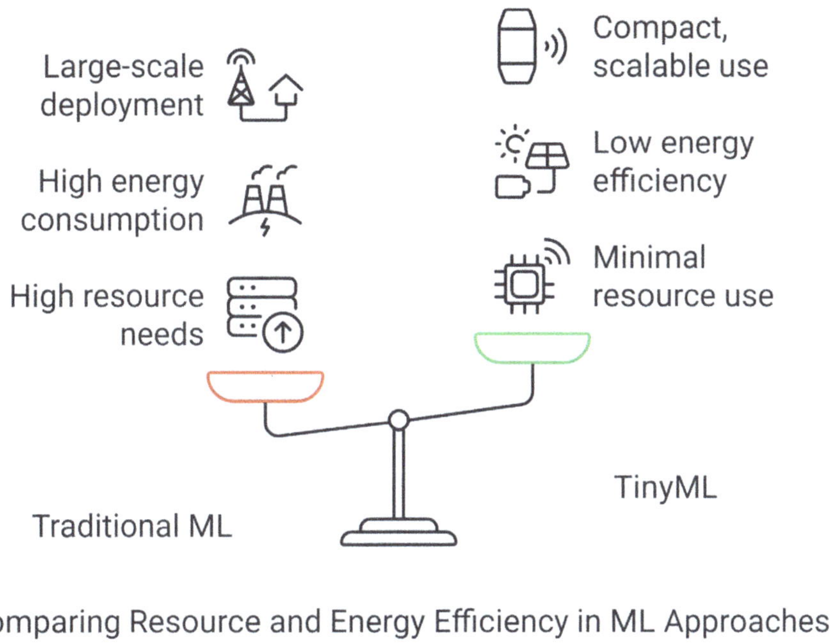

TinyML refers to the deployment of machine learning models on low-power, resource-constrained devices like microcontrollers. These devices, often no larger than a coin, can now perform tasks that were once the domain of powerful servers or cloud platforms. By enabling on-device AI, TinyML reduces latency, enhances privacy, and minimizes energy consumption, making it a game-changer for industries ranging from healthcare to environmental monitoring.

TinyML vs. Traditional ML and Edge AI

To understand TinyML, it's helpful to compare it with traditional machine learning and edge AI. Traditional ML relies on cloud-based servers to train and deploy models, requiring significant computational resources and energy. Edge AI, on the other hand, involves deploying ML models on edge devices like smartphones or gateways, which have more resources than microcontrollers but less than cloud servers.

TinyML takes this a step further by enabling ML on devices with kilobytes of memory and milliwatts of power. For example, while a smartphone might have gigabytes of RAM and a powerful processor, a microcontroller used in TinyML applications might have only 256 KB of memory and run on a battery for months or even years. This makes TinyML ideal for applications where energy efficiency, cost, and size are critical.

Why TinyML Matters

The importance of TinyML lies in its ability to bring intelligence to resource-constrained environments. Consider a scenario where you need to monitor air quality in a remote village with no reliable internet connection. Traditional cloud-based solutions would be impractical, but a TinyML-enabled sensor can analyze data locally and transmit only the most relevant information, conserving energy and bandwidth.

TinyML is also transforming industries by enabling real-time decision-making at the edge. For example, in industrial automation, TinyML-powered sensors can predict equipment failures before they occur, reducing downtime and maintenance costs. In healthcare, wearable devices with TinyML can monitor vital signs and provide real-time diagnostics, improving patient outcomes.

The emergence of TinyML as a key enabler in edge computing has received growing attention, particularly for enabling on-device intelligence in ultra-low-power scenarios. Baresi, Mendonça, and Garriga (2017) introduced an early architecture empowering low-latency applications through serverless edge computing, demonstrating foundational principles for time-critical systems.

Han and Siebert (2022) provide a comprehensive review of TinyML, mapping its evolution since 2019. Their systematic synthesis identifies the key characteristics of TinyML, such as energy efficiency, low latency, and on-device processing, establishing its relevance for embedded machine learning. Alajlan and Ibrahim (2022) expand on this by examining inference architectures and hardware-software co-design strategies that empower deep learning on microcontrollers. Further exploring algorithmic advancements, Capogrosso et al. (2023) present a machine learning-oriented survey, covering supervised, unsupervised, and reinforcement learning techniques applied in TinyML. They emphasize the need for model compression, quantization, and adaptation for real-time edge inference.

Real-world applications underscore TinyML's potential. Andrade et al. (2021) proposed an unsupervised TinyML pipeline for detecting pavement anomalies in intelligent vehicles. This work leverages embedded accelerometers and microcontrollers, illustrating robust anomaly detection without cloud dependencies. In the smart energy domain, Hayajneh et al. (2024) utilize TinyML for solar forecasting, demonstrating how compact ML models can boost solar yield predictions even on constrained devices. Gawate and Rane (2023) investigate the use of edge platforms such as NVIDIA Jetson Nano for intelligent manufacturing. Their work shows how such systems help optimize processes and maintain low latency in industrial automation.

Sharma et al. (2025) propose HamQ, a Hamming-weight-based quantization technique that boosts energy efficiency in analog compute-in-memory accelerators. This technique illustrates the importance of hardware-aware TinyML model design for intelligent sensor networks. Gong et al. (2023) provide a survey on edge intelligence in intelligent transportation systems, showcasing how TinyML can address latency, reliability, and bandwidth limitations. Meanwhile, Afachao and Abu-Mahfouz (2022) discuss the capabilities of intelligent IoT edge devices, stressing the role of embedded ML in enhancing situational awareness and autonomy at the network edge.

Collectively, these studies portray TinyML as a transformative technology capable of bridging resource constraints and intelligent processing, enabling a new class of privacy-preserving, low-power, and real-time edge AI solutions.

APPLICATIONS OF TINYML IN ENGINEERING

Smart Home and IoT Devices

The integration of Tiny Machine Learning (TinyML) into smart home and IoT devices is redefining the way people interact with their living spaces. Unlike traditional smart home systems that rely heavily on cloud computing, TinyML enables real-time, on-device intelligence, reducing latency, enhancing privacy, and improving energy efficiency. This allows devices to process data locally, making them faster, more reliable, and capable of operating even in environments with limited internet connectivity.

One of the most prominent applications of TinyML in smart homes is in intelligent climate control systems. Devices like the Google Nest Thermostat leverage TinyML to analyze user behavior, learn temperature preferences, and optimize energy usage by adjusting settings accordingly. By processing sensor data in real-time, these devices minimize unnecessary heating or cooling, leading to significant energy savings. Similarly, smart lighting systems equipped with TinyML can detect occupancy patterns, adjusting brightness and color temperature based on user activity without requiring manual intervention.

Security and surveillance systems are another area where TinyML is making a significant impact. Traditional security cameras often rely on cloud-based processing, leading to privacy concerns and potential delays in threat detection. With TinyML, smart security cameras and motion detectors can process video feeds and recognize patterns locally, distinguishing between routine movements (such as a passing car) and potential threats (like an unknown person approaching the door).

This ensures that alerts are generated in real time without transmitting sensitive footage to external servers, enhancing both security and privacy.

Voice assistants and smart speakers, such as Amazon Alexa and Google Assistant, also benefit from TinyML. By processing basic voice commands locally, these devices reduce response times and enhance user experience while ensuring that sensitive voice data remains private. This is especially crucial in applications where quick responses are necessary, such as controlling appliances, setting reminders, or triggering emergency alerts.Beyond comfort and convenience, TinyML is also improving home automation by integrating predictive analytics into daily household operations. Smart appliances, such as washing machines and refrigerators, can analyze usage patterns and suggest optimal operation schedules, reducing energy consumption and improving efficiency. Additionally, TinyML-powered health monitoring devices embedded in smart homes can track vital signs, detect anomalies, and provide early warnings for medical conditions, making home environments safer for elderly or at-risk individuals. The ability of TinyML to bring AI-driven decision-making to low-power, resource-constrained devices makes it a game-changer for the smart home industry. By enabling real-time processing, enhancing privacy, and optimizing energy use, TinyML is paving the way for truly intelligent homes that can adapt to user needs with minimal human intervention. As technology continues to evolve, the role of TinyML in smart home automation will only become more integral, leading to smarter, safer, and more efficient living environments.

Healthcare: Real-Time Monitoring and Diagnostics

The integration of Tiny Machine Learning (TinyML) into healthcare is transforming how medical diagnostics and patient monitoring are conducted. Traditional healthcare monitoring systems often rely on cloud-based processing, requiring continuous internet connectivity and raising concerns about latency, data privacy, and energy consumption. TinyML addresses these challenges by enabling real-time, on-device intelligence, allowing wearable and portable medical devices to analyze physiological signals locally. This advancement not only enhances the efficiency of healthcare systems but also empowers individuals with real-time insights into their health conditions. One of the most promising applications of TinyML in healthcare is continuous cardiac monitoring. TinyML-powered electrocardiogram (ECG) devices can continuously track heart rhythms and detect irregularities such as atrial fibrillation (AFib) or arrhythmias. Unlike traditional ECG monitors that require data to be transmitted to cloud servers for analysis, TinyML algorithms process the signals locally, identifying abnormalities instantly and alerting users or healthcare providers when necessary. This early warning system can be life-saving, enabling timely medical intervention and reducing the risk of severe cardiovascular events.

Additionally, because the processing happens on the device itself, power consumption is minimized, allowing for prolonged use in battery-powered wearables.

Another significant application of TinyML in healthcare is in diabetes management through Continuous Glucose Monitoring (CGM) systems. Traditional CGMs require frequent data uploads to cloud platforms, but with TinyML, these devices can analyze blood glucose trends in real-time and provide personalized insights without the need for constant internet access. By recognizing patterns in glucose fluctuations, TinyML models can predict potential hypo- or hyperglycemic events, alert users in advance, and suggest dietary or insulin adjustments. This enhances diabetes self-management, improves patient outcomes, and reduces dependency on manual monitoring.

In addition to cardiovascular and glucose monitoring, TinyML is being used in respiratory health monitoring, particularly for individuals with chronic conditions such as asthma or chronic obstructive pulmonary disease (COPD). TinyML-enabled smart inhalers and wearable respiratory sensors can track breathing patterns, detect anomalies such as sudden drops in oxygen levels, and provide personalized recommendations based on usage history. This capability allows for more precise treatment adjustments and helps prevent emergency situations by providing early warnings before symptoms worsen.

Beyond chronic disease management, TinyML is playing a crucial role in early disease detection. Wearable biosensors equipped with TinyML algorithms can analyze sweat composition, skin temperature, and hydration levels to detect potential infections, dehydration, or electrolyte imbalances. Such devices are particularly beneficial for remote patient monitoring, reducing hospital visits and enabling proactive healthcare interventions. In the field of neurological health, TinyML-based seizure detection systems are being developed to analyze EEG signals locally, allowing epilepsy patients to receive real-time alerts before a seizure occurs. The integration of TinyML in healthcare is paving the way for a new era of personalized, real-time, and data-driven medicine. By reducing the dependency on cloud-based processing, TinyML ensures faster response times, enhanced privacy, and prolonged battery life in wearable health devices. As research and development in TinyML continue to advance, its role in healthcare will expand, leading to more intelligent, efficient, and accessible medical solutions that improve patient care and overall well-being.

Industrial Automation and Predictive Maintenance

Tiny Machine Learning (TinyML) is revolutionizing industrial automation by enabling real-time predictive maintenance, reducing operational costs, and improving equipment reliability. In traditional industrial settings, machinery maintenance is often reactive or scheduled at fixed intervals, leading to unexpected breakdowns or unnecessary servicing. TinyML addresses these inefficiencies by allowing on-device machine learning models to continuously monitor equipment health, detect anomalies, and predict failures before they occur. By minimizing downtime and optimizing maintenance schedules, TinyML significantly enhances industrial productivity and cost-effectiveness.

One of the key applications of TinyML in predictive maintenance is in vibration and acoustic analysis for rotating machinery such as motors, turbines, and pumps. Sensors embedded with TinyML algorithms can analyze subtle changes in vibration patterns, detect misalignments, and identify early signs of wear and tear. Unlike traditional sensor-based monitoring systems that require data to be sent to centralized servers for analysis, TinyML processes the data locally, ensuring real-time fault detection and immediate response. This is particularly beneficial for remote industrial sites, manufacturing plants, and high-risk environments, where timely maintenance can prevent catastrophic failures.

Another important use case is temperature and pressure monitoring in industrial systems such as boilers, pipelines, and heavy machinery. TinyML-powered sensors continuously analyze fluctuations in temperature, pressure, and other environmental parameters. If an abnormal trend is detected, the system can generate alerts for maintenance teams, allowing them to address potential issues before they escalate. This proactive approach prevents costly repairs, enhances worker safety, and extends the lifespan of critical equipment.

A real-world example of TinyML-powered predictive maintenance is its implementation by Siemens in industrial IoT solutions. Siemens uses TinyML-enabled sensors to monitor the condition of industrial machines, analyzing vibration, temperature, and energy consumption data in real time. By identifying deviations from normal operating conditions, these systems can predict potential failures and schedule maintenance only when necessary, optimizing both resources and production efficiency. Beyond predictive maintenance, TinyML is also improving quality control in manufacturing. Sensors embedded in production lines can analyze sound, visual data, and operational parameters to detect defects or irregularities in real-time. For example, TinyML-based computer vision systems can inspect products for defects without the need for cloud-based image processing, making the inspection process faster and more reliable.

The advantages of TinyML in industrial automation extend beyond cost savings and efficiency. By reducing reliance on cloud processing, TinyML enhances data privacy and security, ensuring that sensitive operational data remains within the industrial facility. Additionally, the low power consumption of TinyML models makes them ideal for deployment in battery-operated or energy-harvesting sensors, enabling sustainable and scalable industrial solutions. As industries continue to adopt smart manufacturing and Industry 4.0 technologies, the role of TinyML in predictive maintenance and automation will become increasingly critical. By empowering machines with real-time intelligence, TinyML is reshaping industrial operations, making them more efficient, cost-effective, and resilient to unexpected failures.

Environmental Monitoring: Wildlife Tracking and Air Quality Sensors

Tiny Machine Learning (TinyML) is playing a crucial role in environmental monitoring, offering innovative solutions for tracking wildlife, assessing air quality, and detecting ecological changes in real time. Traditional environmental monitoring systems often rely on cloud-based data processing, which can be impractical in remote locations due to limited connectivity and high power consumption. TinyML overcomes these challenges by enabling on-device intelligence, allowing sensors to process and analyze data locally before transmitting only essential information. This approach significantly reduces energy consumption, enhances efficiency, and makes long-term environmental monitoring more sustainable.

One of the most impactful applications of TinyML in wildlife conservation is in tracking and studying animal behavior. Researchers use TinyML-powered wildlife tracking collars and sensor tags to monitor animal movements, detect behavioral anomalies, and gain insights into migration patterns. These lightweight, low-power devices analyze movement data in real time, identifying unusual activity that may indicate a nearby predator, illness, or environmental threats. For example, if a tracking collar detects an abrupt increase in an animal's movement speed or erratic behavior, it can flag potential danger and send alerts to conservationists. This helps in protecting endangered species, understanding habitat changes, and implementing more effective wildlife conservation strategies.

Another crucial area where TinyML is revolutionizing environmental monitoring is air quality assessment. Traditional air pollution monitoring systems often rely on centralized data collection, which can be inefficient for large-scale deployment. TinyML-powered air quality sensors can analyze pollutant levels (such as PM2.5, CO_2, and NO_x) on-site and send only relevant alerts when thresholds are exceeded. These battery-operated sensors can be deployed in urban areas, forests, and industrial zones, providing real-time pollution insights without requiring constant cloud connectivity. A real-world example of TinyML in environmental monitoring is the use of sensor networks in wildfire detection. TinyML-enabled sensors deployed in remote forests can continuously monitor temperature, humidity, and smoke levels. Instead of transmitting raw data continuously, the device processes the information locally and sends an alert only when early signs of a wildfire—such as sudden spikes in temperature or unusual smoke patterns—are detected. This approach helps firefighters respond faster, preventing large-scale forest destruction and reducing response times significantly.

Figure 3. Tiny ML applications in various areas

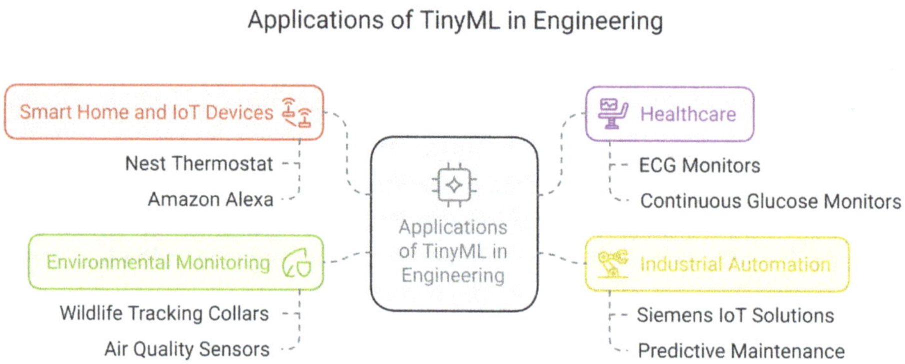

Beyond wildlife tracking and air quality monitoring, TinyML is being used in ocean conservation as well. Smart buoys equipped with TinyML sensors can analyze water quality parameters such as pH levels, dissolved oxygen, and temperature fluctuations to detect pollution or harmful algal blooms. By processing data on-site, these devices reduce power consumption and improve real-time decision-making for marine conservation efforts.The key advantages of TinyML in environmental monitoring include its low energy footprint, reduced reliance on cloud infrastructure, and ability to function in remote, off-grid locations. By making environmental sensors smarter and more efficient, TinyML is transforming the way researchers, conservationists, and policymakers gather and analyze ecological data. As technology advances, the integration of TinyML into smart environmental monitoring systems will continue to enhance sustainability efforts, protect wildlife, and mitigate the impact of climate change.

TOOLS AND FRAMEWORKS FOR TINYML DEVELOPMENT

TensorFlow Lite for Microcontrollers (TFLM)

TensorFlow Lite for Microcontrollers (TFLM) is one of the most popular frameworks for TinyML development. It is a lightweight version of TensorFlow designed specifically for microcontrollers. TFLM supports model quantization and optimization, making it suitable for TinyML applications.

For example, TFLM has been used to develop a TinyML-powered gesture recognition system for smart home devices. By analyzing accelerometer data, the system can recognize gestures like a wave or a swipe, enabling touchless control of devices.

Edge Impulse for Model Deployment

Edge Impulse is a development platform that simplifies the creation and deployment of TinyML models. It provides tools for data collection, model training, and deployment, making it accessible even to beginners.

For instance, Edge Impulse has been used to develop a TinyML-powered anomaly detection system for industrial equipment. By analyzing vibration data, the system can detect anomalies and predict equipment failures, reducing downtime and maintenance costs.

Arduino TinyML Kit for Beginners

The Arduino TinyML Kit is designed for hobbyists and beginners. It includes hardware, software, and tutorials to help users get started with TinyML projects, such as gesture recognition and voice control.

For example, the Arduino TinyML Kit has been used to develop a TinyML-powered voice-controlled robot. By analyzing audio data, the robot can recognize voice commands and perform tasks like moving forward or turning left.

STMicroelectronics' X-Cube-AI for Industrial Applications

X-Cube-AI is a software expansion pack for STM32 microcontrollers. It enables the deployment of AI models on STM32 devices, making it ideal for industrial applications like predictive maintenance and anomaly detection.

Figure 4. Empowering TinyML innovations with key frameworks and tools

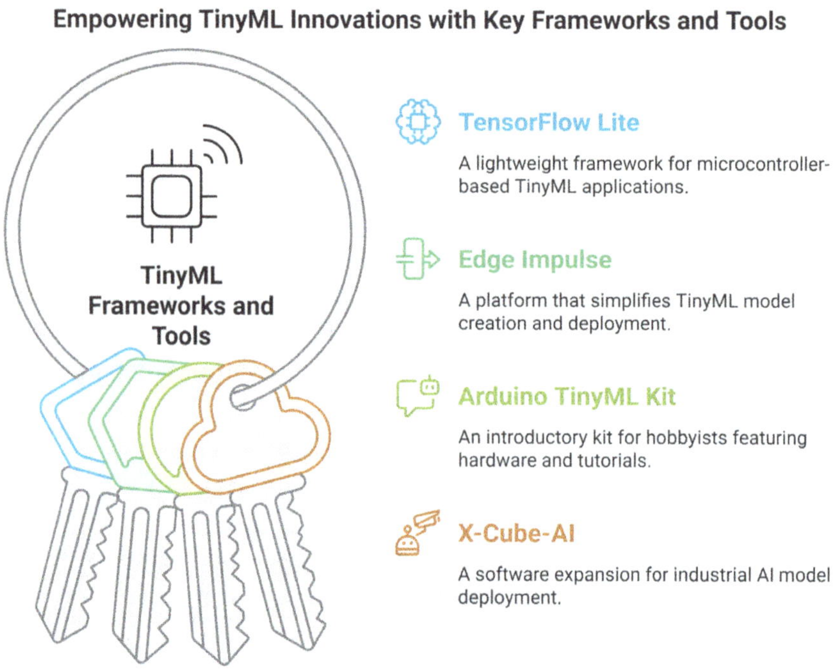

For instance, X-Cube-AI has been used to develop a TinyML-powered predictive maintenance system for wind turbines. By analyzing vibration and temperature data, the system can predict equipment failures and schedule maintenance before a breakdown occurs.

CHALLENGES IN TINYML IMPLEMENTATION

Optimizing Models for Low Memory and Computational Power

One of the biggest challenges in TinyML is optimizing ML models to fit within the limited memory and computational power of microcontrollers. Techniques like model quantization, pruning, and knowledge distillation are often employed to address this challenge.

For example, model quantization reduces the precision of model weights from 32-bit floating-point numbers to 8-bit integers, significantly reducing memory usage without sacrificing accuracy.

Energy Efficiency in Battery-Powered Devices

Figure 5. How to address TinyML challenges

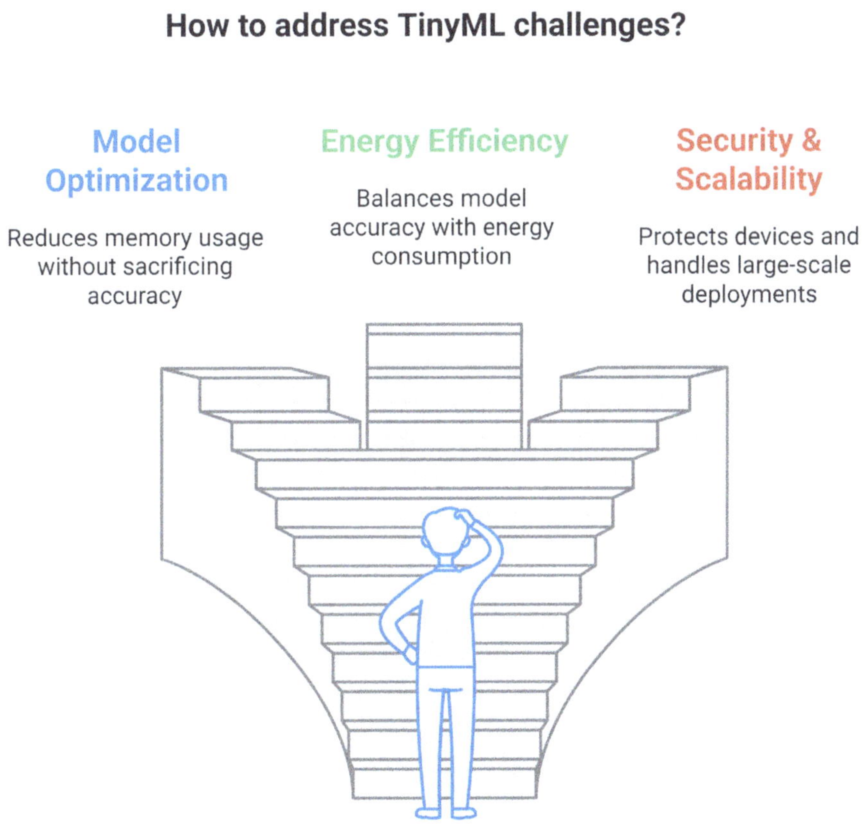

Energy efficiency is critical for battery-powered TinyML devices. Developers must balance model accuracy with energy consumption, often using techniques like duty cycling and low-power hardware design.

For instance, duty cycling involves periodically turning off the device to conserve energy. This technique is commonly used in environmental monitoring applications, where devices need to operate for extended periods without maintenance.

Security and Scalability in Large-Scale Deployments

Ensuring the security and scalability of TinyML deployments is another challenge. Devices must be protected against cyber threats, and systems must be designed to handle large-scale deployments without compromising performance.

For example, encryption and secure boot mechanisms can be used to protect TinyML devices from cyberattacks. Additionally, distributed computing techniques can be used to scale TinyML deployments across large networks.

FUTURE TRENDS IN TINYML

Federated Learning for Collaborative Training

Federated learning allows multiple devices to collaboratively train a shared model without sharing raw data. This approach enhances privacy and reduces the need for centralized data storage, making it ideal for TinyML applications.

Figure 6. Innovative advances shaping the future of TinyML technology

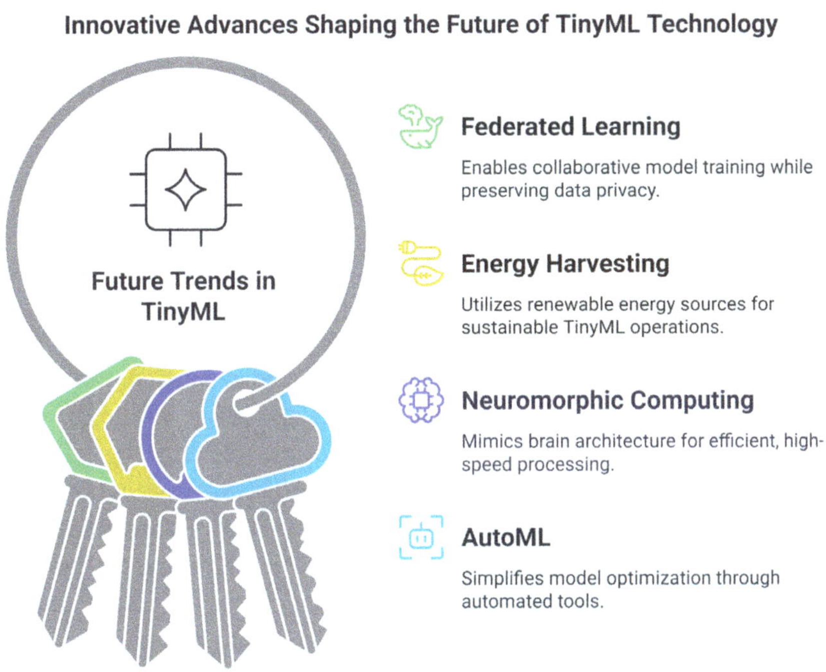

For example, federated learning can be used to train a TinyML model for speech recognition across multiple devices, ensuring that user data remains private.

Energy Harvesting for Sustainable Operations

Energy harvesting technologies, such as solar or kinetic energy, can enable TinyML devices to operate sustainably in remote or inaccessible locations.

For instance, solar-powered TinyML devices can be used to monitor environmental conditions in remote areas, providing valuable data without the need for frequent battery replacements.

Neuromorphic Computing for Enhanced Processing

Neuromorphic computing mimics the human brain's architecture, offering significant improvements in energy efficiency and processing speed. This technology has the potential to revolutionize TinyML by enabling more complex models on low-power devices.

For example, neuromorphic chips can be used to develop TinyML-powered vision systems that can recognize objects in real-time, even in low-light conditions.

AutoML for Simplified Model Optimization

Automated Machine Learning for TinyML: Enabling Intelligent Edge Solutions with Minimal Effort

The convergence of embedded systems and artificial intelligence has given rise to an exciting field known as TinyML—Machine Learning deployed on ultra-low-power microcontrollers and edge devices. These systems, which include sensors, wearables, and various IoT nodes, are increasingly capable of performing intelligent tasks such as speech recognition, anomaly detection, and gesture control. However, developing efficient machine learning models that can operate within the stringent resource constraints of such hardware presents a formidable challenge. Traditional approaches to TinyML development demand substantial expertise in both machine learning and embedded systems. Developers must carefully engineer features, experiment with various model architectures, perform hyperparameter tuning, and then compress the models without sacrificing too much accuracy. These manual tasks are not only labor-intensive but also time-consuming, often stretching development cycles to several weeks or even months. Moreover, as new use cases emerge and data continues to evolve, developers are required to revisit and re-optimize their models frequently.

Automated Machine Learning (AutoML) has emerged as a transformative solution to this challenge. It brings the power of automation to various stages of the machine learning pipeline, enabling developers to create robust, optimized models with minimal manual intervention. When integrated with TinyML workflows, AutoML offers the promise of rapid prototyping and deployment of intelligent edge applications. AutoML leverages algorithms to automatically explore and evaluate multiple combinations of data preprocessing strategies, model types, and hyperparameter settings. In the context of TinyML, it also considers additional constraints such as model size, memory footprint, and inference time, which are critical for deployment on devices with limited resources. The result is a model that is not only accurate but also lightweight and efficient—ready to be deployed on microcontrollers that may have as little as a few kilobytes of RAM.

One of the key areas where AutoML has demonstrated its value in TinyML is in the development of gesture recognition systems. Consider the example of a wearable fitness tracker or a smart remote control, both of which rely on interpreting human gestures using onboard sensors such as accelerometers and gyroscopes. To build a model capable of distinguishing between gestures like tapping, swiping, or shaking, a developer must collect labeled time-series data, engineer features or allow the system to extract them, train several candidate models, and then select the one that best balances performance and efficiency.

With AutoML, much of this complexity is abstracted away. The developer can focus on collecting quality data and defining the problem, while the AutoML engine takes care of exploring model architectures, adjusting parameters such as learning rates and batch sizes, evaluating performance using cross-validation, and applying quantization techniques to reduce the final model size. Tools such as Edge Impulse, Google's TensorFlow Lite Model Maker, Microsoft's Azure Percept Studio, and Qeexo AutoML provide user-friendly interfaces that allow developers—even those with limited backgrounds in machine learning—to create and deploy gesture recognition models in a matter of hours. These tools often include hardware-aware optimization steps, where models are evaluated not only on accuracy but also on their suitability for specific target devices. For instance, a model trained for gesture recognition on an ARM Cortex-M4 processor can be benchmarked for memory consumption, latency, and energy usage before being deployed. The entire optimization loop can be conducted with minimal code, empowering engineers to focus on system integration and user experience.

In addition to accelerating development, AutoML helps standardize the modeling process, reducing the variability introduced by manual experimentation. This consistency is particularly valuable in industrial and commercial applications, where reproducibility and reliability are paramount. AutoML also fosters rapid iteration: as new data becomes available or as system requirements change, models can be retrained and re-optimized with just a few modifications, ensuring that applications remain adaptive and responsive. Nevertheless, it is important to acknowledge the limitations of current AutoML systems in TinyML. One concern is the "black-box" nature of many AutoML workflows, which can obscure the decision-making process behind model selection and tuning. For mission-critical applications, transparency and explainability are often as important as performance. Additionally, some AutoML frameworks rely on cloud-based training, raising questions about data privacy and energy consumption during training. As these tools continue to evolve, future developments may focus on making AutoML more transparent, privacy-preserving, and sustainable.

AutoML represents a significant leap forward in making TinyML development more accessible, efficient, and scalable. By automating key aspects of the machine learning lifecycle, AutoML reduces the technical burden on developers and opens up new possibilities for deploying intelligent features on small, power-constrained devices. Whether it's recognizing a user's gesture, detecting anomalies in a factory machine, or interpreting spoken commands in a smart home, AutoML equips developers with the tools they need to build sophisticated models that meet the demands of real-world edge computing environments. As the field matures, AutoML is poised to become an indispensable component of the TinyML toolkit, driving innovation across industries and use cases.

CASE STUDY: TINYML IN ACTION

Use Case: ML-Powered Agricultural Robotics

One of the most compelling applications of TinyML is in the field of agricultural robotics, where it is revolutionizing crop trials and precision farming. A UK-based robotics startup, in collaboration with Adiuvo, has developed an agricultural robot called **MARV-bot** (Multi-purpose Agricultural Robotics Vehicle). This robot is designed to support crop trials by accurately recording crop growth and providing precise irrigation levels, helping researchers quantify the best-growing conditions for various crops.

Crop trials are inherently intensive operations. They require meticulous recording of crop growth data and controlled irrigation to determine optimal growing conditions. These trials are often conducted in remote locations, making access difficult for researchers and increasing operational costs. MARV-bot addresses these challenges by automating data collection and irrigation, reducing human intervention, and improving the accuracy and efficiency of crop trials.

Implementation Steps and Challenges

The implementation of the MARV-bot is a compelling example of integrating robotics with precision agriculture to achieve sustainable and intelligent field operations. Designed to operate in structured crop trial environments, the MARV-bot is constructed with a high-clearance frame that straddles rows of crops without causing physical disruption. This elevated configuration allows most of the robot's operational hardware—such as imaging modules, sprayers, and control electronics—to remain above the canopy, thereby ensuring that sensitive plant structures are undisturbed during field navigation.

The robot is engineered to perform two core agricultural functions: spraying and imaging with analysis. These functions, although unified in the robotic platform, demand contrasting operational parameters, particularly in terms of speed, control logic, and hardware usage. The careful orchestration of these operations requires a multi-layered system of mechanical, electrical, and computational subsystems, all coordinated by a real-time control framework.

In the spraying mode, the MARV-bot is calibrated to travel at a constant speed of approximately three miles per hour. This speed has been selected based on field trials to ensure optimal coverage and penetration of spray droplets while maintaining uniformity across the treatment area. The spraying system consists of precision nozzles arranged along a lateral bar, controlled by electronically actuated valves. These valves are governed by a control algorithm that adjusts the flow rate based on movement speed, nozzle pressure, and ambient environmental conditions such as wind speed and direction. The consistency of droplet size and coverage area is essential, particularly in agricultural trials where accurate and reproducible application is critical for experimental integrity.

During spraying, navigation is driven by a combination of GPS positioning and inertial measurements. The robot uses pre-mapped coordinates to traverse between rows, with minimal lateral deviation to avoid drift and double coverage. Given that the spraying process is sensitive to misalignment, terrain irregularities can pose challenges. Uneven surfaces may lead to fluctuations in nozzle height or pitch angle, potentially compromising uniform application. Furthermore, environmental factors such as sudden gusts of wind can distort the spray pattern, leading to unintended dispersion or drift. These challenges necessitate the use of closed-loop feedback mechanisms that continuously monitor environmental parameters and adjust spraying parameters in real time.

The second major function of the MARV-bot involves imaging and analytical data acquisition. Unlike spraying, which demands a relatively swift traversal, imaging requires the robot to slow down considerably, typically to a speed of one to one-and-a-half miles per hour. This reduction in velocity is crucial to ensure the acquisition of high-resolution, motion-free images suitable for downstream analysis. The imaging system includes both RGB and multispectral cameras, mounted in fixed downward-facing positions to capture detailed views of the plant canopy and soil surface.

As the robot moves along the trial plot, images are captured at fixed spatial intervals and automatically tagged with geolocation and time-stamp metadata. These images are either stored locally or transmitted wirelessly for real-time or offline processing. Onboard edge computing units, often powered by embedded GPUs or neural accelerators, perform preliminary analysis to extract relevant crop health indicators. Such analysis may include detection of foliar diseases, estimation of leaf area index, identification of weed patches, or assessment of plant vigor using normalized vegetation indices. This capability allows for rapid feedback to researchers and agronomists, enabling them to take timely corrective actions.

Despite its capabilities, the implementation of the MARV-bot is not without challenges. Lighting variability poses a significant obstacle in field imaging. Fluctuations in sunlight due to cloud movement, shadows from structural elements, or the time of day can all affect image consistency. Advanced image normalization and adaptive thresholding techniques are often required to mitigate these variations. In terms of image processing, the computational burden on the robot's onboard systems can be substantial, especially when dealing with high-resolution video streams or performing complex inference tasks. As a result, model optimization techniques such as quantization, pruning, and reduced-precision arithmetic are employed to maintain real-time performance.

Additionally, synchronization between hardware components—such as aligning image capture with position tracking and sensor data—is a non-trivial task. Misalignment can lead to errors in geo-referencing, which in turn affects the accuracy of spatial analyses. Power management is another crucial aspect, as the simultaneous operation of pumps, motors, cameras, and computing units can quickly deplete onboard energy reserves. This is particularly problematic in large-scale fields or during prolonged trials, requiring either battery swap protocols or hybrid energy solutions that incorporate solar charging systems.

At the system level, transitioning between operational modes (from spraying to imaging, or vice versa) requires intelligent task scheduling and state management. These transitions are managed by control software that monitors task queues, environmental triggers, and user commands, ensuring that operational overlap or mechanical conflict does not occur.MARV-bot's deployment demonstrates the potential of robotics to enhance the precision, efficiency, and intelligence of agricultural trial management. Its dual-mode operation, although complex in execution, provides a unified platform for performing repetitive yet critical tasks with high accuracy. The implementation process demands a careful balance between mechanical design, control logic, environmental adaptability, and system integration. The challenges faced and overcome in this context offer valuable insights for future iterations of autonomous agricultural systems, particularly in adapting such technologies to broader and more variable field conditions.

This design allows the robot to perform two primary functions:

1. **Spraying**: The robot travels at a speed of 3 MPH to deliver precise irrigation and spraying.
2. **Imaging and Analysis**: The robot moves at a slower speed to capture high-resolution images of the crops and perform real-time analysis.

Figure 7. Development and impact of MARV-bot

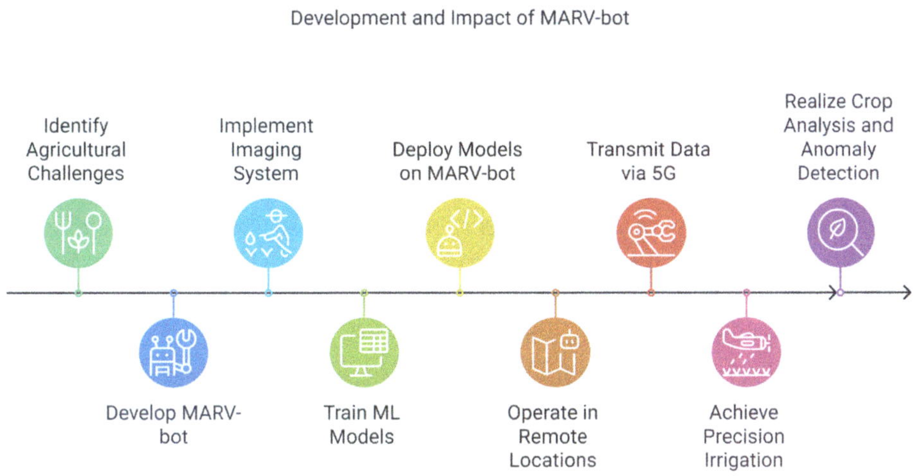

Key Features of MARV-bot

Imaging System

A core component of the MARV-bot's functionality is its sophisticated imaging system, designed to perform visual data acquisition and analysis directly from the field environment. Unlike conventional agricultural monitoring, which often depends on manual observations or aerial surveillance, the MARV-bot offers a ground-level, plant-by-plant imaging perspective. This fine-grained approach is particularly well-suited for crop trials, where precise phenotyping, weed monitoring, and anomaly detection are crucial.

The imaging system is composed of three tightly integrated subsystems: illumination, image capture, and edge processing. The illumination unit includes LED-based active lighting sources that provide consistent brightness and spectrum, allowing the system to operate effectively under varying natural light conditions such as shadows, overcast skies, or early morning light. This artificial illumination ensures that image quality remains consistent, a key factor for the reliability of machine learning-based analyses.

The image capture subsystem includes high-resolution RGB cameras, and optionally, multispectral or near-infrared (NIR) sensors. These sensors are rigidly mounted on the robot's elevated frame and oriented for nadir (downward-facing) imaging.

The fixed height and consistent velocity of the robot help to standardize the image acquisition geometry, reducing distortion and facilitating easier downstream analysis.

What makes MARV-bot's imaging system truly intelligent is its embedded TinyML-based analytical layer. These models, pre-trained and deployed on low-power microcontrollers or edge AI accelerators, are capable of performing real-time inference on each image. Key tasks include the detection and classification of individual plants, measurement of morphological traits (such as leaf area or canopy coverage), and the identification of anomalies such as pest-infested zones, nutrient deficiency patterns, or unauthorized plant species such as weeds. The use of lightweight convolutional neural networks (CNNs) or decision tree models ensures that this analysis is both fast and resource-efficient, avoiding the need for bulk data transmission or cloud dependency for inference.

This localized intelligence allows MARV-bot to respond autonomously to field conditions. For example, upon detecting an anomalous weed patch, the robot could flag the area for additional spraying, mark it for human inspection, or transmit a geo-tagged alert to the trial supervisor. This integration of imaging and analytics transforms the robot into not just a passive observer, but an active, decision-capable field assistant.

Communication and Navigation

Operating in the remote and sometimes infrastructure-poor environments typical of crop trials, MARV-bot must be capable of robust communication and precise navigation. These capabilities form the operational backbone of the system, supporting everything from real-time telemetry to data geo-referencing and autonomous path planning.

For communication, MARV-bot employs hybrid networking strategies, depending on the geographical location and connectivity infrastructure available. In well-connected rural areas, terrestrial 5G networks provide low-latency communication channels, enabling real-time streaming of telemetry and processed data to cloud-based dashboards. In more isolated trial plots, the robot leverages satellite-based 5G services or long-range radio frequency (LoRa) networks. These systems allow the robot to maintain continuous connectivity with control centers or research institutions, even in environments where conventional mobile networks are unavailable.

On the navigation front, Global Navigation Satellite Systems (GNSS) play a pivotal role. The robot is equipped with high-precision GNSS receivers, potentially augmented with Real-Time Kinematic (RTK) correction services to achieve centimeter-level accuracy. This degree of precision is essential for consistently tagging plant locations, aligning image datasets, and returning to the same crop rows in successive missions. GNSS data is also used in coordination with onboard inertial measurement units (IMUs) and wheel encoders to improve localization in areas where satellite signals may temporarily degrade, such as near dense vegetation or metallic structures.

The seamless integration of communication and navigation subsystems allows MARV-bot to operate with a high degree of autonomy while remaining tethered to supervisory systems for mission planning, data synchronization, and operational monitoring. Furthermore, real-time location tagging ensures that all data collected—whether image-based, environmental, or analytical—is spatially contextualized, enhancing the value of the insights derived.

Power Efficiency

One of the most critical aspects of autonomous field robots is energy management. Unlike stationary devices, mobile robots like MARV-bot must allocate power across multiple demanding subsystems, including mobility, sensing, computation, and communication. In the context of agricultural robotics, where access to recharging stations may be limited and field operations can span several hours or days, power efficiency becomes a determining factor in system viability.

The MARV-bot is engineered to function under a constrained power budget, and its subsystems are optimized accordingly. Among the highest consumers of energy are the navigation drives and communication modules. Motors powering the wheels or tracks must be energy-efficient and often include regenerative braking systems to recover energy during deceleration. Communication modules, especially those transmitting over long distances or through satellite uplinks, are designed to operate in duty-cycled modes to minimize active transmission time.

Given these constraints, the choice of computational framework is particularly important. Instead of relying on traditional machine learning models that require substantial computational and memory resources, MARV-bot employs TinyML models—machine learning models specifically optimized for execution on micro-controllers and embedded processors. These models are quantized and pruned to reduce size and computational complexity, allowing them to operate within the limited memory (often less than 256 KB RAM) and power envelope of typical edge devices.

The use of TinyML ensures that critical analytical functions such as object detection, anomaly classification, and image feature extraction can be performed locally without the need for high-power CPUs or continuous cloud connectivity. Additionally, power-aware scheduling algorithms are implemented to prioritize or delay tasks based on current battery levels, solar input (if panels are present), and task criticality.

This design philosophy enables MARV-bot to perform extended missions in the field—sometimes lasting several days—with only minimal recharging requirements.

This self-sufficiency is vital for real-world agricultural applications, where deployment environments may be spread across acres of land and logistical support is limited.

Development Process

Imaging and Analysis System: Embedded Intelligence through Edge Impulse and Sony Spresense

A pivotal element in the MARV-bot's architecture is its imaging and analysis system, which serves as the robot's perceptual core. This subsystem enables the robot not only to "see" the crop field but also to interpret it intelligently in real-time. The design and deployment of this system required careful consideration of both hardware and software components, particularly with respect to energy efficiency, processing power, and deployment flexibility—challenges inherent to mobile robotic systems operating in rural, off-grid environments.

To facilitate rapid prototyping and to minimize the development risk associated with custom hardware design, the MARV-bot team selected off-the-shelf embedded development boards. Among the hardware platforms considered, the Sony Spresense emerged as a highly suitable choice due to its unique combination of capabilities. The board offers an energy-efficient multicore architecture, a built-in GPS module, and native support for a high-resolution 5-megapixel camera module. Its compact form factor allows it to be deployed in multiple locations on the MARV-bot's frame—such as front, side, and downward-facing orientations—providing a comprehensive visual field without breaching the overall power envelope of the system.

One of the standout features of the Spresense is its ability to run advanced computer vision workloads at the edge using TinyML models. These models, by design, are lightweight and optimized for execution on low-power microcontrollers. To streamline the training and deployment of such models, the development team integrated Edge Impulse, a widely used platform for end-to-end TinyML model development. Edge Impulse provided not only the machine learning toolchain but also seamless integration with the Sony Spresense board via its C++ deployment SDK.

The development of the imaging model followed a structured workflow:

The first phase was dataset collection and preparation. Since field deployment was in its early stages, the team sourced an initial image dataset from Kaggle, focusing on three major crops: wheat, maize, and sugarcane. This dataset included a diverse range of crop images captured under different lighting, growth stages, and environmental conditions. The images were curated and annotated, including samples containing common field anomalies, such as the presence of weeds, inter-crop contamination, or signs of fungal infections. This annotated dataset was uploaded into Edge Impulse's data ingestion portal, where it was automatically segmented for training, testing, and validation.

The next phase was model training and optimization. Within the Edge Impulse environment, a compact convolutional neural network (CNN) architecture was selected, suitable for classification tasks on edge devices. The model was trained to classify the three crops based on leaf and canopy features and to detect anomalous patterns indicating the presence of unwanted plant species or disease symptoms. During training, Edge Impulse provided real-time metrics on model accuracy, confusion matrices, and memory usage estimates, allowing the team to iteratively refine the model architecture to strike a balance between accuracy and computational cost.

After satisfactory performance was achieved—typically an overall classification accuracy exceeding 90% on the test dataset—the model was exported using Edge Impulse's C++ SDK, which supports embedded deployment on the Sony Spresense platform. This deployment phase involved compiling the model into the robot's firmware stack and integrating it with the camera control system. Once deployed, the Spresense board became a self-contained vision node, capable of acquiring images from the 5MP camera, processing them locally using the TinyML model, and transmitting classification outputs to the MARV-bot's central controller.

This approach enabled real-time, distributed image analysis across the robot. Each Spresense module could independently classify crops and flag anomalies without requiring high-bandwidth communication or cloud-based inference. The inclusion of the built-in GPS module also allowed each image classification to be geo-tagged at the point of capture, ensuring spatially coherent analysis that could later be visualized on farm management dashboards or GIS platforms.

From a power consumption standpoint, the Sony Spresense and Edge Impulse combination was highly effective. Despite performing real-time image classification, the entire vision module operated within a modest power budget, which was essential for ensuring that the robot could conduct extended missions in the field without frequent recharging.

The development of this subsystem illustrates a paradigm shift in agricultural robotics—from centralized processing to distributed, intelligent edge systems. By equipping the MARV-bot with embedded intelligence, the imaging and analysis system transcended its traditional role as a passive data collection tool, evolving into an active decision-making unit capable of influencing the robot's behavior and contributing to adaptive precision agriculture.The integration of **Edge Impulse**, a leading TinyML development platform, was critical for training and deploying machine learning models on the Spresense board. The process involved:

1. **Dataset Collection**: A dataset of crop images (wheat, maize, and sugarcane) was obtained from Kaggle and imported into Edge Impulse.
2. **Model Training**: A machine learning model was trained to classify the crops and detect anomalies such as weeds.

3. **Deployment**: The trained model was deployed on the Sony Spresense using Edge Impulse's C++ library, enabling real-time crop classification and analysis.

Results and Impact

Outcomes and Impact of the MARV-bot Prototype

The deployment of the MARV-bot prototype has marked a significant milestone in the application of autonomous systems to agricultural research, particularly in the context of crop trial management. Developed with a focus on modularity, affordability, and real-time intelligence, the MARV-bot integrates mechanical precision with embedded machine learning, offering a practical and scalable solution for modern agronomic practices.

At the core of its success is the robot's ability to merge real-time data acquisition with localized analysis. This integration has yielded multiple functional advantages, each contributing to the overarching goals of increased accuracy, reduced human dependency, and enhanced decision-making capabilities in field trials.

One of the most transformative features of the MARV-bot is its real-time crop analysis system, enabled by TinyML-powered embedded vision. The onboard imaging system, equipped with lightweight neural networks trained via the Edge Impulse platform and deployed on Sony Spresense boards, is capable of classifying crops such as wheat, maize, and sugarcane with high accuracy. Beyond classification, the system can identify field anomalies—such as the presence of weeds, stunted growth patterns, or disease symptoms—at the point of image capture. These detections are not merely logged; they are analyzed and interpreted into actionable insights for researchers, who can use them to make timely interventions or annotate their field trials with greater precision.

Complementing its visual intelligence is the robot's capability for precision irrigation, a feature that aligns well with sustainable agriculture goals. Unlike conventional irrigation systems that treat entire plots uniformly, the MARV-bot can adjust water delivery levels based on location-specific requirements. Its motion-controlled spraying apparatus, guided by GPS and real-time environmental sensors, ensures that each crop row receives the precise amount of water needed, minimizing both over-irrigation and under-irrigation. This targeted approach supports optimal crop growth conditions, enhances experimental reliability in trials, and conserves critical natural resources such as water and fertilizer.

Another notable outcome of the MARV-bot prototype is its demonstrated capacity for remote and autonomous operation. Designed to function in geographically isolated or infrastructure-poor environments, the robot leverages a flexible communication architecture that includes both terrestrial 5G and satellite connectivity. This infrastructure allows for near-real-time data telemetry, enabling field researchers and agricultural institutions to remotely monitor the robot's status, visualize incoming crop data, and initiate commands without requiring a physical presence at the field site. This capability not only reduces operational overhead but also opens the possibility of centralized monitoring of multiple robots deployed across different trial locations.

Equally important is the scalability and adaptability of the MARV-bot system. Unlike highly specialized agricultural robots that are limited in scope, MARV-bot has been constructed using off-the-shelf components—most notably the Sony Spresense board for embedded processing, and Edge Impulse for streamlined model development and deployment. This design choice makes the prototype affordable and replicable. It also allows the robot to be easily customized for different crops, terrains, or trial protocols. For instance, adapting the system to monitor a different crop type merely involves retraining the existing model with a new dataset—something that can be accomplished rapidly within the Edge Impulse ecosystem.

MARV-bot prototype has successfully demonstrated how intelligent robotics, when thoughtfully integrated with edge AI and modular hardware, can revolutionize the way crop trials are conducted. It provides a blueprint for next-generation agricultural systems that are autonomous, scalable, resource-efficient, and capable of delivering meaningful insights in real time. As the demand for data-driven agriculture continues to grow, systems like MARV-bot are poised to play an increasingly pivotal role in shaping the future of precision farming and sustainable food production.. Key outcomes include:

- **Real-Time Crop Analysis**: The TinyML-powered imaging system can classify crops and detect anomalies with high accuracy, providing researchers with actionable insights.
- **Precision Irrigation**: By delivering precise irrigation levels, MARV-bot helps optimize crop growth conditions, leading to better yields and resource efficiency.
- **Remote Operation**: The robot's ability to operate in remote locations and transmit data via 5G or satellite communication reduces the need for on-site human intervention, lowering operational costs.
- **Scalability**: The use of off-the-shelf components like the Sony Spresense and Edge Impulse's development tools makes the solution scalable and adaptable to different crops and environments.

Lessons Learned and Future Directions

The journey of developing the MARV-bot—from concept to working prototype—was as much a technological challenge as it was a learning experience. Like many field-deployed robotic systems, the success of the MARV-bot relied not only on innovation but also on iteration, adaptability, and careful consideration of real-world constraints. The lessons drawn from this experience have been instrumental in shaping both the refinement of the current system and the vision for future versions.

One of the most profound takeaways was the critical importance of power efficiency. In theory, machine learning models can deliver high accuracy with complex architectures and large parameter sets. But in the real-world setting of crop trials—often miles from the nearest power outlet—every milliwatt matters. The robot had to operate for long durations, sometimes under harsh weather conditions, and with limited opportunity for recharging. This reality brought the focus squarely onto TinyML and its ability to deliver high-performance inference on microcontrollers with modest energy demands. The team spent considerable time optimizing models for inference efficiency, employing techniques such as quantization, model pruning, and memory-aware scheduling. These optimizations were not just about squeezing performance from the hardware—they were what made autonomous operation truly viable.

Equally influential was the decision to embrace a modular design philosophy. By building the MARV-bot around off-the-shelf components—including the Sony Spresense board, standard camera modules, and existing development platforms like Edge Impulse—the team was able to dramatically reduce both development time and risk. Components could be swapped out or upgraded as needed, and troubleshooting became more manageable. Perhaps most importantly, this modularity made the platform accessible: developers and researchers from different backgrounds could replicate, modify, or build upon the MARV-bot with relatively low entry barriers. It allowed innovation to be more open, inclusive, and collaborative.

Another defining insight came from the system's adaptability. Crop trials are inherently diverse—varying in crop types, geographies, disease challenges, and growth stages. From the beginning, the team designed the MARV-bot to be a general-purpose platform, capable of adjusting its sensing and analytical capabilities to new agricultural contexts. Switching from wheat to maize or sugarcane did not require redesigning the robot; it simply meant collecting new training data and retraining the machine learning model using Edge Impulse. This ability to adapt rapidly to new domains without overhauling the underlying hardware represents a future-proofing strategy that is likely to define next-generation agricultural automation.

Looking ahead, the development team envisions a second-generation MARV-bot—one that pushes the boundaries of autonomy, collaboration, and sensing intelligence. Several new features are being actively explored, each aimed at enhancing performance and scalability.

One such feature is the integration of federated learning, a technique that allows multiple robots to collaboratively train a shared model without exchanging raw data. Instead of uploading images or sensor data to a central server, each MARV-bot trains locally on its own dataset and only shares model updates (such as gradients or weights) with the network. This approach preserves data privacy—an increasingly important concern in agriculture, especially when dealing with proprietary crop genetics or experimental trials. Moreover, it facilitates scalable, decentralized learning, enabling models to evolve over time through shared field experiences across multiple geographic regions.

Another promising enhancement involves energy harvesting. In version one, MARV-bot relied entirely on battery power, which limited continuous operation. Future iterations are expected to include solar panels or other energy-harvesting technologies—such as kinetic charging or thermoelectric generators—that allow the robot to extend its operating time substantially. The idea is not just to prolong battery life, but to inch closer to energy autonomy, where the robot can support day-long or even multi-day missions without human intervention.

Finally, the next MARV-bot is set to include an enhanced imaging suite. While the prototype successfully used RGB imagery for crop classification and anomaly detection, there are limitations to what can be inferred from visible light alone. By incorporating multispectral or thermal imaging capabilities, the robot will be able to detect deeper physiological traits—such as water stress, nutrient deficiencies, or hidden pest infestations—that are invisible to the naked eye. This addition could transform MARV-bot from a reactive tool into a predictive system, capable of spotting early signs of stress before visible symptoms appear.

The development of MARV-bot has not only yielded a functional robotic platform but also offered rich insights into the principles and practicalities of building intelligent systems for the field. It serves as a case study in how machine learning, embedded systems, and robotics can converge to support precision agriculture in a sustainable, scalable, and impactful way. As the team prepares to move toward version two, the foundation laid by the prototype—its efficiency, flexibility, and collaborative spirit—will continue to guide its evolution toward a smarter, more responsive future in agricultural automation.

The development of MARV-bot highlighted several key lessons:

- **Power Efficiency is Critical**: Optimizing TinyML models for low power consumption is essential for agricultural robots operating in remote locations.
- **Modular Design**: Using modular, off-the-shelf components accelerates development and reduces risk.

- **Adaptability**: The ability to switch out ML models for different crops makes the system versatile and future-proof.

Looking ahead, the team plans to develop **version two** of MARV-bot, incorporating advanced features such as:

- **Federated Learning**: Enabling multiple robots to collaboratively train ML models without sharing raw data, enhancing privacy and scalability.
- **Energy Harvesting**: Integrating solar panels or other energy-harvesting technologies to further extend the robot's operational lifespan.
- **Enhanced Imaging**: Adding multispectral or thermal imaging capabilities to provide deeper insights into crop health.

CONCLUSION

Summary of TinyML's Impact on Engineering Solutions

The MARV-bot case study exemplifies the transformative potential of TinyML in engineering solutions. By enabling intelligent decision-making at the edge, TinyML is addressing critical challenges in agriculture, from precision irrigation to real-time crop analysis. Its ability to operate in resource-constrained environments makes it a powerful tool for industries seeking to optimize efficiency and reduce costs.

Future Outlook

As technologies like federated learning, energy harvesting, and advanced imaging continue to evolve, TinyML will play an increasingly important role in creating intelligent, autonomous systems. The future of TinyML is bright, with endless possibilities for innovation and impact across industries. This updated case study provides a detailed, real-world example of TinyML in action, showcasing its potential to revolutionize agriculture and other industries.

REFERENCES

Afachao, K., & Abu-Mahfouz, A. (2022). A review of intelligent IoT devices at the edge. *Proceedings of the 2022 International Conference on Artificial Intelligence of Things*. DOI: 10.1109/ICAIoT57170.2022.10121824

Alajlan, N. N., & Ibrahim, D. M. (2022). TinyML: Enabling of inference deep learning models on ultra-low-power IoT edge devices for AI applications. *Micromachines*, *13*(4), 851. DOI: 10.3390/mi13060851 PMID: 35744466

Andrade, P., Silva, I., Signoretti, G., Da Silva, M. B. D., Dias, J., Marques, L., & Costa, D. G. (2021). An unsupervised TinyML approach applied for pavement anomalies detection under the Internet of Intelligent Vehicles. *2021 IEEE International Workshop on Metrology for Industry*. DOI: 10.1109/MetroInd4.0IoT51437.2021.9488546

Baresi, L., Mendonça, D. F., & Garriga, M. (2017). Empowering low-latency applications through a serverless edge computing architecture.

Capogrosso, L., Cunico, F., Cheng, D. S., Fummi, F., & Cristani, M. (2023). A machine learning-oriented survey on tiny machine learning. *arXiv preprint arXiv:2301.12345*. https://arxiv.org/abs/2301.12345

Gawate, E., & Rane, P. (2023). Empowering intelligent manufacturing with the potential of edge computing with NVIDIA's Jetson Nano. *2023 International Conference on Computing, Communication and Automation*. DOI: 10.1109/ICCCIS60361.2023.10425278

Gong, T., Zhu, L., Yu, F., & Tang, T. (2023). Edge intelligence in intelligent transportation systems: A survey. *IEEE Transactions on Intelligent Transportation Systems*, *24*(1). DOI: 10.1109/TITS.2023.3275741

Han, H., & Siebert, J. (2022). TinyML: A systematic review and synthesis of existing research. *Proceedings of the International Conference on Artificial Intelligence*. DOI: 10.1109/ICAIIC54071.2022.9722636

Hayajneh, A., Alasali, F., Salama, A., & Holderbaum, W. (2024). Intelligent solar forecasts: Modern machine learning models and TinyML role for improved solar energy yield predictions. *IEEE Access : Practical Innovations, Open Solutions*, *12*, 10846–10864. DOI: 10.1109/ACCESS.2024.3354703

Sharma, S., Kang, B., Kidambi, N. V., & Mukhopadhyay, S. (2025). HamQ: Hamming weight-based energy-aware quantization for analog compute-in-memory accelerator in intelligent sensors. *IEEE Sensors Journal*, *25*(2). DOI: 10.1109/JSEN.2024.3382479

Chapter 4
Support Vector Machine in Predictive Maintenance of Belt-Drive System

Saahil Jamad
https://orcid.org/0009-0001-3597-7036
SRM Institute of Science and Technology, India

Anitha Kumari S.
https://orcid.org/0000-0001-6185-5238
SRM Institute of Science and Technology, India

Josika Meenal Alamelu Muthu
https://orcid.org/0009-0000-3743-7933
SRM Institute of Science and Technology, India

ABSTRACT

Belt-drives are widely used as power transmission elements in light as well as heavy-duty industrial applications. Maintenance activities are imperative to ensure the performance and reliability of a belt-drive system. The vibrational signal patterns from the system can reveal failures such as belt misalignment and tension irregularities. These anomalies can be predicted and mitigated before they lead to costly downtime. The work presents Support Vector Machine (SVM) based predictive maintenance of the belt-drive system. The reason for choosing SVM for predictive maintenance of the belt-drive system is that it addresses the challenges posed by the sensor data, including the noise, high dimensionality and complex temporal dependencies. Also, SVM being a discriminant machine learning technique, returns

DOI: 10.4018/979-8-3373-0746-6.ch004

the same optimal hyperplane parameter. This contrasts generative approaches such as Naive Bayes or Gaussian process classification. The SVM model developed has shown high accuracy in classifying healthy and faulty belt-drive systems.

INTRODUCTION

The advent of Industry 4.0 has ushered in a new era of industrial automation and data exchange, profoundly impacting manufacturing systems worldwide. Central to this transformation are belt-drive systems, which play a pivotal role in transmitting power across various machinery components (Pollak et al., 2021; Malaka et al., 2020). Their widespread application in production lines and supply chains underscores their significance in ensuring seamless operations. However, the efficiency and longevity of these systems are contingent upon robust maintenance strategies. This discourse delves into the evolution of maintenance methodologies, the integration of advanced technologies in predictive maintenance, and the challenges and prospects associated with implementing such systems in belt-drive mechanisms (Achouch et al., 2022; Sakib et al., 2018; Gerike et al., 2017).

Over the years, the strategies in maintenance practices have undergone significant evolution, transitioning from reactive approaches to more proactive methodologies (Shehadeh et al., 2024; Fazli et al.,2023; Fitch et al. 2013):

1. Reactive Maintenance: Traditionally, industries operated on a "run-to-failure" model, wherein equipment was repaired or replaced only after a malfunction has occurred. While straightforward, this approach often resulted in unplanned downtimes, escalating operational costs, and potential safety hazards.
2. Preventive Maintenance: To mitigate the pitfalls of reactive maintenance, industries adopted scheduled interventions based on predefined time intervals or usage metrics. This maintenance strategy requires complete knowledge of the degraded equipment. Although this method reduced unexpected failures, it did not account for the actual condition of the equipment, sometimes leading to unnecessary maintenance activities and associated costs (Gerike et al., 2017).
3. Predictive Maintenance: Emerging as a sophisticated alternative, predictive maintenance leverages real-time data and analytical tools to anticipate equipment failures before they occur. By monitoring the actual condition of machinery, maintenance activities are performed only when necessary, optimizing resource utilization and minimizing downtime. Predictive maintenance is also known as condition monitoring and it is based on the ground that till the equipment condition degrades no reactive maintenance is required. However, this maintenance

strategy will not say the exact time at which the equipment fails. Instead, it helps the maintenance expert to schedule the repairs and hence reduce downtime.

Belt-drive systems, integral to numerous industrial applications (Ashworth, 1971), have benefited immensely from predictive maintenance methodologies. This includes vibration analysis, energy consumption monitoring, and acoustic monitoring. Continuous monitoring of vibration patterns helps in detecting anomalies that may indicate misalignment, wear, or other issues in belt-drive systems. Early detection through vibration analysis allows for timely interventions, preventing more severe damage. Analyzing the energy consumption of belt-drive systems can reveal inefficiencies or impending failures. Anomalies in energy usage patterns often precede mechanical failures, making this a valuable predictive maintenance tool. Sound analysis can detect irregular noises produced by belt-drive systems, which may indicate issues such as belt slippage or bearing faults. Acoustic monitoring serves as a non-invasive diagnostic tool for early fault detection.

The implementation of predictive maintenance has been significantly enhanced by the integration of advanced technologies such as Internet of Things (IoT), Machine Learning (ML), Artificial Intelligence (AI), and digital twins (Pulcini and Modoni, 2024). IoT devices facilitate continuous monitoring of equipment by collecting and transmitting data related to operational parameters such as temperature, vibration, and load. This real-time data acquisition is fundamental to predictive maintenance strategies (Lee, Bagheri, & Kao, 2015). Advanced algorithms such as ML and AI analyze the data collected from IoT devices to identify patterns indicative of potential failures. For instance, machine learning models can predict when specific maintenance tasks, such as belt replacements or component inspections, should be scheduled, thereby preventing unexpected breakdowns (Zonta et al., 2020). These technologies help shift maintenance from reactive to proactive strategies, improving overall asset utilization and reducing operational downtime (Carvalho et al., 2019). Digital Twin technology involves creating a virtual replica of physical assets, enabling simulation and analysis of equipment behavior under various conditions. Digital twins facilitate proactive decision-making by predicting how equipment will respond to different scenarios, thus informing maintenance schedules (Tao et al., 2018). The integration of these technologies not only improves failure detection accuracy but also supports resource optimization and enhances maintenance planning across industrial systems (Zhang et al., 2021).

There are several innovative approaches in data analysis and decision-making frameworks that have been previously published and have had a substantial impact on recent developments in predictive maintenance for belt-drive systems (Wu et al., 2023). The productivity and maintenance procedures are enhanced by te introduction of such monitoring systems (Kumari and Srinivasan, 2019; Yu, 2019; Kiangala and

Wang, 2018). A number of machine learning methodologies for predictive maintenance of industrial machines have been proposed (Hosamo et al., 2022; Zarkovic and Stojkovic, 2019; Gutschi et al., 2019; Mathew et al., 2018). The algorithms predominantly employ artificial intelligence based algorithms on belt-drive systems (Szrek et al., 2020; Kozhushko et al., 2020; Al-Kahwati et al., 2022; Gupta et al., 2023). However the key factors that highly impact the conveyor performance are not known. The convolution neural network model proposed by Wang et al. (2020) for fault detection in conveyors was interpreted to overfit which influences the precision of the model. Automatic fault classification in rotary machines using similarity-based models, a non-machine learning model, is performed by F. Ribeiro et al (2017). The accuracy of these models was 96.43%. The MAFAULDA database for machine faults was used in this study. Multiple failure states in a system and the transition probabilities among the states were generated by the hidden Markov model proposed by Liu et al. (2019). Ouerghemmi et al. (2023) utilized spectral analysis methods such as the Extended Park Vector approach and Motor Current Signature to identify faults in V-belt transmission systems. It has been inferred that Extended Park Vector approach has high sensitivity in fault detection. The importance of fault detection based on vibrations for predictive maintenance has been suggested by Pollak et al. (2021). This method also helps in estimating the remaining service life of belt-drive systems. Rumin et al. (2023) have proposed machine learning algorithms for predicting conveyor belt flip over.

To summarize, the integration of predictive maintenance strategies within the framework of Industry 4.0 has revolutionized the management of belt-drive systems in industrial manufacturing. By transitioning from reactive to predictive methodologies and leveraging advanced technologies, industries can achieve optimized performance, reduced downtime, and substantial cost savings. This chapter proposes an intelligent predictive maintenance system using Support Vector Machine to address the challenges in industrial belt-drive systems. The standout feature of the model is its predictive capability empowering the maintenance teams to schedule maintenance at optimal intervals.

METHODOLOGY

Belt-Drive Systems

Belt-drive systems are among the most commonly employed mechanical power transmission systems in industrial settings. Their widespread use stems from their simplicity, cost-effectiveness, and versatility across a wide range of operating conditions. From machine tools and industrial fans to conveyor systems and agricultural equipment, belt drives play a critical role in transferring rotational power between components. These systems are particularly well-suited for situations where components are spaced apart or require speed variation (Zhu et al., 2021).

A belt-drive system operates by connecting two pulleys, a driving pulley and a driven pulley with a continuous loop known as the belt. The driving pulley is attached to the power source, typically an electric motor, and transmits motion to the driven pulley, which is connected to the machinery or load that needs to be operated. These pulleys are mounted on their respective shafts and work together as a unit to transfer mechanical energy through rotational motion. The tensioning mechanism maintains appropriate tension in the belt to ensure consistent power transmission and reduce slippage. The belt wraps tightly around both pulleys, and this setup enables the efficient transmission of torque from the motor to the driven system. It is a flexible loop made from materials like rubber, neoprene, or polyurethane. It connects the two pulleys and transfers motion through friction.

The core principle behind a belt-drive system is this frictional contact. As the driving pulley rotates, it creates friction with the belt at the contact surface. This frictional force, coupled with belt pretension, allows the belt to grip the pulley and transfer motion to the driven pulley. The performance and efficiency of the system depend heavily on this frictional resistance, as noted by Chowdhury et al. (2016). If the friction is insufficient due to slippage, misalignment, or wear, the system will not perform optimally.

The belt-drive systems can be classified based on the type of belt used. Based on the application requirements, such as distance between shafts, power capacity, speed ratios, and synchronization needs, different types of belts are used. Each belt type offers specific advantages and is suited for particular operating environments and design constraints.

- Flat Belts: Flat belts are among the earliest forms of belt drives used in machinery. These belts are typically made from flexible materials such as rubber, leather, synthetic polymers, or fabric-reinforced composites. The defining feature is their flat surface which runs over flat or slightly crowned pulleys. One of the key advantages of flat belts is their efficiency. Due to minimal bending and low internal stress, they can operate with high efficiency, sometimes exceeding 98% in well-aligned systems. This makes them suitable for high-speed applications where long-distance transmission is necessary (Shigley et al., 2011). Despite their simplicity, flat belts do have certain operational limitations. Because they rely heavily on friction, they are more prone to slippage, especially when subjected to sudden load changes or contamination by oil or water. This can make them less ideal in precision-dependent environments unless tensioning and alignment are perfectly maintained.

Applications for flat belts include textile machinery, conveyor systems, woodworking equipment, and agricultural machinery (Khurmi & Gupta, 2005), especially where high-speed, low-to-moderate torque transmission is required.

- V-Belts: V-belts are named for their trapezoidal cross-section, which allows them to sit deeply in the V-shaped grooves of pulleys. This geometry significantly improves the belt's grip by increasing the contact area and producing a wedging action, which helps to minimize slippage even under high load. V-belts are generally constructed from rubber or synthetic materials reinforced with fibers like polyester, Kevlar, or steel cords for added strength and flexibility. They are available in various sizes, designated by standard profiles such as A, B, C, D, and E (based on their cross-sectional dimensions). One of the main advantages of V-belts is their ability to transmit a high amount of power relative to their size, making them a go-to solution in many industrial applications (Rao, 2011). They are also known for their quiet operation, ease of installation, and the ability to operate in compact spaces due to the smaller pulley diameters they can accommodate. There are several variants within the V-belt category, including Classical V-belts that are widely used in older machines, Narrow V-belts which offer better performance due to higher power transmission per unit width, and cogged V-belts that have notches or grooves that provide more flexibility and better heat dissipation (Mott, 2017). Typical applications include HVAC systems, pumps, lathes, drilling machines, compressors, and agricultural equipment.
- Timing Belts (Toothed Belts): Timing belts, also known as synchronous belts, feature teeth on their inner surface that match the grooves on the mating pulley. This unique design allows for a positive engagement between the belt and the pulley, ensuring no relative motion between them. As a result, timing belts eliminate slippage, making them suitable for applications where precise timing is critical (Norton, 2020). These belts are commonly manufactured from polyurethane or rubber and are reinforced with cords made from fiberglass, Kevlar, or steel. They are available in different tooth profiles (e.g., trapezoidal, curvilinear) to suit various applications and load conditions. One of the standout features of timing belts is their ability to maintain exact angular relationships between shafts, which is essential in applications like internal combustion engines, where valve timing must be meticulously controlled. Timing belts are also used in robotics, automated assembly lines, CNC machines, printers, and packaging machinery. Their efficiency and maintenance-free operation over long periods make them ideal for closed systems where lubrication is either undesirable or infeasible (Shigley et al., 2011). However, timing belts must be kept under proper tension, and they require pulleys with matching tooth profiles. If not replaced at recommended

intervals, worn teeth may shear, leading to loss of synchronization and possible equipment damage.

- Round Belts and Multi-Ribbed Belts: Round belts, also called O-ring belts, are typically made from elastomeric materials and have a circular cross-section. These belts are flexible, easy to install, and well-suited for light-duty power transmission applications. Their flexibility allows them to twist and turn through complex routing paths, making them popular in serpentine configurations. They are often used in small appliances, live roller conveyors, and various office machines such as printers and copiers. One of the advantages of round belts is their ability to be welded or spliced on-site, which makes replacement and customization convenient. However, their grip is relatively weak compared to V-belts or timing belts, which limits their use to low-torque applications. The possibility of slippage also increases in high-load or high-speed scenarios unless a grooved pulley is used.
- Multi-ribbed belts combine the features of flat and V-belts. They have multiple small V-shaped ribs running along the width of the belt. These ribs allow for excellent grip and enable the belt to bend easily around small pulley diameters, distributing stress more evenly and reducing belt wear. They are generally constructed from high-strength rubber and reinforced with fibers for increased durability. The design allows them to transmit more power than standard V-belts in a more compact form factor. These belts are commonly used in automotive engines, washing machines, compressors, and fitness equipment. Their ability to handle high-speed operations and relatively high power in limited spaces makes them especially suitable for modern applications where efficiency and compactness are critical.

Each belt type has its advantages and is chosen based on load, speed, torque, environmental conditions, and maintenance requirements. While belt drives are efficient and easy to use, their performance depends on several key factors:

Belt Tension

Maintaining the correct belt tension is critical for the efficient functioning of the system. If the belt is too loose, it may slip, leading to loss of power transmission, heat buildup, and premature wear. On the other hand, if the belt is too tight, it may strain the pulleys, bearings, and shafts, resulting in mechanical damage. A proper tensioning mechanism ensures that the belt remains taut enough to generate sufficient friction without causing excessive stress.

Alignment

Proper pulley alignment ensures that the belt tracks correctly and wears evenly. Misaligned pulleys cause the belt to wear unevenly, slip off the pulley, or develop cracks and frays. Laser alignment tools and periodic inspections are recommended to maintain alignment and ensure optimal operation.

Belt Wear and Replacement

Belts are subject to wear due to continuous operation, friction, and environmental exposure. Worn belts may show signs of glazing, cracking, fraying, or stretching. These conditions reduce grip and increase the risk of slippage or breakage. Regular inspection and timely replacement are crucial to avoid unexpected downtime and system failure.

Environmental Conditions

Temperature, humidity, dust, and chemical exposure can all impact belt performance. For example, extreme heat can cause rubber belts to harden and crack, while oily or greasy environments may reduce friction. Choosing the right belt material—such as heat-resistant, oil-resistant, or anti-static belts—ensures reliability under various operating conditions.

Load Fluctuations

Sudden load changes can cause belt slippage or stretching. In systems where load variations are frequent, it may be necessary to use belts with higher tensile strength or incorporate tensioners and shock-absorbing elements.

Belt drives are used in a wide array of industries and machinery types, including:

- Manufacturing and Automation: In conveyor belts, robotics, and assembly line equipment.
- Agriculture: For driving equipment such as harvesters, threshers, and irrigation pumps.
- Automotive: Timing belts and serpentine belts in engines.
- HVAC Systems: To drive fans, compressors, and blowers.
- Textile Industry: In looms and spinning machines.
- Food Processing: Where non-metallic, hygienic belts are required for handling and packaging.

Each application may demand a specific belt design, material, and operating configuration tailored to the task and working environment.

Common Maintenance Issues in Belt-Drive Systems

Mechanical systems cannot withstand wear, structural failure, and jamming. Wearing in mechanical systems occurs due to overstressing of a surface. Wear in a system degrades the performance and causes functional impairment (Gwidon, 2010). The structural failures may be due to distortion or fractures. It occurs when any of the subsystems can no longer withstand the applied load, perform its function, or when a specific component in the system inhibits the movement of another component in the system. This failure is irreversible (Mobley, 2002). Jamming occurs due to interference in the motion obstruction of a mechanical or fluid contaminant. However, this failure is reversible. Belt-drive systems are susceptible to various anomalies or failures, including:

Misalignment

One of the most common problems in belt-driven systems is pulley misalignment. Misalignment occurs when the pulleys are not positioned in the same plane or if they are angled with respect to each other. This misalignment can be of several types, angular, parallel, or a combination of both. Angular misalignment happens when the pulley shafts are not parallel, while parallel misalignment occurs when the pulleys are offset from each other (Neale, 1995). Causes of misalignment include improper installation, foundation settling, vibration, or thermal expansion of machine parts. In belt systems, even a slight deviation in alignment can cause uneven wear on the belt and reduce the contact surface between the belt and pulley (Bloch & Geitner, 1999).

The consequences of misalignment are severe. The belt may twist, stretch, or even slip off the pulley. Increased friction on one side of the belt causes accelerated wear, heat buildup, and potential belt failure. Misalignment also reduces power transmission efficiency and can lead to frequent maintenance requirements. To prevent misalignment, it's crucial to use laser alignment tools during installation and conduct regular maintenance checks (Lipson & Juran, 1991).

Slippage

Slippage refers to the condition where the belt does not grip the pulley tightly enough, causing it to slide over the pulley surface rather than turning it effectively. While a small amount of slippage is inevitable in many systems, excessive slippage is problematic (Childs, 2013). Slippage typically occurs when the friction between the belt and the pulley surface decreases. This reduction in friction can result from belt wear, contamination by oil or debris, or improper tensioning. For example, a belt that is too loose will not have sufficient grip, while one that is too tight may cause unnecessary strain on the pulley and bearings (White, 2004).

Effects of slippage include reduced power transmission, inefficient operation, overheating, and noise. Slippage can also cause wear marks or glazing on the belt and pulley surface. Over time, this can lead to a vicious cycle where decreased friction leads to more slippage, which in turn accelerates wear. Preventing slippage involves proper belt tensioning, routine inspections, and ensuring the pulley surfaces are clean and free from oil or debris. In some applications, using belts with high-friction materials or cogged belts can help improve grip and reduce slippage (Thompson, 2000).

Belt Wear

Another critical issue in belt and pulley systems is belt wear. Over time, belts are subjected to constant stress, friction, and environmental exposure. These factors lead to gradual degradation of the belt material. Worn belts lose their strength, flexibility, and ability to transmit power efficiently (Collacott, 1979). Common types of wear include abrasion, cracking, fraying, hardening, and stretching. Abrasion typically results from misalignment or debris in the pulley groove. Cracks and fraying indicate material fatigue or thermal degradation, while stretching occurs due to prolonged tension and mechanical loading (Harris & Kotzalas, 2006).

Consequences of belt wear include reduced power transmission, increased noise, and the risk of belt failure during operation. A failing belt can cause sudden machinery stoppage, which may lead to production losses and equipment damage. To mitigate belt wear, it's important to perform regular inspections for signs of cracking, glazing, or uneven wear. Replacing belts before they reach critical wear levels can help avoid unscheduled downtime. Additionally, using high-quality belts designed for specific environmental conditions (e.g., heat-resistant, oil-resistant) can extend their service life (Bishop & Sherrington, 2000).

Pulley Deformation

Though pulleys are typically made of robust materials such as steel or cast iron, they are not immune to deformation or wear. Pulley wear may occur due to continuous contact with the belt, environmental corrosion, or improper mounting. Deformation refers to the change in shape of the pulley, which may result in grooves becoming shallow, uneven surfaces, or out-of-round conditions (Peterson & Winer, 1980). Causes of pulley deformation include excessive load, improper tension, prolonged use without maintenance, and the use of incompatible belt-pulley combinations. Over-tightened belts can exert too much force on the pulley, causing the shaft or hub to deform. In some cases, corrosion or rust can erode the pulley surface (Juvinall & Marshek, 2011).

Impacts of pulley deformation are significant. An uneven pulley surface can cause erratic belt movement, slippage, and noise. It also accelerates belt wear and may lead to misalignment. In the worst cases, it may cause mechanical failure or damage to the shaft and bearings. Routine inspection and maintenance are essential to identify early signs of pulley deformation. Worn or damaged pulleys should be repaired or replaced promptly. Additionally, using pulleys made from corrosion-resistant materials and maintaining proper belt tension can help prevent deformation (Krar et al., 2005).

Vibration and Noise

Unusual vibration and noise in a belt and pulley system often indicate underlying mechanical issues. While some level of vibration is normal in rotating systems, excessive or irregular vibrations are problematic. Noise, such as squealing or chirping sounds, typically results from belt slippage, misalignment, or pulley defects (Heisler, 2002). Sources of vibration and noise include imbalanced pulleys, worn belts, misalignment, loose components, or foreign objects trapped in the pulley grooves. A belt that is too tight can also induce excessive stress on the shaft and bearings, leading to vibrations. Likewise, a loose belt may flap during operation, creating noise (Tlusty, 2000). If left unaddressed, vibration and noise can lead to secondary failures such as bearing wear, fatigue cracks in machine parts, and eventual system breakdown. In environments where machinery operates near personnel, persistent noise can also lead to occupational health concerns.

Diagnosing vibration and noise issues often requires a combination of physical inspection, vibration analysis tools, and sound level measurements. Corrective actions may involve re-aligning the pulleys, re-tensioning the belt, replacing worn components, and ensuring proper lubrication of rotating parts. Current maintenance methods either miss these early warning signs or detect them when it is too late to avoid costly repairs. The challenge is to develop a predictive maintenance solution that can continuously monitor the belt-pulley system's health and provide accurate failure predictions (Ahmad & Kamaruddin, 2012).

SUPPORT VECTOR MACHINE FOR CLASSIFICATION

Support Vector Machine (SVM) is a highly influential and widely used machine learning technique that is particularly effective for both classification and regression tasks. Initially proposed by Vladimir Vapnik and his collaborators in 1995 and further developed over subsequent years (Vapnik, 1999), SVM has since become a foundational model in the field of machine learning, known for its robustness in high-dimensional spaces and strong theoretical foundation grounded in statistical learning theory. SVM can handle high-dimensional data, maximize margins for better generalization, and incorporate kernels for non-linear classification makes them especially valuable across domains like image recognition, bioinformatics, text classification, and financial forecasting. From its mathematical elegance to its practical utility, SVM remains a benchmark algorithm that continues to influence modern machine learning research and applications. Whether applied to small, linearly separable datasets or large, high-dimensional, and non-linear problems, SVM provides a robust framework capable of delivering high accuracy and interpretability.

At its core, SVM is a non-parametric learning method, meaning it does not assume a specific probability distribution for the underlying data. Instead, it seeks to determine the optimal function that captures the relationship between the predictor variables (also known as features or inputs) and the output variable (also called the target or label). This property allows SVMs to flexibly adapt to the structure of the data, making them especially suitable for complex, real-world applications.

Consider a binary classification problem where we are given a training data set represented as:

$$\{(X_i, y_i), i=1,\ldots,N\}, X_i \in R^n, y_i \in \{1,-1\}^N$$

In this mathematical formulation, X_i represents the ith feature vector in an n-dimensional real space. y_i is the class label associated with X_i, which takes values of either +1 or -1 and N is the total number of training examples. The objective of SVM is to find a function or decision boundary that best classifies new input data into one of the two categories. The decision function can be written as:

$$y = g(X;\alpha)$$

where α denotes the parameters of the classifier that need to be learned from the data. These parameters include the weights associated with the features and a bias term that collectively define the decision surface—more specifically, a hyperplane.

A hyperplane in an n-dimensional space is an $n-1$-dimensional flat affine subspace. In SVM, the hyperplane is the decision boundary that separates the feature space into two distinct regions, each corresponding to one of the classes y = +1 or y = -1. The equation of a hyperplane is given by $w^T.X + b = 0$. w is an n×1 vector (column vector) that is orthogonal to the hyperplane. X represents any point lying on the hyperplane and b is a scalar representing the bias or offset, which determines the position of the hyperplane relative to the origin.

The sign of the expression $w^T.X + b$ determines on which side of the hyperplane a given data point lies. If $w^T.X + b > 0$, the point belongs to class +1. If $w^T.X + b < 0$, the point belongs to class -1 or if $w^T.X + b = 0$, the point lies exactly on the hyperplane.

The SVM algorithm aims to find the hyperplane that not only separates the two classes but also maximizes the margin—the distance between the hyperplane and the nearest data points from either class.

The margin is a crucial concept in SVM. It refers to the shortest distance between the hyperplane and any point in the dataset. More formally, it is the perpendicular distance from the hyperplane to the nearest data point. SVM seeks to maximize this margin, as a larger margin is typically associated with better generalization and robustness to noise in unseen data. The data points that lie exactly on the margin boundaries (i.e., the closest points to the hyperplane) are known as support vectors. These points are critical because the position of the optimal hyperplane is entirely determined by them. Removing any other point from the dataset would not affect the solution, but altering or removing a support vector could shift the hyperplane significantly. Maximizing the margin is equivalent to minimizing the norm of the weight vector w, subject to constraints ensuring that all data points are correctly classified. This leads us to the formulation of an optimization problem. The optimization problem is thus

$$\min_{w,b} \frac{1}{2} \|w\|^2$$

subject to the constraints $y_i(w \bullet X+b) - 1 \geq 1$ for all i. For a perfect linearly separable dataset, the constraints can be written as $y_i(w \bullet X+b) - 1 \geq 1$ for i=1,…,N. This ensures that all data points are on the correct side of the margin.

This is a convex optimization problem, where the objective is to find the weight vector w and bias b that minimize the squared norm of w while ensuring that the constraints are satisfied. Minimizing $||w||^2$ effectively maximizes the margin.

DATASET COLLECTION AND PREPARATION

The backbone of predictive maintenance is accurate data collection. Sensors like accelerometers, temperature sensors, proximity sensors, and current transducers are integrated into machinery to monitor real-time conditions. Vibration sensors detect imbalances, misalignments, or loose belts by analyzing oscillations in the drive system. Excess heat often indicates friction or overload. An abnormal rise can suggest bearing failure or improper belt tension. This information is obtained by using temperature sensors. Acoustic sensors such as microphones and sound sensors can capture anomalies in operating noise, which may indicate slippage or component fatigue. Current sensors monitor the electrical current consumed by motors and it can provide indirect insights into mechanical strain or misalignment.

Effective data preparation is a cornerstone of developing robust and accurate machine learning models, particularly when utilizing algorithms like Support Vector Machines (SVMs). The process encompasses several critical steps: data preprocessing, feature selection, and data splitting. Each of these stages plays a pivotal role in ensuring that the model not only learns effectively from the data but also generalizes well to unseen instances. The following comprehensive guide delves into these steps, offering insights and best practices to optimize the model development process.

Data Preprocessing

Data preprocessing involves transforming raw data into a clean and structured format suitable for analysis. This step is vital for enhancing the quality of the data, which directly impacts the performance of machine learning models. The key components of data preprocessing include:

Handling Missing Values

Missing data can lead to biased estimations and reduced model accuracy. Addressing missing values is essential to maintain the integrity of the dataset. Common strategies include removal of the data, imputation, and predictive imputation. Removal of data involves elimination of records with missing values, which is feasible only when such records constitute a small fraction of the dataset. Imputation is replacing missing values with statistical measures like mean, median, or mode. For example, if a dataset contains missing entries in a numerical feature, imputing the mean value ensures that the overall distribution remains unaffected. Whereas predictive imputation utilizes regression or other predictive models to estimate missing values based on other features. Choosing the appropriate method depends on the nature and extent of the missing data, as well as the specific requirements of the analysis.

Identifying and Handling Outliers

Outliers are data points that significantly deviate from the majority of the data. They can distort statistical analyses and adversely affect model performance. Detecting outliers can be achieved through visual inspection or statistical methods. visual Inspection use box plots, scatter plots, or histograms to visually identify anomalies. Statistical approaches apply techniques such as the Z-score or the IQR (Interquartile Range) method to flag outliers.

Once the outliers are identified, they can be addressed by deleting the records, especially if they result from data entry errors or applying mathematical transformations (e.g., logarithmic) to reduce the impact of outliers. Extreme values can also be grouped to bins to minimize their influence. The chosen approach should align with the overall data analysis strategy and the specific context of the dataset.

Encoding Categorical Variables

Machine learning algorithms typically require numerical input. Therefore, categorical variables must be converted into numerical formats. Common encoding techniques include:

- Label Encoding: Assigning a unique integer to each category. While straightforward, this method may inadvertently introduce ordinal relationships where none exist.
- One-Hot Encoding: Creating binary columns for each category, indicating the presence or absence of a particular category. This method avoids the pitfalls of implied ordinality and is widely used in practice.

The choice of encoding technique should consider the nature of the categorical data and the requirements of the specific machine learning algorithm being employed.

Feature Scaling

Feature scaling ensures that all numerical features contribute equally to the model by standardizing their ranges. This is particularly crucial for algorithms like SVMs, which are sensitive to the magnitudes of feature values. Common scaling methods include:

- Min-Max Scaling (Normalization): Rescales features to a specific range, typically [0, 1], The mathematical formulation for Min-Max scaling is $x' = \frac{X - \bar{X}}{\sigma}$.

- Standardization (Z-score Normalization): Centers the feature values around zero with a standard deviation of one. The mathematical formulation for Z-score normalization is $x' = \frac{x - \bar{x}}{\sigma}$

  where $\bar{x}$ is the mean and σ is the standard deviation of the feature.

Implementing feature scaling can significantly enhance the efficiency of SVMs by ensuring that each feature contributes proportionately to the computation of distances between data points.

Feature Selection

Effective feature selection plays a vital role in the success of machine learning models, especially in predictive maintenance, where data from sensors and signal sources can be voluminous and noisy. It leads to more efficient models by eliminating redundant or irrelevant features. Techniques like Univariate Selection, Recursive Feature Elimination, and Principal Component Analysis offer powerful tools to extract meaningful information, reduce computational burden, and enhance model generalization. When applied appropriately, these methods not only boost performance but also provide deeper insights into system behavior, enabling more accurate, interpretable, and actionable predictions. Feature selection involves identifying and retaining the most relevant features for model training, thereby reducing dimensionality, enhancing model performance, and improving interpretability. Selecting the right features is critical for several reasons, such as increase in accuracy, reduced overfitting, enhanced interpretability, and reduced computational cost. Eliminating irrelevant features can enhance the model's predictive performance. It also simplifies the model and decreases the risk of overfitting. Models with fewer features are easier to understand and interpret. Also, Fewer features lead to faster training times and lower computational resource requirements.

In machine learning pipeline, when dealing with high-dimensional datasets commonly encountered in predictive maintenance, fault diagnosis, and various industrial applications, feature selection is a critical step. The objective of feature selection is to identify and retain the most relevant variables (features) that contribute significantly to the predictive power of the model while removing irrelevant, redundant, or noisy data that could lead to overfitting, reduced accuracy, or increased computational complexity. Feature selection not only improves model performance but also enhances the interpretability of the model, making it easier to understand the influence of each variable on the output. This is particularly valuable in industrial systems, where engineers and maintenance personnel must make informed decisions based on model outputs.

Several sophisticated techniques exist to perform feature selection, each with its unique advantages, methodologies, and appropriate use cases. The most commonly used feature selection methods include Univariate Selection, Recursive Feature Elimination (RFE), and Principal Component Analysis (PCA). A detailed discussion of each method is given below.

Univariate Selection

Univariate selection is a filter-based feature selection technique in which each feature is evaluated individually to determine its statistical relationship with the target variable. In this method, features are ranked according to the strength of their association with the output variable, and only the top-ranked features are selected for model training. Univariate methods apply statistical hypothesis testing for each feature independently. For classification tasks, common tests include the chi-squared test, ANOVA F-test, or mutual information. For regression tasks, Pearson correlation or F-regression scores may be used. These tests yield a score for each feature, and features with the highest scores are considered most relevant.

Recursive Feature Elimination (RFE)

Recursive Feature Elimination is a wrapper-based method that uses a machine learning model to identify and remove the least significant features recursively. The model is trained multiple times, each time eliminating the weakest features until the optimal subset is selected. RFE starts by training a model (e.g., logistic regression, SVM, or decision tree) on the complete set of features. Based on the model's internal feature importance scores (such as coefficients or weights), it eliminates the least important feature. This process is repeated recursively until the desired number of features is reached.

In industrial applications, RFE is particularly useful when the relationships among features are complex and when domain knowledge suggests that the interaction among sensors or signal components is critical to identifying faults.

Principal Component Analysis (PCA)

Principal Component Analysis is not a feature selection method in the traditional sense but a dimensionality reduction technique. It transforms the original correlated features into a set of uncorrelated variables called principal components. These components are ordered in such a way that the first few retain most of the variation present in the original dataset.

PCA works by computing the eigenvalues and eigenvectors of the covariance matrix of the features. The eigenvectors define the new axes (principal components), and the corresponding eigenvalues represent the amount of variance captured by each component. By selecting the top N components, one can reduce the dimensionality while retaining most of the important information. In fault detection systems where vibration signals produce high-dimensional data, PCA can significantly reduce complexity while preserving the essential patterns of signal variation. This makes it suitable for real-time predictive maintenance applications.

The selection of an appropriate feature selection technique depends on several factors such as the size of the dataset and the type of the model. Recursive Feature Elimination or univariate selection may work best for small to medium-sized datasets. For large datasets with high dimensionality, PCA is more computationally efficient. Recursive Feature Elimination is tightly coupled with the model used; hence, choosing a model with a meaningful feature importance measure is critical. If interpretability is crucial, univariate or RFE is preferable. If performance and dimensionality reduction are more important, PCA is ideal. PCA and univariate selection are generally faster than RFE, which can become slow with many features.

STATISTICAL ANALYSIS OF DATA

In the realm of vibration analysis, extracting and interpreting statistical features from vibrational signals is fundamental for diagnosing mechanical conditions and identifying potential faults. These features provide quantitative metrics that encapsulate the behavior of machinery, facilitating the detection of anomalies and the implementation of predictive maintenance strategies. Among the myriads of statistical features, five stand out for their efficacy and widespread application: Variance, Mean Absolute Deviation (MAD), Zero Crossing Rate (ZCR), Autocorrelation, and Energy. This comprehensive exploration delves into each of these features, elucidating their theoretical foundations, computational methodologies, and practical applications in vibration analysis.

Variance

Variance quantifies the dispersion of a set of data points around their mean value. In the context of vibrational signals, it measures the extent to which individual vibration measurements deviate from the average vibration level. A higher variance indicates greater variability, which may suggest irregularities or instabilities in the machinery's operation. For a discrete vibrational signal represented as x[n], where n denotes the time index, the variance σ^2 is computed as:

$$\sigma 2 = \frac{1}{N} \sum_{n=1}^{N} (x[n] - \mu)^2$$

Here, N is the total number of samples, and μ represents the mean of the signal.

$$\mu = \frac{1}{N} \sum_{n=1}^{N} x[n]$$

In vibration analysis, variance serves as an indicator of the overall energy content of the signal. Elevated variance levels can be symptomatic of mechanical issues such as imbalance, misalignment, or looseness. By monitoring variance over time, maintenance professionals can detect deviations from normal operating conditions and initiate timely interventions.

Mean Absolute Deviation (MAD)

Mean Absolute Deviation provides a measure of the average absolute deviations of data points from their mean. Unlike variance, MAD is less sensitive to extreme values, offering a robust assessment of variability. In vibrational analysis, MAD is instrumental in understanding the consistency of vibration amplitudes. For a vibrational signal x[n], the MAD is calculated as:

$$MAD = \frac{1}{N} \sum_{n=1}^{N} |x[n] - \mu|$$

MAD is particularly useful in scenarios where outliers may skew the analysis. For instance, in detecting gradual wear or degradation in machine components, MAD can reveal subtle changes in vibration patterns that might be overlooked when relying solely on variance.

Zero Crossing Rate (ZCR)

Zero Crossing Rate refers to the frequency at which a signal changes its sign, crossing the zero amplitude axis. In vibration analysis, ZCR is indicative of the signal's frequency content, with higher rates suggesting the presence of high-frequency components often associated with certain types of mechanical faults. For a discrete signal x[n], ZCR is computed by counting the number of times consecutive samples have different signs:

$$ZCR = \frac{1}{2N} \sum_{n=1}^{N-1} |sgn(x[n]) - sgn(x[n+1])|$$

where sgn is the sign function. An increased ZCR in a vibrational signal can point to issues such as bearing defects or gear tooth faults, which introduce high-frequency oscillations. By analyzing ZCR, engineers can identify and address such problems before they escalate.

Autocorrelation

Autocorrelation measures the similarity between a signal and a delayed version of itself over varying time lags. It reveals repeating patterns within the signal, making it invaluable for detecting periodicities and resonant frequencies in vibrational data.

The autocorrelation function Rxx(τ) for a signal x[n] is given by:

$$Rxx(\tau) = \sum_{n=1}^{N-\tau} x[n]\, x[n+\tau]$$

where τ represents the time lag. In vibration analysis, autocorrelation assists in identifying the presence of periodic forces or motions, such as those caused by rotating components. A pronounced peak in the autocorrelation function at a specific lag indicates a strong periodic component, which could be linked to specific machine operations or faults.

Energy

Energy in the context of vibrational signals refers to the total magnitude of the signal squared over a given duration. It reflects the power and intensity of the vibrations, serving as a direct measure of the mechanical activity within the system. For a signal x[n], the energy E is calculated as:

$$E= \sum_{n=1}^{N} x[n]^2$$

Monitoring the energy of vibrational signals enables the detection of changes in the operational state of machinery. Sudden increases in energy may indicate the onset of mechanical faults such as imbalance, misalignment, bearing defects, or increased friction due to wear and tear. These anomalies often lead to elevated vibrational amplitudes, resulting in a spike in the overall energy of the signal. Monitoring such changes helps in early fault detection and prevents potential machinery breakdowns.

RESULTS AND DISCUSSION

The belt-pulley system is modeled and data related to vibrations, temperature, tension, and rotational speed under various operating conditions are collected. This data will be analyzed to identify patterns indicative of wear, misalignment, or impending failures. Machine learning algorithms will then be trained to predict potential issues, enabling timely interventions. This study utilizes the vibration data in the condition monitoring of the belt-drive system that was previously published (Khalifa et al., 2023). The vibration data was collected from two accelerometers mounted near driver and pulley using USB data acquisition cards. The data was collected in three scenarios, including healthy operating conditions, the presence of unbalanced weight, and faulty operation. The dataset comprises belt pretensions at 70 and 150 N. A speed controller allows for the controllable variation of the motor's speed, providing a broad range of operating rotational velocities between 500-2000 RPM. In this work, two operational cases, viz healthy and faulty, with belt pretension of 150 N and specific operational speed of 500 RPM have been chosen. The healthy and faulty states are labeled as 'Maintenance Needed' or 'No Maintenance Needed' to indicate the system's condition. This data can be used for condition monitoring to detect faults or other anomalies in the belt-drive system. This data is used for SVM model development for predictive maintenance framework.

An in-depth examination of the system under various operational conditions, including changes in speed, pretension levels, and states (both normal and faulty), was performed. Advanced statistical features are derived from vibration data for enhanced analysis. Five representative features, such as variance, Mean Absolute Deviation, Zero Crossing Rate, Autocorrelation, and signal energy, were derived from vibration data, and learning was conducted using these features and three external inputs, such as temperature, tension, and rotational speed. These features play a vital role in accurately identifying vibrational patterns across different operational scenarios. These metrics are essential in predictive maintenance for evaluating equipment conditions and forecasting potential failures effectively. Also, these metrics offer a thorough understanding of the equipment performance, enabling early detection of issues and optimizing maintenance schedules to prevent unexpected breakdowns.

Variance measures the spread of data points around the mean, helping to identify fluctuations or instability in machine behavior. A higher variance often indicates irregularities, such as wear or misalignment, which may require maintenance. Figure. 1 compares the variance between a “Faulty” and a “Healthy” system. The higher variance in the faulty system shows that the signal is more spread out, with greater fluctuations, compared to the more consistent behavior seen in the healthy system. This is another useful metric in fault detection, as increased variance can signify abnormal behavior or noise in the system.

Figure 1. Variance of the vibrational signal at 500 RPM and 150 N belt tension

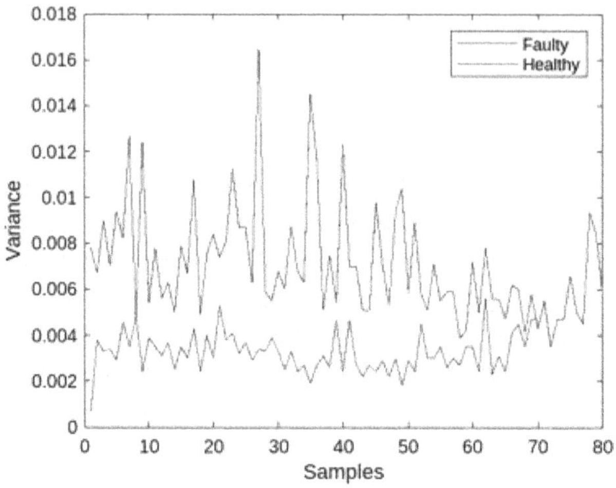

Mean Absolute Deviation, which calculates the average magnitude of deviations from the median, is robust to outliers and can detect gradual changes in equipment performance, signaling early signs of degradation. It measures the average deviation of values from the mean of a data set, indicating the variability or dispersion. Figure 2 compares the Mean Absolute Deviation of "Faulty" and "Healthy" systems, where the faulty system shows more variability in the signal compared to the healthy one.

Figure 2. Mean absolute deviation of the vibrational signal at 500 RPM and 150 N belt tension

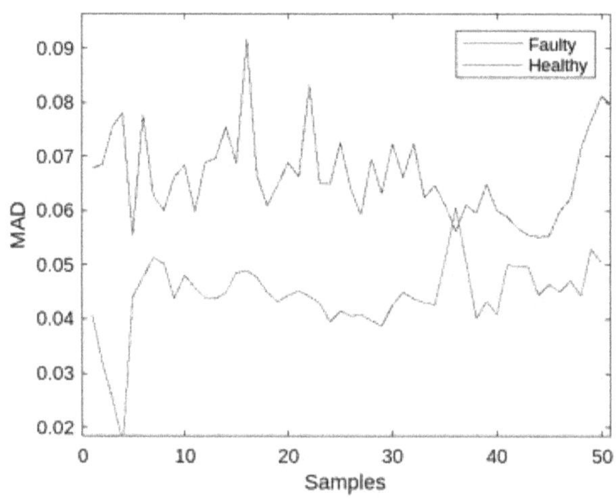

Zero Crossing Rate is a 3×1 feature vector with mean, standard deviation, and skew as its components. It refers to the number of times the signal crosses the zero axis, providing insight into the frequency content and detecting mechanical irregularities or vibrations that may indicate malfunctioning components. It is commonly used in signal processing, particularly in fault detection and vibration analysis. Figure 3 shows the comparison between "Faulty" and "Healthy" systems over multiple samples.

Figure 3. Zero crossing rate of the vibrational signal of the vibrational signal at 500 RPM and 150 N belt tension

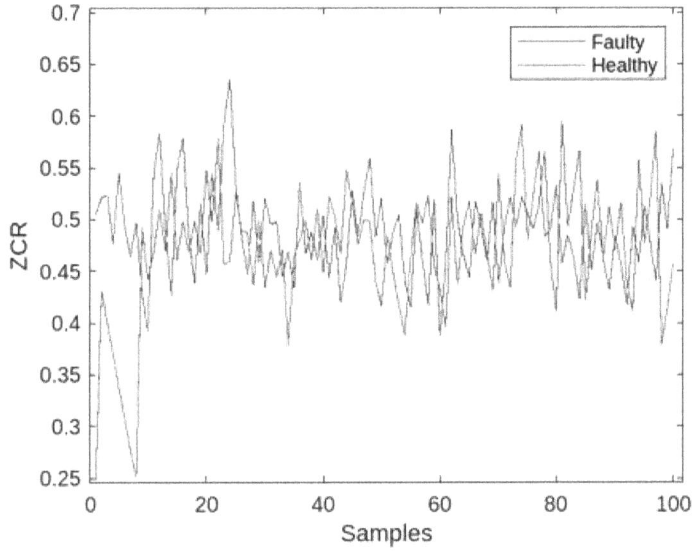

Autocorrelation measures the correlation of a signal with a lagged version of itself, helping to identify periodic patterns or repetitive faults in machinery behavior. Autocorrelation measures the similarity between a signal and its time-lagged version. Figure 4 shows the autocorrelation coefficient over multiple samples, comparing how "Faulty" and "Healthy" systems behave in terms of signal consistency and periodicity.

Figure 4. Autocorrelation of the vibrational signal of the vibrational signal at 500 RPM and 150 N belt tension

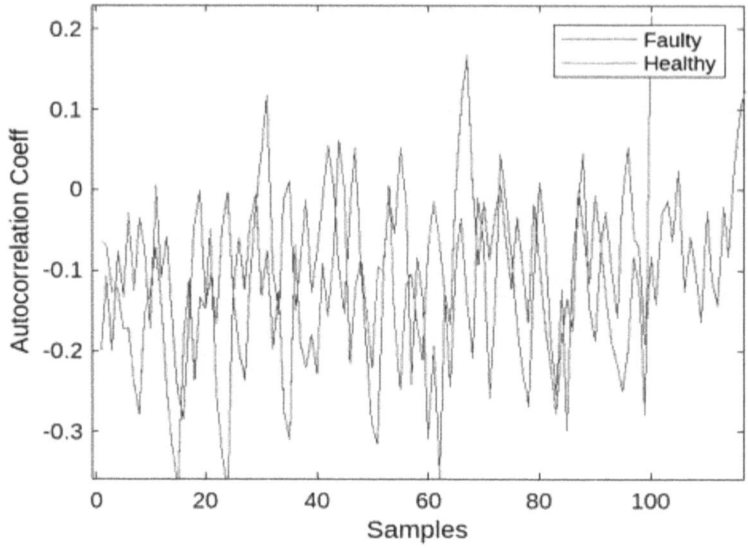

Energy quantifies the total signal strength over time, with drops in energy potentially indicating inefficiencies, wear, or the onset of failures. Figure 5 represents the energy of the signal, calculated by summing the squared values of the amplitude over time. Higher energy often indicates a more significant disturbance or anomaly, which can be observed in the "Faulty" system compared to the "Healthy" system.

Figure 5. Energy of the vibrational signal of the vibrational signal at 500 RPM and 150 N belt tension

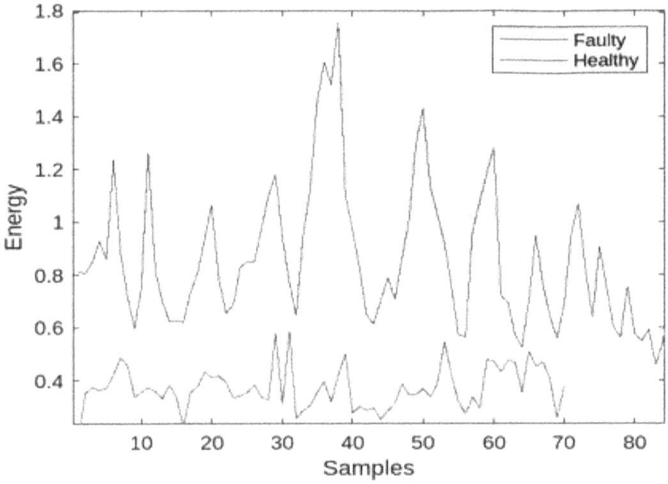

Table 1 shows the efficacy metrics of the proposed SVM model. The SVM model shows the best classification with an accuracy of 95% in predicting maintenance needs, demonstrating its reliability in distinguishing between healthy and faulty system states. Precision (true positive rate) was approximately 93%, indicating that most detected faults were indeed correct. Recall (sensitivity) reached 90%, showing the model's effectiveness at identifying actual faults. F1 Score (harmonic mean of precision and recall) was 91.5%, signifying a balance between detecting all failures and less false alarms. ROC-AUC Score: The model's Area Under the ROC curve (AUC) was 0.9, reflecting strong performance in classifying between healthy and faulty conditions. Also, as the threshold changes, the model maintains a strong balance between true positive and false positive rates, showing reliable performance in binary classification tasks for maintenance needs.

Table 1. Efficacy metrics of proposed SVM model

Proposed model	Train/Test set (%)	Total input features	Accuracy	AUC
SVM Model	80/20	5+5	95%	0.9

Figure 6. Confusion matrix

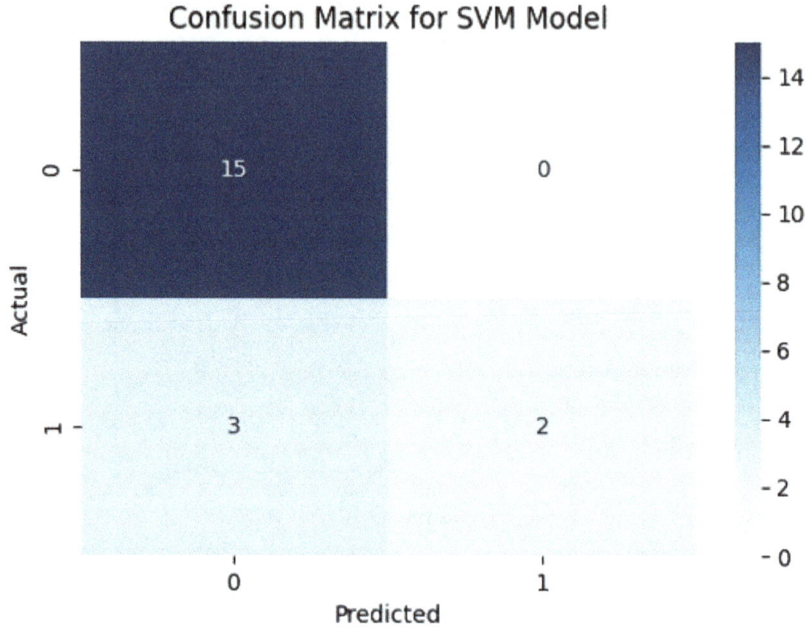

The confusion matrix shown in Figure 6 illustrates the SVM model's performance in classifying belt-pulley system health. Out of the total test samples, 15 instances were correctly classified as "No Maintenance Needed" (True Negatives). 2 instances were correctly classified as "Maintenance Needed" (True Positives). 3 instances of "Maintenance Needed" were misclassified as "No Maintenance Needed" (False Negatives). This matrix indicates the model's accuracy and error rate. The SVM model successfully predicts the majority of cases, but some misclassifications highlight potential areas for improving the model's sensitivity to failure indicators.

Figure 7 illustrates the separation between two classes: "No Maintenance Needed" (blue) and "Maintenance Needed" (red) in a reduced feature space using Principal Component Analysis (PCA). The decision boundary provides a linear separation between the two classes, suggesting that the underlying classification model assumes a linear relationship in the PCA-transformed data. Most points in the "No Maintenance Needed" class are clustered below or around the boundary, while the "Maintenance Needed" points are predominantly above it. However, there is some overlap near the boundary, indicating potential misclassifications or areas where the model struggles to distinctly separate the two classes. Overall, the visualization highlights the model's ability to differentiate the two classes effectively, with some limitations in ambiguous regions.

Figure 7. Classification between maintenance required and not required

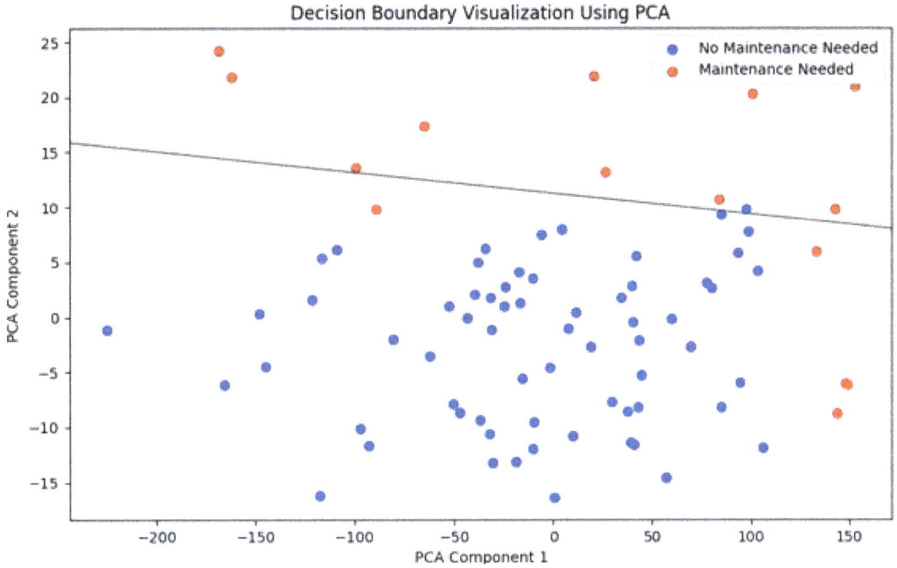

CONCLUSION

By utilizing vibrational data and machine learning techniques, the system achieves reliable fault detection and classification, significantly reducing the risk of unplanned downtime.

1. Sophisticated statistical features, including Variance, Mean Absolute Deviation, Zero Crossing Rate, Autocorrelation Coefficient, and energy of the signal, play a vital role in accurately identifying vibrational patterns across different operational scenarios. These metrics are essential in predictive maintenance for evaluating equipment condition and forecasting potential failures effectively. Together, these metrics offer a thorough understanding of the equipment performance, enabling early detection of issues and optimizing maintenance schedules to prevent unexpected breakdowns. In this proposed predictive maintenance approach, these features were extracted from the data collected under varying operational conditions, including changes in belt tension, rotational speed (RPM), and load scenarios.

2. To classify the condition of the belt-drive system based on these features, the Support Vector Machine algorithm was employed. The reason for choosing this algorithm is that it is a robust and widely used machine learning method, particularly effective in binary and multi-class classification problems. The SVM model was trained to categorize the system's health into distinct classes, viz healthy and faulty states. Due to its capacity to construct optimal separating hyperplanes and its resistance to overfitting in high-dimensional spaces, SVM proved to be exceptionally accurate and dependable for the fault classification task.
3. During testing, the SVM model exhibited high classification accuracy, even when presented with data collected under challenging and variable operational environments. This demonstrates the model's generalization ability and highlights its practical applicability for real-world predictive maintenance tasks. Also, the model consistently demonstrated its reliability under diverse operating conditions, including varying belt tensions and RPM levels. Moreover, the system's computational efficiency ensures that it can operate in near-real-time, enabling industries to shift from reactive or scheduled maintenance to proactive strategies. This transition drastically reduces unnecessary machine downtimes and extends the service life of critical components. The system's fast computational performance ensures its suitability for real-time industrial applications, enabling proactive maintenance strategies.
4. Future enhancements to the system can be made by incorporating additional sensor modalities. For example, temperature data can provide insights into overheating components or lubrication issues, while acoustic signals may capture early-stage faults that are not yet detectable through vibration alone. The inclusion of such multisensory data could further improve the accuracy and reliability of fault detection systems and allow for the identification of more complex failure patterns.
5. Furthermore, while SVM was highly effective in this current framework, other machine learning models such as Random Forests, Gradient Boosting Machines, or even Deep Learning models like Convolutional Neural Networks (CNNs) can be explored for performance comparison. These models may provide enhanced classification abilities when dealing with more complex or noisy datasets.
6. Another promising direction for future work is the application of the Taguchi method, a statistical approach used for optimizing process parameters and conducting robust design experiments. In the context of predictive maintenance, the Taguchi method can be employed for life prediction assessments, helping industries to estimate the remaining useful life of machine components more accurately. This would further strengthen decision-making processes regarding component replacement, maintenance scheduling, and inventory management.

REFERENCES

Achouch, M., Dimitrova, M., Ziane, K., Sattarpanah Karganroudi, S., Dhouib, R., Ibrahim, H., & Adda, M. (2022). On predictive maintenance in industry 4.0: Overview, models, and challenges. *Applied Sciences (Basel, Switzerland), 12*(16), 8081. DOI: 10.3390/app12168081

Ahmad, R., & Kamaruddin, S. (2012). An overview of time-based and condition-based maintenance in industrial application. *Computers & Industrial Engineering, 63*(1), 135–149. DOI: 10.1016/j.cie.2012.02.002

Al-Kahwati, K., Birk, W., Nilsfors, E. F., & Nilsen, R. (2022). Experiences of a digital twin based predictive maintenance solution for belt conveyor systems. *PHM Soc. Eur. Conf.*, *7*(1), 1–8. DOI: 10.36001/phme.2022.v7i1.3355

Ashworth, D. H. (1971). Belt Drives. In *Mechanical Power Transmission* (pp. 103–110). Palgrave Macmillan UK. DOI: 10.1007/978-1-349-01197-1_11

Bishop, R. E. D., & Sherrington, I. (2000). *Mechanical wear prediction and analysis*. Arnold Publishers.

Bloch, H. P., & Geitner, F. K. (1999). *Practical machinery management for process plants: Machinery failure analysis and troubleshooting* (Vol. 2). Gulf Professional Publishing.

Carvalho, T. P., Soares, F. A. A. M. N., Vita, R., Francisco, R. P., Basto, J. P., & Alcalá, S. G. (2019). A systematic literature review of machine learning methods applied to predictive maintenance. *Computers & Industrial Engineering, 137*, 106024. DOI: 10.1016/j.cie.2019.106024

Childs, P. R. N. (2013). *Mechanical design* (2nd ed.). Butterworth-Heinemann.

Chowdhury, S., & Yedavalli, R. K. (2016). Dynamics of belt-pulley-shaft systems. *Mechanism and Machine Theory*, *98*, 199–215. DOI: 10.1016/j.mechmachtheory.2015.11.011

Collacott, R. A. (1979). *Mechanical fault diagnosis and condition monitoring*. Springer.

Fazli, H., Farooq, S., Yang, C., & Wæhrens, B. V. (2023). Proactive and reactive approaches towards sustainable practices in manufacturing companies: Emerging economies perspective. *Sustainability (Basel)*, *15*(17), 12796. DOI: 10.3390/su151712796

Fitch, E. C. (2013). *Proactive maintenance for mechanical systems* (No. 5). Elsevier.

Gerike, B., Panachev, I., & Kuzin, E. (2017). Development of the preventive maintenance system for belt conveyors reducers. In *E3S Web of Conferences* (Vol. 15, p. 03008). EDP Sciences. DOI: 10.1051/e3sconf/20171503008
Gerike, B., Panachev, I., & Kuzin, E. (2017). Development of the preventive maintenance system for belt conveyors reducers. In *E3S Web of Conferences* (Vol. 15, p. 03008). EDP Sciences. DOI: 10.1051/e3sconf/20171503008
Gupta, V., Mitra, R., Koenig, F., Kumar, M., & Tiwari, M. K. (2023). Predictive maintenance of baggage handling conveyors using IoT. *Computers & Industrial Engineering*, *177*, 109033. DOI: 10.1016/j.cie.2023.109033

Gutschi, C., Furian, N., Suschnigg, J., Neubacher, D., & Voessner, S. (2019). Log-based predictive maintenance in discrete parts manufacturing. *Procedia CIRP*, *79*, 528–533. DOI: 10.1016/j.procir.2019.02.098

Gwidon, W. S. (2010). *Mechanical engineering handbook*. CRC Press.

Harris, T. A., & Kotzalas, M. N. (2006). *Rolling bearing analysis* (5th ed.). CRC Press. DOI: 10.1201/9781482275148

Heisler, H. (2002). *Advanced vehicle technology* (2nd ed.). Butterworth-Heinemann.

Hosamo, H. H., Svennevig, P. R., Svidt, K., Han, D., & Nielsen, H. K. (2022). A Digital Twin predictive maintenance framework of air handling units based on automatic fault detection and diagnostics. *Energy and Building*, *261*, 111988. DOI: 10.1016/j.enbuild.2022.111988

Juvinall, R. C., & Marshek, K. M. (2011). *Fundamentals of machine component design* (5th ed.). Wiley.

Khalifa, R. M., Yacout, S., Bassetto, S., & Shaban, Y. (2023). Experimental vibration data collected for a belt-drive system under different operating conditions. *Data in Brief*, *48*, 109175. DOI: 10.1016/j.dib.2023.109175 PMID: 37383798

Khurmi, R. S., & Gupta, J. K. (2005). *A Textbook of Machine Design. S.* Chand Publishing.

Kiangala, K. S., & Wang, Z. (2018). Initiating predictive maintenance for a conveyor motor in a bottling plant using industry 4.0 concepts. *International Journal of Advanced Manufacturing Technology*, *97*(9–12), 3251–3271. DOI: 10.1007/s00170-018-2093-8

Kozhushko, G. G., Yampolsky, D. A., & Lukashuk, M. D. (2020). Analysis of forced vibrations of conveyor belts. IOP Conf. Ser. Mater. Sci. Eng. 709 (3), 033022. DOI: 10.1088/1757-899X/709/3/033022

Krar, S., Gill, A., & Smid, P. (2005). *Technology of machine tools* (6th ed.). McGraw-Hill.

Kumari, S. A., & Srinivasan, S. (2019). Ash fouling monitoring and soot-blow optimization for reheater in thermal power plant. *Applied Thermal Engineering*, *149*, 62–72. DOI: 10.1016/j.applthermaleng.2018.12.031

Lee, J., Bagheri, B., & Kao, H. A. (2015). A cyber-physical systems architecture for industry 4.0-based manufacturing systems. *Manufacturing Letters*, *3*, 18–23. DOI: 10.1016/j.mfglet.2014.12.001

Lipson, C., & Juran, I. (1991). *Quality by design: Taguchi methods and U.S. industry*. McGraw-Hill.

Liu, X., He, D., Lodewijks, G., Pang, Y., & Mei, J. (2019). Integrated decision making for predictive maintenance of belt conveyor systems. *Reliability Engineering & System Safety*, *188*, 347–351. DOI: 10.1016/j.ress.2019.03.047

Malaka, J., & Hetmańczyk, M. (2020). Intelligent Drive in Industry 4.0–Protection of Toothed Belt Transmission on the Basis of Its Digital Twin. In *Sympozjon Modelowanie w Mechanice* (pp. 201–210). Springer International Publishing.

Mathew, V., Toby, T., Singh, V., Rao, B. M., & Kumar, M. G. (2018). "Prediction of Remaining Useful Lifetime (RUL) of turbofan engine using machine learning," in IEEE international conference on circuits and systems. *ICCS*, *2017*, 306–311. DOI: 10.1109/ICCS1.2017.8326010

Mobley, R. K. (2002). *An introduction to predictive maintenance* (2nd ed.). Butterworth-Heinemann.

Mott, R. L. (2017). *Machine Elements in Mechanical Design* (6th ed.). Pearson.

Neale, M. J. (1995). *The tribology handbook* (2nd ed.). Butterworth-Heinemann.

Norton, R. L. (2020). *Design of Machinery* (6th ed.). McGraw-Hill Education.

Ouerghemmi, N., Salem, S. B., Salah, M., Bacha, K., & Chaari, A. (2023). V-belt fault detection using Extended Park Vector Approach (EPVA) in centrifugal fan driven by an induction motor. *Mechanical Systems and Signal Processing*, *200*, 110566. DOI: 10.1016/j.ymssp.2023.110566

Peterson, M. B., & Winer, W. O. (1980). *Wear control handbook*. ASME.

Pollak, A., Temich, S., Ptasiński, W., Kucharczyk, J., & Gąsiorek, D. (2021). Prediction of belt-drive faults in case of predictive maintenance in industry 4.0 platform. *Applied Sciences (Basel, Switzerland)*, *11*(21), 10307. DOI: 10.3390/app112110307

Pulcini, V., & Modoni, G. (2024). Machine learning-based digital twin of a conveyor belt for predictive maintenance. *International Journal of Advanced Manufacturing Technology, 133*(11), 6095–6110. DOI: 10.1007/s00170-024-14097-3

Rao, J. S. (2011). *Mechanical Vibrations.* New Age International Publishers.

Ribeiro, F., Marins, M., Netto, S., & da Silva, E. A. (2017). Rotating machinery fault diagnosis using similarity-based models. XXXV Simpósio Brasileiro de Telecomunicações e Processamento de Sinais-sbrt2017.

Rumin, P., Kotowicz, J., Hogg, D., & Zastawna-Rumin, A. (2023). Utilization of measurements, machine learning, and analytical calculation for preventing belt flip over on conveyor belts. *Measurement, 218,* 113157. DOI: 10.1016/j.measurement.2023.113157

Sakib, N., & Wuest, T. (2018). Challenges and opportunities of condition-based predictive maintenance: A review. *Procedia CIRP, 78,* 267–272. DOI: 10.1016/j.procir.2018.08.318

Shehadeh, H. K. (2024). Evaluating the Impact of Transitioning Maintenance Strategy from Reactive to Proactive in Power Generation Companies: An Empirical Analysis. *WSEAS Transactions on Business And Economics, 21,* 820–838. DOI: 10.37394/23207.2024.21.69

Shigley, J. E., Mischke, C. R., Budynas, R. G., & Nisbett, K. (2011). *Mechanical Engineering Design* (9th ed.). McGraw-Hill Education.

Szrek, J., Wodecki, J., Błazej, R., & Zimroz, R. (2020). An inspection robot for belt conveyor maintenance in underground mine-infrared thermography for overheated idlers detection. *Applied Sciences (Basel, Switzerland), 10*(14), 4984–5017. DOI: 10.3390/app10144984

Tao, F., Cheng, Y., Qi, Q., Zhang, M., Zhang, H., & Sui, F. (2018). Digital twin-driven product design, manufacturing and service with big data. *International Journal of Advanced Manufacturing Technology, 94*(9-12), 3563–3576. DOI: 10.1007/s00170-017-0233-1

Thompson, R. (2000). *Power transmission: Drives and belts.* Industrial Press.

Tlusty, J. (2000). *Manufacturing processes and equipment.* Prentice Hall.

Vapnik, V. (1999). *The nature of statistical learning theory.* Springer science & business media.

Wang, P. Y., Liu, W. C., Liu, N., & You, Y. P. (2020). Digital twin-driven system for roller conveyor line: Design and control. *Journal of Ambient Intelligence and Humanized Computing*, *11*(11), 5419–5431. DOI: 10.1007/s12652-020-01898-z

White, G. (2004). *Industrial belts and drives: Selection and applications*. Newnes.

Wu, M., Goh, K. W., Chaw, K. H., Koh, Y. S., Dares, M., Yeong, C. F., Su, E. L. M., William, H., & Zhang, Y. (2024). An intelligent predictive maintenance system based on random forest for addressing industrial conveyor belt challenges. *Frontiers of Mechanical Engineering*, *10*, 1383202. DOI: 10.3389/fmech.2024.1383202

Wu, M., Yeong, C. F., Su, E. L. M., Holderbaum, W., & Yang, C. (2023). A review on energy efficiency in autonomous mobile robots. Robotic Intell. *Automation*, *43*(6), 648–668. DOI: 10.1108/RIA-05-2023-0060

Yu, X. (2019). Machine learning based predictive maintenance and regression based control of a servo motor for industrial conveyor belts. Berlin: Technische Universität. Master Thesis.

Žarković, M., & Stojković, Z. (2019). Artificial intelligence SF6 circuit breaker health assessment. *Electric Power Systems Research*, *175*, 105912. DOI: 10.1016/j.epsr.2019.105912

Zhang, Y., Sun, S., Wang, X., & Wang, Z. (2021). A digital twin-based approach for smart manufacturing system design and evaluation. *Journal of Manufacturing Systems*, *58*, 168–179. DOI: 10.1016/j.jmsy.2020.12.003

Zhu, H., Zhu, W. D., & Fan, W. (2021). Dynamic modeling, simulation and experiment of power transmission belt drives: A systematic review. *Journal of Sound and Vibration*, *491*, 115759. DOI: 10.1016/j.jsv.2020.115759

Zonta, T., da Costa, C. A., da Rosa Righi, R., de Lima, M. J., da Trindade, E. S., & Li, G. P. (2020). Predictive maintenance in the Industry 4.0: A systematic literature review. *Computers & Industrial Engineering*, *150*, 106889. DOI: 10.1016/j.cie.2020.106889

Chapter 5
Exploring the Magnitude–Shape Plot Framework for Anomaly Detection in Crowded Video Scenes

Zuzheng Wang
Computer, Electrical and Mathematical Sciences and Engineering (CEMSE) Division, King Abdullah University of Science and Technology, Thuwal, Saudi Arabia

Fouzi Harrou
https://orcid.org/0000-0002-2138-319X
Computer, Electrical and Mathematical Sciences and Engineering (CEMSE) Division, King Abdullah University of Science and Technology, Thuwal, Saudi Arabia

Ying Sun
Computer, Electrical and Mathematical Sciences and Engineering (CEMSE) Division, King Abdullah University of Science and Technology, Thuwal, Saudi Arabia

Marc G. Genton
Computer, Electrical and Mathematical Sciences and Engineering (CEMSE) Division, King Abdullah University of Science and Technology, Thuwal, Saudi Arabia

ABSTRACT

This study explores video anomaly detection within a Functional Data Analysis framework, focusing on the application of the Magnitude-Shape (MS) Plot. Autoencoders

DOI: 10.4018/979-8-3373-0746-6.ch005

are used to learn and reconstruct normal behavioral patterns from anomaly-free training data, resulting in low reconstruction errors for normal frames and higher errors for frames with potential anomalies. The reconstruction error matrix for each frame is treated as multivariate functional data, with the MS-Plot applied to analyze both magnitude and shape deviations, enhancing the accuracy of anomaly detection. Using its capacity to evaluate the magnitude and shape of deviations, the MS-Plot offers a statistically principled and interpretable framework for anomaly detection. The proposed methodology is evaluated on two widely used benchmark datasets, UCSD Ped2 and CUHK Avenue, demonstrating promising performance. It performs better than traditional univariate functional detectors (e.g., FBPlot, TVDMSS, Extremal Depth, and Outliergram) and several state-of-the-art methods.

INTRODUCTION

Anomaly detection in video sequences is a critical task in computer vision, particularly in scenarios involving crowded environments such as public spaces, events, transportation hubs, and marketplaces (Santhosh et al., 2020). The goal is to identify unusual or potentially dangerous events that deviate from expected crowd behavior, such as sudden dispersals, confrontations, or individuals engaging in atypical activities (Nayak et al., 2021, Mu et al. 2021). Early detection of such anomalies is crucial for ensuring public safety, managing emergencies, and preventing incidents. In crowded areas, the task of anomaly detection becomes especially challenging due to the complexity and high dimensionality of spatiotemporal data. Video sequences in these environments consist of continuous frames with numerous individuals and objects interacting, creating intricate patterns of motion and behavior (Duong et al., 2023). This complexity requires the use of advanced methods capable of capturing not only spatial features within individual frames, but also temporal and contextual dependencies that characterize typical crowd dynamics (Zhu et al. 2012, Pawar and Attar, 2019).

Over the past two decades, researchers have developed numerous methods to effectively identify abnormal events in video data (Morris and Trivedi, 2008). Trajectory-based approaches track moving objects to detect deviations from typical patterns (Kim and Grauman, 2009, Mehran et al., 2009, Basharat et al., 2008). For instance, the study in (Piciarelli et al., 2008) explored anomaly detection through trajectory analysis using a single-class support vector machine (SVM) clustering approach, effectively identifying unusual trajectories without prior knowledge of outlier distributions. In (Cosar et al., 2016), snapped trajectories are introduced as a high-level representation that reduces computational load and identifies key scene regions. This combined approach detects anomalies in speed, direction, and finer motion details with fewer false alarms. Trajectory-based methods rely on accurate detection and tracking but are limited by crowd density, resolution, motion, and occlusion, making them more suitable for sparse crowds. However, in crowded scenes, frequent occlusions, intersecting paths, and the challenge of distinguishing similar individuals (Ma et al., 2014) reduce their effectiveness, and they lack contextual awareness for identifying anomalies.

Alternatively, other techniques use spatiotemporal features to represent events in videos without requiring trajectory analysis. These methods capture motion dynamics by analyzing changes in pixel intensities and patterns over time (Liet al., 2013). Techniques such as spatiotemporal gradients (Lu et al., 2013), Histograms of Oriented Gradients (HOG) (Dalal and Triggs, 2005), 3D spatiotemporal gradient (Kratz and Nishino, 2009), and Histograms of Optical Flow (HOF) (Dalal et al., 2006) help detect collective crowd behaviors and abnormalities at the pixel or region level, enhancing anomaly detection in dense and dynamic scenes. For example, in (Roy and Om, 2018), an SVM classifier trained on HOG features is proposed to automatically detect violent activities in surveillance videos, distinguishing actions like kicking and punching. This system triggers alerts for detected violence and monitors loitering duration to flag suspicious behavior in real-time, thus enhancing traditional surveillance effectiveness. In (Ahad et al., 2018), a method is introduced for recognizing overlapping and multi-dimensional actions using a spatiotemporal representation and enhanced Motion History Image (MHI). It employs Speeded-Up Robust Features (SURF) to capture key motion features, gradient-based optical flow for motion representation, and RANSAC (Random Sample Consensus) for outlier removal. The nearest neighbour classifier with leave-one-out cross-validation achieves improved recognition rates for complex actions in outdoor scenes compared to traditional MHI approaches. The study in (Cheng et al., 2015) introduces a hierarchical framework for video anomaly detection that integrates multi-level features, including 3D-SIFT (scale-invariant feature transform), HOF, and HOG, with Gaussian process regression. However, these low-level visual features (e.g., motion or texture) may still struggle to fully capture higher-level semantic information or complex contextual interactions in crowded scenes, limiting their effectiveness in more challenging anomaly detection scenarios (Duong et al., 2023).

To address these limitations, deep learning-based methods have emerged as advanced alternatives for video anomaly detection, leveraging their capacity to capture high-level semantic information and complex interactions in crowded and dynamic environments Duong et al. [2023]. Unlike traditional techniques, deep learning models such as Convolutional Neural Networks (CNNs) (Sabokrou et al., 2018, Mansour et al., 2021), Recurrent Neural Networks(RNNs), and Long Short-Term Memory networks (LSTMs) can automatically learn and extract relevant spatial and temporal features directly from the data (Ullah et al., 2021). These models are particularly effective in modeling nuanced deviations from normal patterns, enabling them to generalize better across diverse scenarios. Moreover, autoencoders and Generative Adversarial Networks (GANs) are commonly used to reconstruct normal patterns in video frames, with deviations in reconstruction used as indicators of potential anomalies (Luo et al., 2021, Micorek et al., 2024, Chen et al., 2022). The ability of deep learning models to handle complex feature representations has significantly improved the performance and adaptability of video anomaly detection systems across various application settings (Nayak et al., 2021, Duong et al., 2023, Pawar and Attar, 2019, Wu et al., 2024). The study in (Sabokrou et al., 2018) presents a method for anomaly detection in crowded scenes using fully convolutional networks (FCNs) with temporal data. The model enhances feature extraction by combining a pre-trained CNN (a modified AlexNet) with a custom convolutional layer tailored to specific video data. In (Zhou et al., 2016), a spatial-temporal CNN model is proposed for anomaly detection and localization using spatial-temporal volumes with motion information from static camera scenes. By capturing both appearance and motion features through spatial-temporal convolutions, the model enhances robustness. Evaluated on four benchmark datasets, it outperforms state-of-the-art methods, especially on challenging pixel-level criteria. In (Hong et al., 2024), a video anomaly detection approach leverages a frame-to-label and motion (F2LM) generator to intentionally reduce the quality of abnormal regions, followed by a Destroyer that transforms these areas into zero vectors, making anomalies more prominent. This technique surpasses state-of-the-art performance on the UCSD Ped2, CUHK Avenue, and Shanghai Tech datasets. The study (Dilek and Dener, 2024) proposes an efficient frame-level video anomaly detection (VAD) method that utilizes transfer learning and fine-tuning on 20 CNN-based deep learning models, including variants of VGG, Xception, MobileNet, and ResNet.

Over the past decade, numerous reconstruction-based methods for video anomaly detection have emerged, leveraging models that learn to reconstruct normal video frames or sequences and using reconstruction errors to identify anomalies (Gong et al., 2019; Le and Kim, 2023). The core idea of these approaches is that the model trained exclusively on anomaly-free data will struggle to accurately reconstruct anomalous events, resulting in higher reconstruction errors for frames with unusual or abnormal content. This discrepancy between normal and abnormal reconstructions provides a basis for detecting anomalies effectively. In (Zhao et al., 2017), the authors tackle the challenge of detecting anomalies in complex video scenes by introducing a spatiotemporal AutoEncoder (STAE) that uses deep neural networks with 3D convolutions to learn spatial and temporal features. It incorporates a weight-decreasing prediction loss for future frame generation, improving motion feature learning beyond standard reconstruction loss. In (Deepak et al., 2021), a residual spatiotemporal autoencoder (STAE) is proposed for video anomaly detection, where anomalies are identified as deviations from normal patterns using reconstruction loss. Residual connections enhance model performance, effectively reconstructing normal frames with low cost while detecting irregularities as abnormal frames. Another study in (Luo et al., 2019) proposed a sparse coding-inspired Deep Neural Network (DNN) for video anomaly detection, known as Temporally-coherent Sparse Coding (TSC). TSC maintains frame similarity using a temporal coherence term optimized via the Sequential Iterative Soft-Thresholding Algorithm (SIATA). The enhanced stacked Recurrent Neural Network Autoencoder (sRNN-AE) model introduces data-dependent similarity, reduced model depth for real-time detection, and temporal pooling for efficiency. The study in (Kumar and Khari, 2023) considered a Residual Variational Autoencoder (RVAE) for unsupervised video anomaly detection, which captures complex patterns and minimizes reconstruction error through low-dimensional latent encoding and decoding. The model incorporates a ConvLSTM layer for improved spatiotemporal learning and uses residual connections to address the vanishing gradient problem. Recently, in (Aslam and Kolekar, 2024), a deep multiplicative attention-based autoencoder (DeMAAE) was introduced for video anomaly detection. DeMAAE applies a global attention mechanism at the decoder to enhance feature learning, leveraging an attention map created from encoder-decoder hidden states to guide decoding via a context vector. In (Le and Kim, 2023), a spatial-temporal network achieved 97.4% AUC on UCSD Ped2. A Multivariate Gaussian Fully Convolutional Adversarial Autoencoder in (Li and Chang, 2019) scored 91.6% AUC on UCSD Ped2. The spatiotemporal 3D Convolutional Auto-Encoder (ST-3DCAE) in (Hu et al., 2022a) achieved 85.3% on UCSD Ped2 and 75.8% on UCF-Crime. OF-ConvAE-LSTM in (Duman and Erdem, 2019) reached 92.9% on UCSD Ped2 and 89.5% on Avenue. Finally, (Mishra and Jabin, 2024 achieved 86.4% on UCSD Peds1 and 88.9% on Avenue using a deep autoencoder with regularity-based thresholding.

While deep learning models have significantly advanced video anomaly detection, statistical methods, particularly functional data analysis (FDA), remain underexplored in this area. FDA offers unique advantages by treating each frame's reconstruction error as a continuous, multivariate function over time, enabling the capture of both temporal and spatial dependencies and allowing for a more effective distinction between typical fluctuations and true anomalies by analyzing the shape and magnitude of deviations (Dai and Genton, 2018). Despite these benefits, only a few studies have investigated statistical approaches in video anomaly detection. For example, Rousseeuw et al. (2018) introduced a measure of directional outlyingness specifically applied to image and video data, using a statistical method for anomaly detection based on directional outlyingness. However, this approach has only been validated on images and non-crowded video sequences, emphasizing the need for further testing in more complex, crowded environments. Another statistical approach for anomaly detection in high-dimensional data is the Depthgram method by Alemán-Gómez et al. (2022) designed for functional data visualization in fMRI. Depthgram uses depth-based 2Drepresentations to identify outliers, variability, and sample composition, supporting the exploration of neuroscientific patterns across individuals and brain regions. However, it is limited to static images, such as fMRI data, and has not been explored for video applications.

This study explores video anomaly detection in crowded scenes within a Functional Data Analysis framework, offering a statistical perspective to enhance the detection process. Many statistical video anomaly detection methods (Naji et al. 2022; Yang et al. 2025; R. Singh et al. 2023; Le and Kim 2023; Cao, Lu, and Zhang 2024; Park et al. 2022; Ristea et al. 2024; Hong et al. 2024; Liu et al. 2023) often rely on thresholds determined in a non-automatic and non-systematic manner, focusing primarily on anomalies in the mean or variance of reconstructed residuals. To mitigate these limitations, this work explores the application of the Magnitude-Shape (MS) Plot, a Multivariate Functional Data Visualization and Outlier Detection approach, for video anomaly detection (Dai and Genton, 2018). The MS-Plot provides a statistically principled and interpretable framework by treating reconstruction errors as multivariate functional data. It effectively captures both magnitude (amplitude deviations) and shape (pattern deviations), allowing for the detection of subtle anomalies and complex patterns indicative of abnormal behavior. By simultaneously monitoring amplitude and shape deviations, this approach provides a comprehensive statistical methodology to analyze and understand reconstruction errors in video anomaly detection. In addition, the approach trains solely on normal data, making it effective in scenarios with scarcely labeled anomalies by learning typical patterns and identifying deviations during evaluation. To illustrate, this study investigates two reconstruction-based models: a simple autoencoder and the advanced MAMA autoencoder (Hong et al., 2024). These models learn and reconstruct normal behavior from anomaly-free datasets, producing low reconstruction errors for normal frames and significantly higher errors for anomalous frames during testing. The integration of the MS-Plot enhances this approach by jointly analyzing magnitude and shape deviations, offering a comprehensive framework for anomaly detection. This approach was evaluated on two publicly available datasets, UCSD Ped2 and CUHK Avenue, demonstrating superior performance compared to traditional univariate functional detectors, including Functional Boxplot (FBplot) (Sun and Genton, 2011), Total Variation Depth with Modified Shape Similarity (TVDMSS), Extremal Depth (ED) (Narisetty and Nair, 2016), and Outliergram (OG) (Arribas-Gil and Romo, 2014). Unlike these univariate approaches, which focus on either shape or mean deviations, MS-Plot captures complex, multivariate outliers, proving more sensitive to both subtle and significant anomalies in video data. Also, the proposed approach surpassing state-of-the-art techniques.

MATERIALS AND METHODS

Problem Statement

Video anomaly detection aims to identify frames with abnormal behavior by learning normal patterns during training and detecting deviations during testing. Frame-level detection evaluates each frame for anomalies, such as unexpected objects, irregular behaviors, or environmental changes, without relying on object- or scene-based approaches. Let X denote a video consisting of multiple frames $X = \{X_1, X_2, \dots, X_T\}$, where T represents the total number of frames. Each frame X_t at time t is represented as:

$$X_t = \{p_{ij}(t) \mid 1 \leq i \leq H,\ 1 \leq j \leq W\}, \tag{1}$$

where H and W are the height and width of the frame, and $p_{ij}(t)$ denotes the intensity of the pixel located at (i,j) in frame t.

The goal is to develop a model f that assigns a binary label y_t to each frame X_t such that:

$$y_t = f(X_t) \text{ where } y_t \in \{0, 1\} \tag{2}$$

Here, $y_t = 0$ indicates a normal frame, while $y_t = 1$ indicates an anomalous frame. This detection is performed at the frame level.

Frame-level anomaly detection in videos presents challenges such as diverse anomaly types and limited labeled data. Anomalies range from sudden lighting changes to unexpected behaviors, complicating model generalization. The scarcity of labeled data necessitates unsupervised or semi-supervised methods focused on normal patterns. This study employs autoencoder models combined with the MS-Plot to enhance video anomaly detection.

MS-Plot for Functional Data Outlier Detection

The MS-Plot, introduced by Dai and Genton (Dai and Genton 2018), detects outliers in functional data by evaluating magnitude and shape deviations from a central region. Extending the concept of functional directional outlyingness, it measures both the extent and direction of deviations. This measure allows the MS-Plot to effectively capture two components of outlyingness: magnitude (distance from the central region) and shape (direction of deviation), making it suitable for both univariate and multivariate functional data.

- Mean Directional Outlyingness (MO): This quantifies the average magnitude of a function's deviation from the central region across its domain, effectively identifying anomalies in overall amplitude or intensity. For a function $X(t)$ with distribution F_X, it is defined as (Dai and Genton, 2018):

$$MO(X,F_X) = \int_I O(X(t),F_{X(t)})w(t)\,dt, \tag{3}$$

where $O(X(t),F_{X(t)})$ represents the directional outlyingness at each point t in the domain I, and $w(t)$ is a weight function.

- Variation of Directional Outlyingness (VO): This measures the variability in directional outlyingness across the domain, capturing deviations in the shape or structure of the function. By assessing this variability, VO enables the detection of shape anomalies or irregular patterns within the data (Dai and Genton 2018).

$$VO(X,F_X) = \int_I \| \ O(X(t),F_{X(t)}) - MO(X,F_X)\|^2 w(t)\,dt. \tag{4}$$

Functional Directional Outlyingness (FO) quantifies the overall outlyingness of a function, defined as (Dai and Genton 2018):

$$FO(X,F_X) = \| \ MO(X,F_X)\|^2 + VO(X,F_X). \tag{5}$$

In Equation (**5**), FO combines magnitude outlyingness $\| \ MO \ \|$, representing deviation extent, and shape outlyingness VO, reflecting structural variability. This decomposition enhances the ability to quantify centrality and identify abnormal patterns in functional data.

The MS-Plot technique classifies observations as normal or anomalous by applying a threshold to the FO measure, which combines magnitude and shape deviations. Dai and Genton (Dai and Genton, 2018) proposed calculating FO using MO and VO. Assuming normal data distribution and using random projection point-wise depth, they compute functional directional outlyingness. The method utilizes the squared robust Mahalanobis distance (SRMD) for $(MO, VO)^T$, where the covariance matrix is determined using the Minimum Covariance Determinant (MCD) algorithm (Rousseeuw and Driessen, 1999). The SRMD's tail distribution is approximated with a Fisher's F distribution (Hardin and Rocke, 2005), and curves with SRMD values exceeding the threshold are flagged as outliers.

Autoencoder-Based Video Anomaly Detection

An autoencoder is a neural network framework that learns to compress input data into a latent representation and then reconstruct it back to its original form (Bengio et al. 2013). This process is facilitated by two key components: an encoder, which reduces the data's dimensionality, and a decoder, which restores it (Hinton and Salakhutdinov, 2006).

- **Encoder:** transforms the input data into a compact latent space, preserving critical features and filtering out noise and unnecessary details.
- **Decoder:** Reconstructs the input data from the latent representation, aiming to reduce the reconstruction error by aligning the output closely with the original input.

Let **X** represent an input frame. The encoder function E transforms **X** into a latent representation **Z**, and the decoder function D reconstructs the original frame as $\widehat{\mathbf{X}}$:

$$\mathbf{Z} = E(\mathbf{X}), \widehat{\mathbf{X}} = D(\mathbf{Z}). \tag{6}$$

During training, the autoencoder is optimized to minimize a reconstruction loss, commonly the Mean Squared Error (MSE) between the input **X** and its reconstruction $\widehat{\mathbf{X}}$:

$$L(\mathbf{X}, \widehat{\mathbf{X}}) = \frac{1}{N}\sum_{i=1}^{N}(\mathbf{X}_i - \widehat{\mathbf{X}}_i)^2, \tag{7}$$

where N is the total number of pixels in the frame.

Figure **1** presents a flowchart of an autoencoder for video anomaly detection. Trained solely on normal frames, the autoencoder minimizes reconstruction errors for normal patterns. Anomalous frames deviating from these patterns result in higher reconstruction errors, forming the basis for anomaly detection.

Figure 1. Basic autoencoder for video anomaly detection

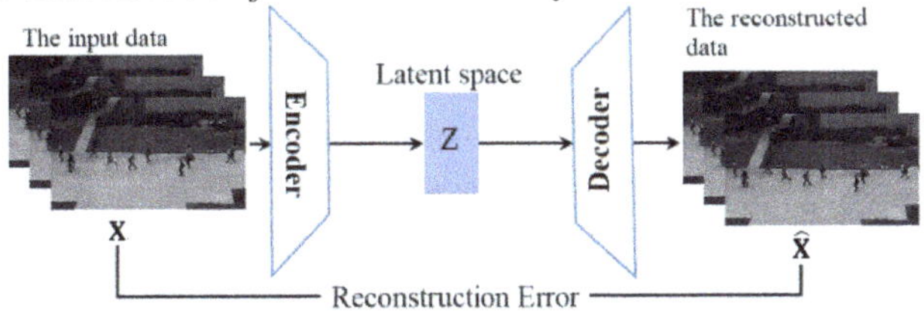

The process of using an autoencoder for video anomaly detection is summarized as follows.

1. **Training Phase:** The autoencoder is trained on a dataset of normal video frames, optimizing it to reconstruct typical, expected frames.
2. **Testing Phase:** For each test frame $\mathbf{X}_t$, the autoencoder reconstructs the frame as $\widehat{\mathbf{X}}_t$ and computes the reconstruction error:

$$e_t = \| \mathbf{X}_t - \widehat{\mathbf{X}}_t \| , \tag{8}$$

where e_t denotes the overall reconstruction error for frame $\mathbf{X}_t$.

3. **Anomaly Detection:** A frame is identified as an anomaly when the reconstruction error e_t surpasses a specified threshold τ.

$$y_t = \begin{cases} 1 & \text{if } e_t > \tau \\ 0 & \text{otherwise.} \end{cases} \tag{9}$$

Here, $y_t = 1$ indicates an anomalous frame, while $y_t = 0$ indicates a normal frame.

Autoencoders effectively detect video anomalies by learning compact representations of normal patterns, highlighting subtle deviations through reconstruction errors (Sabokrouet al., 2016). Trained exclusively on normal data, they excel in unsupervised settings, making them suitable for applications like surveillance and traffic monitoring (Duong et al., 2023). Their ability to flag deviations without labeled anomalies is especially valuable in scenarios with limited anomaly data (Gong et al. 2019). Various enhancements have been proposed to improve their detection capabilities (Hasan et al. 2016; Zhao et al. 2017; Hong et al. 2024; Liu et al. 2023).

MS-PLOT FOR VIDEO ANOMALY DETECTION WITH RECONSTRUCTION MODELS

This study introduces a framework that integrates the MS-Plot with reconstruction-based models for anomaly detection. Residuals (reconstruction errors) generated by these models are treated as multivariate functional data and analyzed using the MS-Plot to identify anomalies. Frame-level anomaly detection involves learning normal behavior patterns, calculating reconstruction errors, and applying the MS-Plot for multivariate functional outlier analysis. This enhances detection accuracy and versatility, enabling the identification of diverse anomalies in video sequences.

For simplicity, the MS-Plot framework is demonstrated with an autoencoder model, but it can be extended to other reconstruction-based models.

This approach consists of three steps: training a reconstruction-based model on anomaly-free data to learn normal patterns, generating residuals by comparing input and reconstructed frames during testing, and analyzing these residuals with the MS-Plot to detect anomalies based on magnitude and shape outlyingness. This integration of reconstruction learning and multivariate analysis ensures effective anomaly detection.

Training Phase - Learning Normal Behavior

During training, only normal frames are provided to the model. The model is typically an autoencoder comprising an encoder function E and a decoder function D. The model aims to reconstruct the input frame $\mathbf{X}_t$ as $\widehat{\mathbf{X}}_t$:

$$\widehat{\mathbf{X}}_t = D(E(\mathbf{X}_t)) \tag{10}$$

The autoencoder is trained to minimize a reconstruction loss function $L(\mathbf{X}_t, \widehat{\mathbf{X}}_t)$, which measures the difference between the original frame $\mathbf{X}_t$ and the reconstructed frame $\widehat{\mathbf{X}}_t$. A common choice for L is the mean squared error (MSE):

$$L(\mathbf{X}_t, \widehat{\mathbf{X}}_t) = \frac{1}{HW}\sum_{i=1}^{H}\sum_{j=1}^{W}\left(p_{ij}(t) - \hat{p}_{ij}(t)\right)^2 \tag{11}$$

where $\hat{p}_{ij}(t)$ represents the intensity of the reconstructed pixel at location (i,j).

Testing Phase - Residual Generation and Anomaly Detection

In the testing phase, the model receives frames from unseen video sequences. For each frame $\mathbf{X}_t$, it computes the reconstruction $\widehat{\mathbf{X}}_t$ and the corresponding reconstruction error matrix:

$$\mathbf{E}_t = \{e_{ij}(t) \mid 1 \leq i \leq H,\ 1 \leq j \leq W\} \tag{12}$$

where $e_{ij}(t) = |p_{ij}(t) - \hat{p}_{ij}(t)|$ denotes the absolute difference between the original and reconstructed pixel intensities.

To enhance anomaly detection, the reconstruction error matrices $\mathbf{E}_t$ are treated as multivariate functional data. For each frame $\mathbf{X}_t$, the MS-Plot calculates MO and VO to detect deviations from normal patterns. The directional outlyingness $O(\mathbf{E}_t$

$,F)$ quantifies the extent to which the error matrix $\mathbf{E}_t$ deviates from a reference distribution F:

$$MO(\mathbf{E}_t,F) = \int_I O(\mathbf{E}_t,F_{\mathbf{E}_t})\,w(t)\,dt,\; VO(\mathbf{E}_t,F) = \int_I \| O(\mathbf{E}_t,F_{\mathbf{E}_t}) - MO(\mathbf{E}_t,F)\|^2 w(t)\,dt.$$

The MS-Plot visualizes $|MO|$ against VO to distinguish normal from anomalous frames, with the relationship defined as:

$$FO = \| MO\|^2 + VO. \tag{14}$$

The MS-Plot applies a threshold on functional outlyingness FO to classify each frame $\mathbf{X}_t$ as normal or anomalous. The decision rule is:

$$y_t = \begin{cases} 1 & \text{if } FO(\mathbf{E}_t) > \tau_{FO} \\ 0 & \text{otherwise.} \end{cases} \tag{15}$$

Algorithm **1** summarizes the steps for using the MS-Plot to analyze residuals and detect frame-level anomalies.

Algorithm 1 MS-Plot framework for anomaly detection based on reconstruction residuals.

Require: Functional data (reconstruction error matrices) $\{\mathbf{E}_1, \mathbf{E}_2, ..., \mathbf{E}_n\}$, Reference distribution $F_{\mathbf{E}}$, Threshold τ_{FO}

Ensure: Anomaly detection result for each $\mathbf{E}_i$ (0: normal, 1: anomaly)

1: Compute the reference distribution $F_{\mathbf{E}}(t)$ (e.g., median function) at each time point t

2: **for** each error matrix $\mathbf{E}_i$ **do**

3: **Compute directional outlyingness** $O(\mathbf{E}_t,F_{\mathbf{E}_i})$ at each time point t:

$$O(E_t,F_{E_i}) = (E_i(t) - F_{E(t)}) \times Sign(E_i(t) - F_{E(t)})$$

4: Calculate Mean Directional Outlyingness (MO) for $\mathbf{E}_i$:

$$MO(\mathbf{E}_i) = \frac{1}{T}\sum_{t=1}^{T} O(\mathbf{E}_i(t), F_{\mathbf{E}(t)})$$

5: Calculate Variation of Directional Outlyingness (VO) for $\mathbf{E}_i$:

$$VO(\mathbf{E}_i) = \frac{1}{T}\sum_{t=1}^{T} O(\mathbf{E}_i(t), F_{\mathbf{E}(t)} - MO(\mathbf{E}_i))^2$$

6: Calculate Functional Outlyingness (FO) for $\mathbf{E}_i$:

$$FO(\mathbf{E}_i) = \| MO(\mathbf{E}_i) \|^2 + VO(\mathbf{E}_i)$$

7: **end for**

8: **Construct MS-Plot**: Plot $|MO(\mathbf{E}_i)|$ on the x-axis and $VO(\mathbf{E}_i)$ on the y-axis for each error matrix $\mathbf{E}_i$

9: **Set Threshold and Detect Outliers:**

10: **for** each error matrix $\mathbf{E}_i$ **do**

11: **if** $FO(\mathbf{E}_i) > \tau_{FO}$ **then**

12: Classify $\mathbf{E}_i$ as anomaly ($y_i = 1$)

13: **else**

14: Classify $\mathbf{E}_i$ as normal ($y_i = 0$)

15: **end if**

16: **end for**

17: **return** Anomaly labels $\{y_1, y_2, ..., y_n\}$ for all error matrices

Evaluation Metrics for Anomaly Detection in Videos

In video anomaly detection, evaluation metrics like True Positive Rate (TPR), False Positive Rate (FPR), Precision, F1 Score, and Accuracy are essential for assessing the effectiveness of identifying anomalies while minimizing false detections. These metrics quantify detection effectiveness by comparing model predictions against ground truth labels. The mathematical definitions of these metrics are summarized in Table **1**, where TP, FP, TN, and FN represent True Positives, False Positives, True Negatives, and False Negatives, respectively.

Table 1. Summary of evaluation metrics for video anomaly detection

Metric	**Equation**
True Positive Rate (TPR)	$\text{TPR} = \frac{\text{TP}}{\text{TP} + \text{FN}}$
False Positive Rate (FPR)	$\text{FPR} = \frac{\text{FP}}{\text{FP} + \text{TN}}$
Precision	$\text{Precision} = \frac{\text{TP}}{\text{TP} + \text{FP}}$
F1 Score	$\text{F1} = 2 \cdot \frac{\text{Precision} \cdot \text{Recall}}{\text{Precision} + \text{Recall}}$
Accuracy	$\text{Accuracy} = \frac{\text{TP} + \text{TN}}{\text{TP} + \text{FP} + \text{TN} + \text{FN}}$

TPR measures the proportion of actual anomalies correctly identified, while FPR indicates the rate of normal frames misclassified as anomalies, a critical factor in reducing false alarms in video data. Precision evaluates the accuracy of predicted anomalies, reflecting the proportion of correctly identified anomalies among all flagged instances. Accuracy captures the overall proportion of correctly classified frames, encompassing both normal and anomalous ones. Finally, the F1-score provides a balanced assessment by combining Precision and Recall, particularly useful in scenarios with class imbalance.

RESULTS AND DISCUSSION

Data Description

In this section, the investigated anomaly detection methods are tested on two widely-used benchmark datasets: the UCSD Pedestrian (Ped2) dataset (Mahadevan et al. 2010) and the CUHK Avenue dataset (Lu et al. 2013), summarized in Table **2**.

Table 2. Details of the UCSD Ped2 and CUHK avenue datasets

Dataset	Total Frames	Training Frames	Testing Frames	Resolution (pixels)
UCSD Ped2	4,560	2,550	2,010	360×240
CUHK Avenue	30,652	15,328	15,324	640×360

UCSD Ped2 dataset: The UCSD Ped2 dataset (Mahadevan et al. 2010) is a widely-used benchmark for video anomaly detection, particularly in surveillance contexts. It captures video sequences of a pedestrian-only walkway, where anomalies such as bicycles, skateboards, and vehicles disrupt typical pedestrian activity. The dataset is divided into a training set with 16 clips (2,550 frames) containing only normal pedestrian behavior and a testing set with 12 clips (2,010 frames) featuring both normal events and anomalies. This structure allows models to learn typical patterns during training and evaluate their ability to detect anomalies. Figure **2** illustrates examples from the dataset, highlighting various anomalies, including cyclists, vehicles, and skaters, which challenge models to accurately differentiate normal and abnormal events.

Figure 2. Sample frames from the UCSD Ped2 dataset showing anomalies: (a) Cyclist, (b) Van and cyclist, (c) Cyclist and skater, and (d) two cyclists. Red boxes highlight anomalous regions

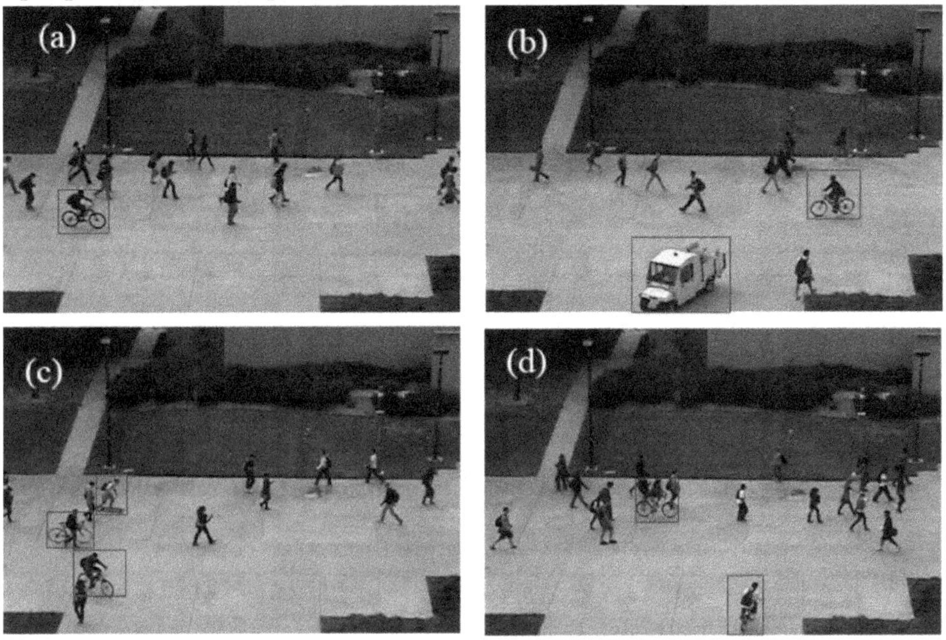

The UCSD Ped2 dataset is recorded at a resolution of 240×360 pixels with a frame rate of 10 frames per second (fps). It includes frame- and pixel-level annotations in the test set, marking anomalies, anomalies timing and location, and facilitating detailed performance evaluation. Frame-level annotations assess a model's ability to detect anomalies within a frame, while pixel-level annotations enable precise localization of anomalous events. Its controlled environment and clear distinction between normal and anomalous events make it a widely accepted standard for evaluating video anomaly detection models. This dataset is essential for benchmarking new methods, assessing model accuracy in anomaly detection, and comparing performance across established metrics. Table **3** outlines the anomalous frames and total video lengths in the UCSD Ped2 dataset, focusing on videos with both normal and anomalous events for MS-Plot analysis.

Table 3. Anomalous frame ranges and total frame count for each test video in the UCSD Ped2 dataset

Test Video	Anomalous Frame Range	Total Frames
Video 1	61–180	180
Video 2	95–180	180
Video 3	1–146	150
Video 4	31–180	180
Video 5	1–129	150
Video 6	1–159	180
Video 7	46–180	180
Video 8	1–180	180
Video 9	1–120	120
Video 10	1–150	150
Video 11	1–180	180
Video 12	88–180	180

CUHK Avenue dataset: The CUHK Avenue dataset (Lu et al. 2013) is a widely used benchmark for video anomaly detection. Captured by a fixed camera overlooking an avenue at the Chinese University of Hong Kong, it offers consistent scene composition, varied lighting conditions, and interactions among multiple pedestrians. The dataset contains 37 video sequences, recorded at 25 frames per second (fps) with a resolution of 640 × 360 pixels, comprising approximately 30,652 frames. It includes 16 training videos featuring normal pedestrian activities, such as walking and entering or exiting the scene, and 21 testing videos containing both normal and abnormal events. Abnormal behaviors include running, throwing objects, loitering, walking in unusual directions, and abandoning items. Figure **3** presents examples from the CUHK Avenue dataset, showcasing a range of anomalies, such as running, walking in unconventional directions, and loitering.

Figure 3. Sample frames from the CUHK Avenue dataset showing anomalies: (a) running, (b) walking in unconventional directions, and (c) loitering

Detection Results Using UCSD2 Ped2 Dataset

In the first experiment, the potential of the MS-Plot technique as an effective detection framework is demonstrated by employing three distinct reconstruction-based models: a simple autoencoder and the MAMA-based model (Hong et al. 2024). Each model is trained exclusively on anomaly-free data to establish a baseline representation of normal behavior. The autoencoder architecture used in this study features an encoder with three convolutional blocks, each comprising a 3×3 Conv2D layer with ReLU activation followed by a 2×2 MaxPooling layer. The encoder's convolutional layers include 32, 16, and 8 filters. The decoder mirrors this structure, using 3×3 Conv2D layers with ReLU activation and 2×2 upsampling, concluding with a 3×3 Conv2D layer (three filters) and a Sigmoid activation function for output generation. Input images are resized to 64×64 pixels, normalized to [0, 1], and processed in batches of 32. The model is trained for 50 epochs using the Adam optimizer and a Mean Squared Error (MSE) loss function to measure reconstruction error.

The second approach combines the MAMA reconstruction-based model (Hong et al. 2024) with the MS-Plot for anomaly detection. The MAMA model operates in two stages: the F2LM generator processes video frames through three parallel streams—raw frames, semantic labels (DeepLabv3), and motion data (FlowNet2)—using a Feature Transform Convolutional (FTC) block to generate high-quality reconstructions for normal events and degraded outputs for anomalies. In the second stage, the Destroyer network enhances these distinctions by suppressing low-quality regions in the F2LM output. To enhance temporal context and motion evolution, the model processes five consecutive frames at each time point. The residuals generated by the two considered reconstruction-based models are analyzed using the MS-Plot, which takes the residual matrix as input and evaluates anomalies through a multidimensional assessment of magnitude and shape outlyingness. By treating the residual matrix as functional data, the MS-Plot detects subtle and complex anomalies, leveraging its nuanced analysis of deviations and the models' ability to handle diverse normal patterns, thereby improving sensitivity and accuracy in video anomaly detection.

Tables **4** and **5** summarize the detection performance of the investigated models on the UCSD Ped2 testing set, using the MS-Plot to analyze reconstruction errors. Table **4** highlights the results for the AE-MS-Plot method, demonstrating varying performance across the testing videos. The method achieves perfect TPR and AUC scores for certain videos (e.g., Videos 9–11), reflecting accurate detection of anomalies. However, its performance falters in scenarios like Video 8, where cyclists and skaters move in different directions, leading to notably low TPR and Accuracy scores. Similarly, Videos 1 and 5, where a bike appears in crowded areas, show higher false positive rates, indicating difficulty in distinguishing anomalies in complex scenes. These results suggest that while the AE-MS-Plot method can effectively detect clear deviations, it faces challenges in scenarios involving subtle anomalies or overlapping movements, emphasizing the need for models capable of handling diverse anomaly patterns.

Table 4. Detection performance of AE-MS-plot on UCSD Ped2 test videos

Video	TPR	FPR	Accuracy	Precision	F1-score	AUC
1	100.00	46.67	84.44	81.08	89.55	78.09
2	18.60	0.00	61.11	100	31.37	77.53
3	95.89	0.00	96.00	100	97.90	98.18
4	95.33	0.00	96.11	100	97.61	97.90
5	100.00	33.33	95.33	94.85	97.36	84.66
6	98.74	9.52	97.78	98.74	98.74	95.06
7	63.70	0.00	72.78	100	77.83	83.13
8	23.33	0.00	23.33	100	37.84	63.41
9	100	0.00	100	100	100	100
10	100	0.00	100	100	100	100
11	100	0.00	100	100	100	100
12	98.92	0.00	99.44	100	99.46	99.52

Table 5. Detection performance of MAMA-MS-plot on UCSD Ped2 test videos

Dataset	TPR	FPR	Accuracy	Precision	F1-score	AUC
1	97.50	1.79	97.73	99.15	98.32	98.08
2	97.67	1.11	98.30	98.82	98.25	98.43
3	98.59	0.00	98.63	100.00	99.29	99.38
4	98.00	0.00	98.30	100.00	98.99	99.24
5	97.60	4.76	97.26	99.19	98.39	96.60
6	97.42	4.76	97.16	99.34	98.37	96.53
7	97.71	0.00	98.30	100.00	98.84	98.97
8	98.86	0.00	98.86	100.00	99.43	99.66
9	97.41	0.00	97.41	100.00	98.69	98.87
10	98.63	0.00	98.63	100.00	99.31	99.47
11	98.30	0.00	98.30	100.00	99.14	99.32
12	95.70	1.20	97.16	98.89	97.27	97.40

The results in Table **5** demonstrate the superior performance of the MAMA-MS-Plot method in detecting different types of anomaly in crowded videos within the UCSD Ped2 dataset. The method consistently achieves high TPR, Precision, F1-scores, and AUC values across most videos, demonstrating its capability to identify both clear and subtle anomalies effectively. Compared to the AE-MS-Plot method, the MAMA-MS-Plot provides enhanced accuracy and maintains low false positive rates, emphasizing its effectiveness in handling diverse testing conditions and crowded environments. This indicates that the sophisticated reconstruction mechanisms in these models generate residuals sensitive enough for the MS-Plot to detect subtle deviations that a simpler autoencoder might miss. Overall, the MS-Plot framework enhances anomaly detection by analyzing magnitude and shape deviations in residuals over time, capturing subtle anomalies, and reducing false positives in complex scenarios.

Comparison of MS-Plot and Univariate Functional Anomaly Detection Methods

The second experiment compares the MS-Plot-based approach with established univariate functional anomaly detection methods, including Functional Boxplot (FBPlot) (Sun and Genton, 2011), Total Variation Depth with Modified Shape Similarity (TVDMSS) (Huang and Sun, 2019), Extremal Depth (ED) (Narisetty and Nair, 2016), and Outliergram (OG) (Arribas-Gil and Romo, 2014). To apply these methods, residual matrices from the autoencoder models were reshaped into one-dimensional vectors. A brief overview of each method is provided for context before comparing their performance with the MS-Plot.

- The FBPlot, introduced by Sun and Genton (Sun and Genton, 2011), extends the traditional boxplot for functional data by using depth measures, such as modified band depth (MBD), to detect outliers. It centers on the median curve and constructs an envelope representing the central 50% of data, identifying outliers as curves that deviate beyond 1.5 times the interquartile range.
- The TVDMSS method (Huang and Sun, 2019), proposed by Huang and Sun, identifies functional outliers using total variation depth (TVD) for magnitude deviations and modified shape similarity (MSS) for shape anomalies. Thresholds on TVD and MSS scores allow detection of outliers in magnitude, shape, or both.
- The ED (Narisetty and Nair, 2016), proposed by Narisetty and Nair, ranks functional data based on "extremeness" to detect both boundary and central outliers. Curves with ED values below a threshold are flagged as outliers.

- The OG (Arribas-Gil and Romo 2014), introduced by Arribas-Gil and Romo, identifies shape outliers in functional data using Modified Band Depth (MBD) for centrality and Half-Region Depth (HRD) for spread. Outliers are visualized as points deviating from the main cluster in an MBD-HRD plot, emphasizing unusual shapes.

Table **6** presents the comparative results of the proposed MS-Plot-based approaches (AE-MS-Plot and MAMA-MS-Plot) against traditional univariate functional anomaly detection methods (FBPlot, TVDMSS, ED, and OG) on the UCSD Ped2 dataset. The analysis highlights notable differences in performance, with MS-Plot-based methods, particularly MAMA-MS-Plot, demonstrating superior results across all evaluated metrics. These findings underscore the effectiveness of MS-Plot in capturing both magnitude and shape deviations, offering a more comprehensive detection framework compared to univariate techniques. For example, MAMA-MS-Plot achieves an AUC of 98.74%, along with high TPR and precision values (97.90% and 99.68%, respectively), demonstrating strong detection accuracy with a minimal false positive rate (FPR of 1.45%). In comparison, AE-MS-Plot achieves a moderate AUC of 88.3%, with a TPR of 82.46%, precision of 94.11%, and an FPR of 23.48%. These results highlight the MS-Plot's effectiveness, with MAMA-MS-Plot excelling in accuracy and sensitivity. In contrast, univariate methods applied to autoencoder residuals perform poorly compared to MS-Plot-based approaches (Table **6**). For instance, AE-FBPlot and AE-TVDMSS achieve an AUC of 52.41% with a TPR of 27.19%, reflecting low sensitivity to anomalies. Even the best univariate method, AE-ED, attains an AUC of 60.16%, far below the performance of MS-Plot-based techniques. Additionally, ED and OG methods applied to vectorized residuals show inconsistent TPR and FPR values, further highlighting their limitations.

Table 6. Comparison of average performance: MS-plot vs. univariate methods on UCSD Ped2 test videos

Approach	TPR	FPR	Accuracy	Precision	F1-score	AUC
AE-MS-Plot	82.46	23.48	81.39	94.11	87.90	88.3
MAMA-MS-Plot	97.90	1.45	98.01	99.68	98.78	98.74
AE-FBPlot	27.19	22.38	36.39	84.48	41.14	52.41
AE-TVDMSS	27.19	22.38	36.39	84.48	41.14	52.41
AE-ED	48.40	28.07	52.50	89.09	62.72	60.16
AE-OG	0.00	0.00	17.60	0.00	0.00	50.00
MAMA-FBPlot	3.09	0.00	19.98	100.00	12.76	51.54

continued on following page

Table 6. Continued

Approach	TPR	FPR	Accuracy	Precision	F1-score	AUC
MAMA-TVDMSS	3.77	0.00	20.54	100.00	15.14	51.88
MAMA-ED	4.69	1.75	21.00	92.68	8.93	51.47
MAMA-OG	0.00	0.00	17.60	0.00	0.00	50.00
DMAD-FBPlot	0.19	0.00	17.58	100.00	0.37	50.09
DMAD-TVDMSS	0.19	0.00	17.58	100.00	0.37	50.09
DMAD-ED	11.05	17.84	23.45	74.58	19.25	46.61
DMAD-OG	49.81	55.56	48.88	80.94	61.67	47.13

This comparison reveals the superiority of MS-Plot-based methods over univariate approaches. By analyzing both magnitude and shape outlyingness, MS-Plot effectively detects anomalies in functional residuals data. Unlike univariate methods like FBPlot and ED, which rely on single-dimensional measures, MS-Plot captures nuanced deviations in reconstruction errors, critical for frame-level video anomaly detection. Its ability to handle variability in residuals and assess both the level and direction of outlyingness enables the detection of subtle shifts missed by univariate methods, improving overall detection accuracy.

MS-Plot Visualization Results based on UCSD Ped2 Testing Set

This study utilizes the UCSD Ped2 testing set to demonstrate the MS-Plot's ability to visually represent frame-level anomaly detection, as illustrated in Figure **4**. The 3D MS-Plot, with ($|MO|$) on the x-axis, (VO) on the y-axis, and frame numbers on the z-axis, highlights anomalies (red dots) and normal frames (blue dots) using residuals from the MAMA-based autoencoder. By excluding videos with solely anomalous frames (Table **3**), the analysis highlights the MS-Plot's ability to distinguish anomalies from normal behavior in mixed-event scenarios and track their progression over time.

Figure **4** illustrates the MS-Plot's capability to distinguish normal frames (blue points) from anomalies (red points) using magnitude outlyingness ($|MO|$) and shape outlyingness (VO). Normal frames cluster at lower $|MO|$ and VO values, while anomalies show higher values, often forming distinct clusters. Elevated VO values highlight unusual shape variations, such as those caused by cyclists or skaters. The z-axis, representing frame numbers, tracks the temporal progression of anomalies, effectively visualizing their persistence and separation from normal behavior.

Figure 4. 3D MS-plot representation for videos in the UCSD Ped2 testing set

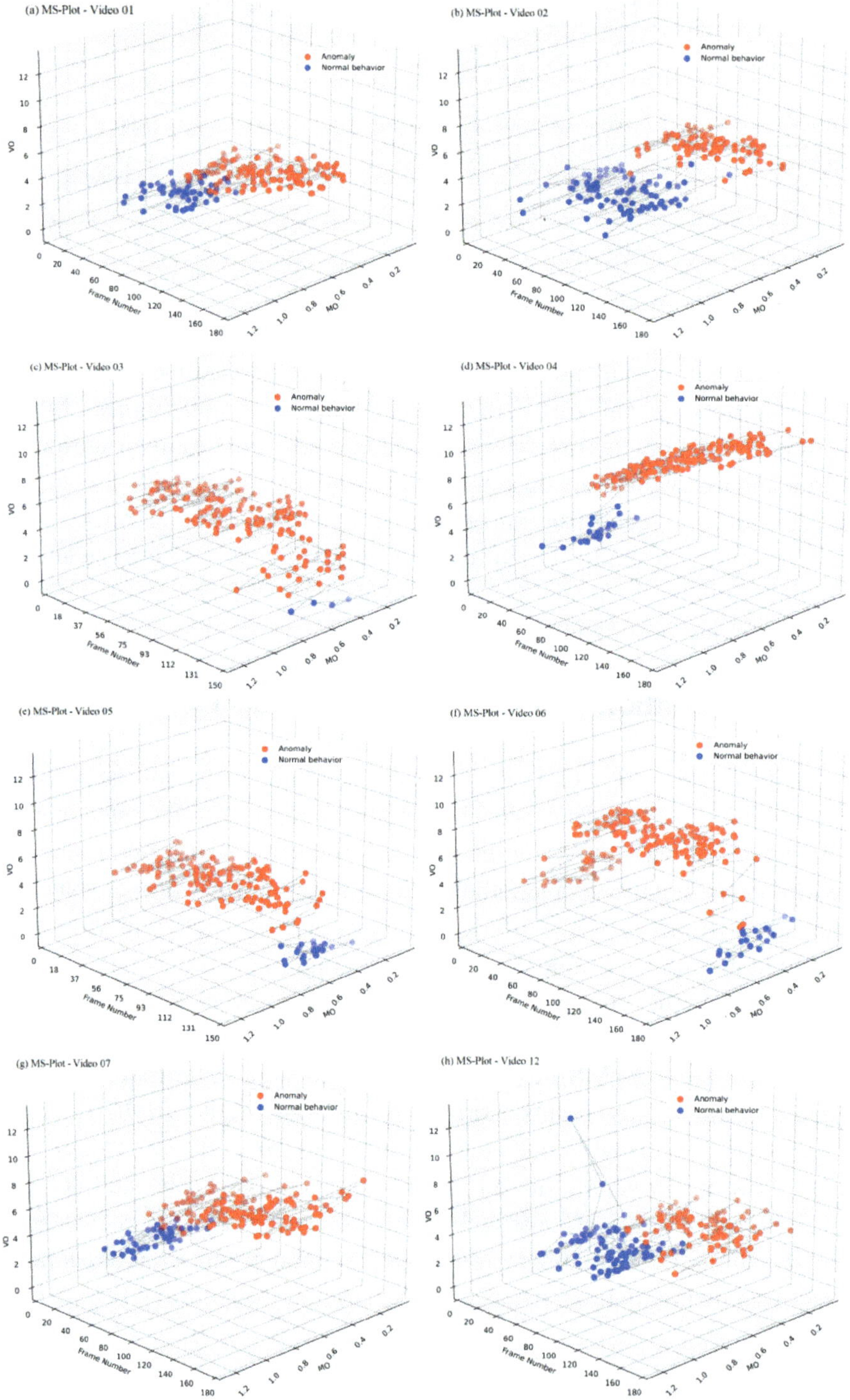

MS-Plot Results Based on CUHK Avenue Data

We extend the evaluation of the two investigated models, the AE-MS-Plot and MAMA-MS-Plot approaches, to the CUHK Avenue dataset to further analyze their performance. The CUHK Avenue dataset is particularly challenging due to its diverse set of anomalous behaviors, such as running, loitering, and abandoning objects, captured in a dynamic and semi-structured environment. By examining the performance of the MS-Plot-based methods on individual videos, the analysis provides insights into the robustness and sensitivity of these approaches to varying anomaly patterns and complexities within the dataset. Following the same protocol as the UCSD Ped2 dataset, both models are trained exclusively on anomaly-free data to learn typical behavioral patterns and subsequently applied to the test set for anomaly detection. This methodology ensures that the models generalize effectively from normal data without prior exposure to anomalous events. The results from univariate functional detectors on the CUHK dataset were excluded due to their poor performance. The performance of the AE-MS-Plot and MAMA-MS-Plot models on the CUHK Avenue dataset is detailed in Tables **7** and **8**. From Tables **7**, the AE-MS-Plot performs well on simpler anomalies (e.g., Videos 1, 4, and 12 with AUC > 97%) but struggles with complex or predominantly anomalous frames (e.g., Videos 3, 10, and 19 with AUC of 50%), highlighting its limitations in handling diverse scenarios. The unsatisfactory AE-based results may stem from dataset imbalances, particularly in CUHK Avenue, where the dominance of normal frames amplifies the impact of false negatives. Moreover, frames containing partial abnormal events, such as a small portion of a bike or an object at the start of a video, further complicate detection, revealing the limitations of the AE-based approach.

Table 7. Performance of AE-MS-plot on CUHK test videos

Video	TPR	FPR	Accuracy	Precision	F1-score	AUC
1	94.90	0.00	98.40	100.00	97.38	97.70
2	40.29	10.01	84.27	34.36	37.09	65.40
3	100.00	100.00	9.35	9.35	17.10	50.00
4	94.62	0.00	99.47	100.00	97.24	97.50
5	92.14	0.00	97.51	100.00	95.91	96.30
6	91.92	0.00	97.27	100.00	95.79	96.20
7	67.62	0.00	94.35	100.00	80.68	84.00
8	100.00	0.00	100.00	100.00	100.00	100.00
9	15.92	0.00	76.11	100.00	27.46	58.20

continued on following page

Table 7. Continued

Video	TPR	FPR	Accuracy	Precision	F1-score	AUC
10	100.00	100.00	19.21	19.21	32.23	50.00
11	100.00	0.00	100.00	100.00	100.00	100.00
12	100.00	0.00	100.00	100.00	100.00	100.00
13	92.59	0.00	98.90	100.00	96.15	96.50
14	72.60	0.00	96.03	100.00	84.13	86.50
15	84.44	20.04	80.36	29.46	43.68	82.40
16	91.92	12.07	88.47	54.17	68.16	90.10
17	79.61	0.00	82.74	100.00	88.65	90.00
18	83.02	0.00	84.54	100.00	90.72	91.70
19	100.00	100.00	53.88	53.88	70.03	50.00
20	44.06	0.00	58.15	100.00	61.17	72.20
21	64.15	10.00	71.23	94.44	76.40	77.30

The MAMA-MS-Plot performs well on the CUHK Avenue dataset, achieving high AUC values above 95% in videos like 1, 4, 8, 10, and 13 (Table **8**). However, it faces challenges in videos with complex anomalies, such as 9 and 18, where AUC scores drop but remain above 88%. These results indicate the method's overall reliability while highlighting areas for potential improvement in detecting anomalies in highly skewed scenarios.

Table 8. Performance of MAMA-MS-plot on CUHK test videos

Video	TPR	FPR	Accuracy	Precision	F1-score	AUC
1	84.92	3.15	93.11	92.51	88.55	91.21
2	84.89	3.18	95.45	77.63	81.10	91.18
3	95.35	3.36	96.52	74.55	83.67	96.32
4	93.55	1.65	97.88	86.14	89.69	96.27
5	82.70	1.02	93.82	97.41	89.46	91.16
6	89.15	1.18	95.55	97.47	93.12	94.30
7	80.95	1.01	95.85	94.44	87.18	90.29
8	90.00	0.00	96.97	100.00	94.74	95.32
9	76.88	0.72	92.92	97.71	86.05	88.40
10	93.79	0.30	98.57	98.69	96.18	97.07
11	81.42	0.00	92.75	100.00	89.76	91.03
12	79.92	0.79	95.43	96.14	87.28	89.89

continued on following page

Table 8. Continued

Video	TPR	FPR	Accuracy	Precision	F1-score	AUC
13	95.06	0.00	99.27	100.00	97.47	97.85
14	90.41	2.09	96.83	88.00	89.19	94.48
15	96.67	2.75	97.19	77.68	86.14	97.28
16	91.92	1.25	97.83	91.92	91.92	95.65
17	77.93	0.00	81.32	100.00	87.60	89.29
18	78.49	0.00	80.41	100.00	87.95	89.57
19	80.30	1.77	88.57	98.15	88.33	89.59
20	78.71	0.00	84.07	100.00	88.09	89.68
21	79.25	0.00	84.93	100.00	88.42	89.94

The overall AUC results show an improvement with the MAMA-MS-Plot (91.43%) over the AE-MS-Plot (81.40%), demonstrating enhanced anomaly detection performance through its advanced residual analysis combined with the MS-Plot framework.

Comparison with SOTA Methods

Table **9** provides an overall comparison of AUC scores for the proposed MS-Plot-based approaches and state-of-the-art (SOTA) methods on the UCSD Ped2 dataset. The MAMA-MS-Plot achieves an impressive AUC of 98.74%, while the AE-MS-Plot records 88.30%, showcasing the overall effectiveness of the MS-Plot framework in accurately identifying video anomalies. Specifically, compared to other detection methods, the MAMA-MS-Plot demonstrated the highest performance, achieving the best AUC score among the evaluated approaches.

Table 9. Comparison of AUC (%) with SOTA methods on the UCSD Ped2 dataset

Method	AUC (%)
MPPCA (Kim and Grauman, 2009)	77.00
Motion-appearance model (Zhang et al., 2016)	90.00
Spatial Temporal CNN (Zhou et al., 2016)	86.00
Conv-AE (Hasan et al., 2016)	90.00
AMDN (Xu et al., 2015)	90.80
GMFC-VAE (Fan et al., 2020)	92.20
StackRNN (Luo et al., 2017a)	92.20
MGFC-AAE (Li and Chang, 2019)	91.60

continued on following page

Table 9. Continued

Method	AUC (%)
ST-CaAE (Li et al., 2020)	92.90
MemAE (Gong et al. 2019)	94.10
DAW (Wang et al. 2018)	96.40
MPN (Lv et al. 2021)	96.90
ConvLSTM (Luo et al., 2017b)	88.10
Spatiotemporal AE (STAE) (Zhao et al., 2017)	91.20
Conv2D (Hasan et al., 2016)	90.00
TMAE (J. Hu et al., 2022)	94.10
MAMC (Ning et al., 2024)	96.70
Cascade Reconstruction (Zhong et al., 2022)	97.70
MAMA (Hong et al., 2024)	98.20
AE-MS-Plot	88.30
MAMA-MS-Plot	98.74

The MAMA-MS-Plot method demonstrates improved performance over the original MAMA-based approach, achieving an AUC of 98.74% compared to 98.20%. This improvement highlights the benefits of using the MS-Plot framework for anomaly detection, particularly in its ability to enhance the interpretability and precision of detection outcomes. In the original MAMA model, anomaly detection is based on empirically determined thresholds primarily based on abnormal changes in the magnitude of residuals. These thresholds, while functional, lack a systematic foundation, which can lead to inconsistencies when addressing diverse anomaly patterns or capturing more nuanced deviations in behavior. The integration of the MS-Plot framework addresses this limitation by treating the residuals of the MAMA model as multivariate functional data. The MS plot evaluates outlyingness (amplitude deviations) and shape (pattern deviations), providing a multidimensional analysis of residual patterns. This capability is particularly beneficial in video anomaly detection, where anomalies often exhibit subtle or complex deviations that may not be effectively captured by simpler thresholding techniques. Using the MS-Plot framework, the MAMA-MS-Plot approach offers a statistically principled method to analyze residuals, enhancing the consistency and reliability of anomaly detection. The framework's capacity to capture both amplitude and shape deviations allows for a more comprehensive understanding of anomalies, ultimately improving detection performance in complex and dynamic video environments.

We compare the proposed MS-Plot-based methods with several state-of-the-art anomaly detection techniques in the CUHK Avenue dataset, as detailed in Table

10. The MAMA-MS-Plot approach achieves an AUC of 91.43%, demonstrating its strong competitiveness among SOTA methods. In particular, the previous highest reported AUC was 91.30%, achieved by SD-MAE (Ristea et al. 2024). The MAMA-MS-Plot exceeds this by a small but meaningful margin of 0.13%, setting a new benchmark for anomaly detection performance in this data set.

Table 10. Comparison of AUC (%) with SOTA methods on the CUHK avenue dataset

Method	**AUC (%)**
Unmasking (Tudor Ionescu et al., 2017)	80.60
StackRNN (Luo et al., 2017a)	81.70
MemAE (Gong et al., 2019)	83.3
FastAno (Park et al., 2022)	85.30
STemGAN (Singh et al., 2023a)	86.00
EVAL (Singh et al., 2023b)	86.02
MESDnet (Fang et al., 2020)	86.3
Any-Shot Sequential (Doshi and Yilmaz, 2020)	86.40
AMMC-Net (Cai et al., 2021)	86.60
ASTNet (Le and Kim, 2023)	86.70
Context Pre (D. Li et al., 2022)	87.10
Learning not to reconstruct (Astrid et al., 2021)	87.10
Siamese Net (Ramachandra et al., 2020)	87.20
Object-centric AE (Ionescu et al., 2019)	87.40
AKD-VAD (Croitoru et al., 2022)	88.30
AnomalyRuler (Yang et al., 2025)	89.70
Two-stream (Cao et al., 2024)	90.80
MAMA (Hong et al., 2024)	91.20
SD-MAE (Ristea et al., 2024)	91.30
AE-MS-Plot	81.40
MAMA-MS-Plot	91.43

This improvement, while incremental, underscores the value of integrating the MS-Plot framework, which provides a systematic and statistically principled approach to anomaly detection. By treating residuals as multivariate functional data, the MAMA-MS-Plot captures both magnitude and shape deviations, enabling more precise differentiation between normal and anomalous frames. In contrast, traditional methods often rely on empirically determined thresholds or univariate analyses, which may miss subtle or multidimensional patterns indicative of anomalies. The MS-Plot's ability to enhance the performance of already strong models like MAMA highlights its potential as a generalizable and effective addition to video anomaly detection pipelines. This advantage is particularly significant given the challenging nature of the CUHK Avenue dataset, which features diverse and complex anomalous behaviors.

CONCLUSION

This study demonstrates the efficacy of the Magnitude-Shape (MS) Plot framework for video anomaly detection, combining statistical functional data analysis with reconstruction-based models. By treating reconstruction errors as multivariate functional data, the MS-Plot captures both magnitude and shape deviations, enabling accurate identification of anomalies in complex, crowded video scenes. The integration of autoencoders with the MS-Plot enhances anomaly detection by leveraging their capacity to model normal behavior and identify deviations with statistical rigor. Experimental results on benchmark datasets, including UCSD Ped2 and CUHK Avenue, highlight the advantages of the proposed framework over traditional univariate functional detectors and several state-of-the-art methods. Specifically, the MAMA-MS-Plot approach achieves consistently high AUC scores, showcasing its capability to generalize across diverse scenarios and detect both subtle and pronounced anomalies effectively.

The findings emphasize the potential of MS-Plot-based frameworks to address the challenges of video anomaly detection, such as the variability of anomalies and limited labeled data. An important direction for future work is to focus on developing unsupervised and robust statistical methods that do not rely on anomaly-free training data. Such methods would further enhance the practicality of video anomaly detection by eliminating the dependence on curated normal datasets and addressing the inherent challenges of real-world applications, such as data scarcity and variability in anomaly types.

REFERENCES

Alemán-Gómez, Y., Arribas-Gil, A., Desco, M., Elías, A., & Romo, J. (2022). Depthgram: Visualizing outliers in high-dimensional functional data with application to fMRI data exploration. *Statistics in Medicine, 41*(11), 2005–2024. DOI: 10.1002/sim.9342 PMID: 35118686

Arribas-Gil, A., & Romo, J. (2014). Shape outlier detection and visualization for functional data: The outliergram. *Biostatistics (Oxford, England), 15*(4), 603–619. DOI: 10.1093/biostatistics/kxu006 PMID: 24622037

Astrid, M., Zaheer, M. Z., Lee, J.-Y., & Lee, S.-I. (2021). Learning not to reconstruct anomalies. arXiv. https://arxiv.org/abs/2110.09742 DOI: 10.5244/C.35.205

Basharat, A., Gritai, A., & Shah, M. (2008). Learning object motion patterns for anomaly detection and improved object detection. In *2008 IEEE Conference on Computer Vision and Pattern Recognition* (pp. 1–8). IEEE. DOI: 10.1109/CVPR.2008.4587510

Bengio, Y., Courville, A., & Vincent, P. (2013). Representation learning: A review and new perspectives. *IEEE Transactions on Pattern Analysis and Machine Intelligence, 35*(8), 1798–1828. DOI: 10.1109/TPAMI.2013.50 PMID: 23787338

Cai, R., Zhang, H., Liu, W., Gao, S., & Hao, Z. (2021). Appearance-motion memory consistency network for video anomaly detection. In Proceedings of the AAAI Conference on Artificial Intelligence (Vol. 35, pp. 938–946). https://ojs.aaai.org/index.php/AAAI/article/view/16163

Cao, C., Lu, Y., & Zhang, Y. (2024). Context recovery and knowledge retrieval: A novel two-stream framework for video anomaly detection. *IEEE Transactions on Image Processing, 33*, 1810–1825. DOI: 10.1109/TIP.2024.3372466 PMID: 38451764

Chen, Z., Duan, J., Kang, L., & Qiu, G. (2022). Supervised anomaly detection via conditional generative adversarial network and ensemble active learning. *IEEE Transactions on Pattern Analysis and Machine Intelligence, 45*(6), 7781–7798. DOI: 10.1109/TPAMI.2022.3225476 PMID: 36445991

Cheng, K.-W., Chen, Y.-T., & Fang, W.-H. (2015). Video anomaly detection and localization using hierarchical feature representation and Gaussian process regression. In *Proceedings of the IEEE Conference on Computer Vision and Pattern Recognition* (pp. 2909–2917). IEEE. DOI: 10.1109/CVPR.2015.7298909

Coşar, S., Donatiello, G., Bogorny, V., Garate, C., Alvares, L. O., & Brémond, F. (2017). Toward abnormal trajectory and event detection in video surveillance. *IEEE Transactions on Circuits and Systems for Video Technology*, *27*(3), 683–695. DOI: 10.1109/TCSVT.2016.2589859

Croitoru, F.-A., Ristea, N.-C., Dascalescu, D., Ionescu, R. T., Khan, F. S., & Shah, M. (2022). Lightning fast video anomaly detection via adversarial knowledge distillation. arXiv. https://arxiv.org/abs/2211.15597

Dai, W., & Genton, M. G. (2018). Multivariate functional data visualization and outlier detection. *Journal of Computational and Graphical Statistics*, *27*(4), 923–934. DOI: 10.1080/10618600.2018.1473781

Dalal, N., & Triggs, B. (2005). Histograms of oriented gradients for human detection. In *2005 IEEE Computer Society Conference on Computer Vision and Pattern Recognition (CVPR'05)* (Vol. 1, pp. 886–893). IEEE. DOI: 10.1109/CVPR.2005.177

Dalal, N., Triggs, B., & Schmid, C. (2006). Human detection using oriented histograms of flow and appearance. In A. Leonardis, H. Bischof, & A. Pinz (Eds.), Computer Vision – ECCV 2006: 9th European Conference on Computer Vision, Graz, Austria, May 7-13, 2006, Proceedings, Part II (pp. 428–441). Springer. DOI: 10.1007/11744047_33

Deepak, K., Chandrakala, S., & Mohan, C. K. (2021). Residual spatiotemporal autoencoder for unsupervised video anomaly detection. *Signal, Image and Video Processing*, *15*(1), 215–222. DOI: 10.1007/s11760-020-01740-1

Dilek, E., & Dener, M. (2024). Enhancement of video anomaly detection performance using transfer learning and fine-tuning. *IEEE Access : Practical Innovations, Open Solutions*, *12*, 123456–123467. DOI: 10.1109/ACCESS.2024.3404553

Doshi, K., & Yilmaz, Y. (2020). Any-shot sequential anomaly detection in surveillance videos. In *Proceedings of the IEEE/CVF Conference on Computer Vision and Pattern Recognition Workshops* (pp. 934–935). IEEE. DOI: 10.1109/CVPRW50498.2020.00475

Duman, E., & Erdem, O. A. (2019). Anomaly detection in videos using optical flow and convolutional autoencoder. *IEEE Access : Practical Innovations, Open Solutions*, *7*, 183914–183923. DOI: 10.1109/ACCESS.2019.2960654

Duong, H.-T., Le, V.-T., & Hoang, V. T. (2023). Deep learning-based anomaly detection in video surveillance: A survey. *Sensors (Basel)*, *23*(11), 5024. DOI: 10.3390/s23115024 PMID: 37299751

Fan, Y., Wen, G., Li, D., Qiu, S., Levine, M. D., & Xiao, F. (2020). Video anomaly detection and localization via Gaussian mixture fully convolutional variational autoencoder. *Computer Vision and Image Understanding*, *195*, 102920. Advance online publication. DOI: 10.1016/j.cviu.2020.102920

Fang, Z., Zhou, J. T., Xiao, Y., Li, Y., & Yang, F. (2020). Multi-encoder towards effective anomaly detection in videos. *IEEE Transactions on Multimedia*, *23*, 4106–4116. DOI: 10.1109/TMM.2020.3037538

Gong, D., Liu, L., Le, V., Saha, B., Mansour, M. R., Venkatesh, S., & van den Hengel, A. (2019). Memorizing normality to detect anomaly: Memory-augmented deep autoencoder for unsupervised anomaly detection. In *Proceedings of the IEEE/CVF International Conference on Computer Vision* (pp. 1705–1714). IEEE. DOI: 10.1109/ICCV.2019.00179

Hardin, J., & Rocke, D. M. (2005). The distribution of robust distances. *Journal of Computational and Graphical Statistics*, *14*(4), 928–946. DOI: 10.1198/106186005X77685

Hasan, M., Choi, J., Neumann, J., Roy-Chowdhury, A. K., & Davis, L. S. (2016). Learning temporal regularity in video sequences. In *Proceedings of the IEEE Conference on Computer Vision and Pattern Recognition* (pp. 733–742). IEEE. DOI: 10.1109/CVPR.2016.85

Hinton, G. E., & Salakhutdinov, R. R. (2006). Reducing the dimensionality of data with neural networks. *Science*, *313*(5786), 504–507. DOI: 10.1126/science.1127647 PMID: 16873662

Hong, S., Ahn, S., Jo, Y., & Park, S. (2024). Making anomalies more anomalous: Video anomaly detection using a novel generator and destroyer. *IEEE Access : Practical Innovations, Open Solutions*, *12*, 98765–98776. DOI: 10.1109/ACCESS.2024.3374383

Hu, J., Yu, G., Wang, S., Zhu, E., Cai, Z., & Zhu, X. (2022). Detecting anomalous events from unlabeled videos via temporal masked auto-encoding. In *2022 IEEE International Conference on Multimedia and Expo (ICME)* (pp. 1–6). IEEE. DOI: 10.1109/ICME52920.2022.9859873

Hu, X., Lian, J., Zhang, D., Gao, X., Jiang, L., & Chen, W. (2022). Video anomaly detection based on 3D convolutional auto-encoder. *Signal, Image and Video Processing*, *16*(7), 1885–1893. DOI: 10.1007/s11760-022-02148-9

Huang, H., & Sun, Y. (2019). A decomposition of total variation depth for understanding functional outliers. *Technometrics*, *61*(4), 445–458. DOI: 10.1080/00401706.2019.1574241

Ionescu, R. T., Khan, F. S., Georgescu, M.-I., & Shao, L. (2019). Object-centric auto-encoders and dummy anomalies for abnormal event detection in video. In *Proceedings of the IEEE/CVF Conference on Computer Vision and Pattern Recognition* (pp. 7842–7851). IEEE. DOI: 10.1109/CVPR.2019.00803

Ionescu, R. T., Smeureanu, S., Alexe, B., & Popescu, M. (2017). Unmasking the abnormal events in video. In *Proceedings of the IEEE International Conference on Computer Vision* (pp. 2895–2903). IEEE. DOI: 10.1109/ICCV.2017.312

Kim, J., & Grauman, K. (2009). Observe locally, infer globally: A space-time MRF for detecting abnormal activities with incremental updates. In *2009 IEEE Conference on Computer Vision and Pattern Recognition* (pp. 2921–2928). IEEE. DOI: 10.1109/CVPR.2009.5206569

Kratz, L., & Nishino, K. (2009). Anomaly detection in extremely crowded scenes using spatio-temporal motion pattern models. In *2009 IEEE Conference on Computer Vision and Pattern Recognition* (pp. 1446–1453). IEEE. DOI: 10.1109/CVPR.2009.5206771

Kumar, A., & Khari, M. (2023). Efficient video anomaly detection using residual variational autoencoder. In *2023 International Conference on Communication System, Computing and IT Applications (CSCITA)* (pp. 50–55). IEEE. DOI: 10.1109/CSCITA55725.2023.10104785

Li, D., Nie, X., Li, X., Zhang, Y., & Yin, Y. (2022). Context-related video anomaly detection via generative adversarial network. *Pattern Recognition Letters, 156*, 183–189. DOI: 10.1016/j.patrec.2022.03.004

Li, N., & Chang, F. (2019). Video anomaly detection and localization via multivariate Gaussian fully convolution adversarial autoencoder. *Neurocomputing, 369*, 92–105. DOI: 10.1016/j.neucom.2019.08.044

Li, N., Chang, F., & Liu, C. (2020). Spatial-temporal cascade autoencoder for video anomaly detection in crowded scenes. *IEEE Transactions on Multimedia, 23*, 203–215. DOI: 10.1109/TMM.2020.2984093

Li, W., Mahadevan, V., & Vasconcelos, N. (2014). Anomaly detection and localization in crowded scenes. *IEEE Transactions on Pattern Analysis and Machine Intelligence, 36*(1), 18–32. DOI: 10.1109/TPAMI.2013.111 PMID: 24231863

Lu, C., Shi, J., & Jia, J. (2013). Abnormal event detection at 150 fps in Matlab. In *Proceedings of the IEEE International Conference on Computer Vision* (pp. 2720–2727). IEEE. DOI: 10.1109/ICCV.2013.338

Luo, W., Liu, W., & Gao, S. (2017a). A revisit of sparse coding based anomaly detection in stacked RNN framework. In *Proceedings of the IEEE International Conference on Computer Vision* (pp. 341–349). IEEE. DOI: 10.1109/ICCV.2017.45

Luo, W., Liu, W., & Gao, S. (2017b). Remembering history with convolutional LSTM for anomaly detection. In *2017 IEEE International Conference on Multimedia and Expo (ICME)* (pp. 439–444). IEEE. DOI: 10.1109/ICME.2017.8019325

Luo, W., Liu, W., Lian, D., & Gao, S. (2019). Video anomaly detection with sparse coding inspired deep neural networks. *IEEE Transactions on Pattern Analysis and Machine Intelligence, 43*(3), 1070–1084. DOI: 10.1109/TPAMI.2019.2944377 PMID: 31567072

Luo, W., Liu, W., Lian, D., Tang, J., Duan, L., Peng, X., & Gao, S. (2021). Future frame prediction network for video anomaly detection. *IEEE Transactions on Pattern Analysis and Machine Intelligence, 44*(11), 7505–7520. DOI: 10.1109/TPAMI.2021.3129349 PMID: 34797762

Lv, H., Chen, C., Cui, Z., Xu, C., Li, Y., & Yang, J. (2021). Learning normal dynamics in videos with meta prototype network. In *Proceedings of the IEEE/CVF Conference on Computer Vision and Pattern Recognition* (pp. 15425–15434). IEEE. DOI: 10.1109/CVPR46437.2021.01517

Ma, X., Wang, H., Xue, B., Zhou, M., Ji, B., & Li, Y. (2014). Depth-based human fall detection via shape features and improved extreme learning machine. *IEEE Journal of Biomedical and Health Informatics, 18*(6), 1915–1922. DOI: 10.1109/JBHI.2014.2304357 PMID: 25375688

Mahadevan, V., Li, W., Bhalodia, V., & Vasconcelos, N. (2010). Anomaly detection in crowded scenes. In *2010 IEEE Computer Society Conference on Computer Vision and Pattern Recognition* (pp. 1975–1981). IEEE. DOI: 10.1109/CVPR.2010.5539872

Mansour, R. F., Escorcia-Gutierrez, J., Gamarra, M., Villanueva, J. A., & Leal, N. (2021). Intelligent video anomaly detection and classification using faster RCNN with deep reinforcement learning model. *Image and Vision Computing, 112*, 104229. DOI: 10.1016/j.imavis.2021.104229

Mehran, R., Oyama, A., & Shah, M. (2009). Abnormal crowd behavior detection using social force model. In *2009 IEEE Conference on Computer Vision and Pattern Recognition* (pp. 935–942). IEEE. DOI: 10.1109/CVPR.2009.5206641

Micorek, J., Possegger, H., Narnhofer, D., Bischof, H., & Kozinski, M. (2024). Mulde: Multiscale log-density estimation via denoising score matching for video anomaly detection. In Proceedings of the IEEE/CVF Conference on Computer Vision and Pattern Recognition (pp. 18868–18877). IEEE. https://openaccess.thecvf.com/content/CVPR2024/html/Micorek_Mulde_Multiscale_Log-Density_Estimation_via_Denoising_Score_Matching_for_Video_CVPR_2024_paper.html

Mishra, S., & Jabin, S. (2024). Anomaly detection in surveillance videos using deep autoencoder. *International Journal of Information Technology : an Official Journal of Bharati Vidyapeeth's Institute of Computer Applications and Management*, *16*(2), 1111–1122. DOI: 10.1007/s41870-023-01659-z

Morris, B. T., & Trivedi, M. M. (2008). A survey of vision-based trajectory learning and analysis for surveillance. *IEEE Transactions on Circuits and Systems for Video Technology*, *18*(8), 1114–1127. DOI: 10.1109/TCSVT.2008.927109

Mu, H., Sun, R., Yuan, G., & Wang, Y. (2021). Abnormal human behavior detection in videos: A review. *Information Technology and Control*, *50*(3), 522–545. DOI: 10.5755/j01.itc.50.3.27864

Naji, Y., Setkov, A., Loesch, A., Gouiffès, M., & Audigier, R. (2022). Spatio-temporal predictive tasks for abnormal event detection in videos. In 2022 18th IEEE International Conference on Advanced Video and Signal Based Surveillance (AVSS) (pp. 1–8). IEEE. DOI: 10.1109/AVSS56176.2022.9959669

Narisetty, N. N., & Nair, V. N. (2016). Extremal depth for functional data and applications. *Journal of the American Statistical Association*, *111*(516), 1705–1714. DOI: 10.1080/01621459.2015.1110033

Nayak, R., Pati, U. C., & Das, S. K. (2021). A comprehensive review on deep learning-based methods for video anomaly detection. *Image and Vision Computing*, *106*, 104078. DOI: 10.1016/j.imavis.2020.104078

Ning, Z., Wang, Z., Liu, Y., Liu, J., & Song, L. (2024). Memory-enhanced appearance-motion consistency framework for video anomaly detection. *Computer Communications*, *216*, 159–167. DOI: 10.1016/j.comcom.2024.01.004

Park, C., Cho, M., Lee, M., & Lee, S. (2022). FastAno: Fast anomaly detection via spatio-temporal patch transformation. In *Proceedings of the IEEE/CVF Winter Conference on Applications of Computer Vision* (pp. 2249–2259). IEEE. DOI: 10.1109/WACV51458.2022.00197

Pawar, K., & Attar, V. (2019). Deep learning approaches for video-based anomalous activity detection. *World Wide Web (Bussum)*, *22*(2), 571–601. DOI: 10.1007/s11280-018-0582-1

Piciarelli, C., Micheloni, C., & Foresti, G. L. (2008). Trajectory-based anomalous event detection. *IEEE Transactions on Circuits and Systems for Video Technology*, *18*(11), 1544–1554. DOI: 10.1109/TCSVT.2008.2005599

Rahman Ahad, M. A., Tan, J. K., Kim, H., & Ishikawa, S. (2018). Activity representation by SURF-based templates. *Computer Methods in Biomechanics and Biomedical Engineering. Imaging & Visualization*, *6*(5), 573–583. DOI: 10.1080/21681163.2017.1298472

Ramachandra, B., Jones, M. J., & Vatsavai, R. (2020). Learning a distance function with a Siamese network to localize anomalies in videos. In *Proceedings of the IEEE/CVF Winter Conference on Applications of Computer Vision* (pp. 2598–2607). IEEE. DOI: 10.1109/WACV45572.2020.9093417

Ristea, N.-C., Croitoru, F.-A., Ionescu, R. T., Popescu, M., Khan, F. S., & Shah, M. (2024). Self-distilled masked auto-encoders are efficient video anomaly detectors. In *Proceedings of the IEEE/CVF Conference on Computer Vision and Pattern Recognition* (pp. 15984–15995). IEEE., https://openaccess.thecvf.com/content/CVPR2024/html/Ristea_Self-Distilled_Masked_Auto-Encoders_are_Efficient_Video_Anomaly_Detectors_CVPR_2024_paper.html DOI: 10.1109/CVPR52733.2024.01513

Rousseeuw, P. J., Raymaekers, J., & Hubert, M. (2018). A measure of directional outlyingness with applications to image data and video. *Journal of Computational and Graphical Statistics*, *27*(2), 345–359. DOI: 10.1080/10618600.2017.1366912

Rousseeuw, P. J., & Van Driessen, K. (1999). A fast algorithm for the minimum covariance determinant estimator. *Technometrics*, *41*(3), 212–223. DOI: 10.1080/00401706.1999.10485670

Roy, P. K., & Om, H. (2018). Suspicious and violent activity detection of humans using HOG features and SVM classifier in surveillance videos. In Satapathy, S. C., Bhateja, V., Le, P. N. H., Nguyen, N., & Rao, B. L. (Eds.), *Advances in soft computing and machine learning in image processing* (pp. 277–294). Springer., DOI: 10.1007/978-3-319-63754-9_13

Sabokrou, M., Fathy, M., & Hoseini, M. (2016). Video anomaly detection and localisation based on the sparsity and reconstruction error of auto-encoder. *Electronics Letters*, *52*(13), 1122–1124. DOI: 10.1049/el.2016.0440

Sabokrou, M., Fayyaz, M., Fathy, M., Moayed, Z., & Klette, R. (2018). Deep-anomaly: Fully convolutional neural network for fast anomaly detection in crowded scenes. *Computer Vision and Image Understanding*, *172*, 88–97. DOI: 10.1016/j.cviu.2018.02.006

Santhosh, K. K., Dogra, D. P., & Roy, P. P. (2020). Anomaly detection in road traffic using visual surveillance: A survey. *ACM Computing Surveys*, *53*(6), 119. DOI: 10.1145/3417988

Singh, A., Jones, M. J., & Learned-Miller, E. G. (2023). Eval: Explainable video anomaly localization. In *Proceedings of the IEEE/CVF Conference on Computer Vision and Pattern Recognition* (pp. 18717–18726). IEEE. DOI: 10.1109/CVPR52729.2023.01793

Singh, R., Saini, K., Sethi, A., Tiwari, A., Saurav, S., & Singh, S. (2023). STEMGAN: Spatio-temporal generative adversarial network for video anomaly detection. *Applied Intelligence*, *53*(23), 28133–28152. DOI: 10.1007/s10489-023-04940-7

Sun, Y., & Genton, M. G. (2011). Functional boxplots. *Journal of Computational and Graphical Statistics*, *20*(2), 316–334. DOI: 10.1198/jcgs.2011.09224

Ullah, W., Ullah, A., Hussain, T., Khan, Z. A., & Baik, S. W. (2021). An efficient anomaly recognition framework using an attention residual LSTM in surveillance videos. *Sensors (Basel)*, *21*(8), 2811. DOI: 10.3390/s21082811 PMID: 33923712

Wang, S., Zeng, Y., Liu, Q., Zhu, C., Zhu, E., & Yin, J. (2018). Detecting abnormality without knowing normality: A two-stage approach for unsupervised video abnormal event detection. In *Proceedings of the 26th ACM International Conference on Multimedia* (pp. 636–644). ACM. DOI: 10.1145/3240508.3240615

Wu, P., Pan, C., Yan, Y., Pang, G., Wang, P., & Zhang, Y. (2024). Deep learning for video anomaly detection: A review. arXiv. https://arxiv.org/abs/2409.05383

Xu, D., Ricci, E., Yan, Y., Song, J., & Sebe, N. (2015). Learning deep representations of appearance and motion for anomalous event detection. arXiv. https://arxiv.org/abs/1510.01553 DOI: 10.5244/C.29.8

Yang, Y., Lee, K., Dariush, B., Cao, Y., & Lo, S.-Y. (2025). Follow the rules: Reasoning for video anomaly detection with large language models. In *European Conference on Computer Vision* (pp. 304–322). Springer. DOI: 10.1007/978-3-031-73004-7_18

Zhang, Y., Lu, H., Zhang, L., & Ruan, X. (2016). Combining motion and appearance cues for anomaly detection. *Pattern Recognition*, *51*, 443–452. DOI: 10.1016/j.patcog.2015.09.005

Zhao, Y., Deng, B., Shen, C., Liu, Y., Lu, H., & Hua, X.-S. (2017). Spatio-temporal autoencoder for video anomaly detection. In *Proceedings of the 25th ACM International Conference on Multimedia* (pp. 1933–1941). ACM. DOI: 10.1145/3123266.3123451

Zhong, Y., Chen, X., Jiang, J., & Ren, F. (2022). A cascade reconstruction model with generalization ability evaluation for anomaly detection in videos. *Pattern Recognition*, *122*, 108336. DOI: 10.1016/j.patcog.2021.108336

Zhou, S., Shen, W., Zeng, D., Fang, M., Wei, Y., & Zhang, Z. (2016). Spatial–temporal convolutional neural networks for anomaly detection and localization in crowded scenes. *Signal Processing Image Communication*, *47*, 358–368. DOI: 10.1016/j.image.2016.06.007

Zhu, Y., Nayak, N. M., & Roy-Chowdhury, A. K. (2013). Context-aware activity recognition and anomaly detection in video. *IEEE Journal of Selected Topics in Signal Processing*, *7*(1), 91–101. DOI: 10.1109/JSTSP.2012.2234722

Chapter 6
Artificial Intelligence-Based Automated System for Building Information Modelling Data Extraction and Analysis

Oluseye Olugboyega
https://orcid.org/0000-0003-4556-6211
Obafemi Awolowo University, Nigeria

ABSTRACT

The emergence of AI enables the use of AI-based automated systems for information extraction from building information models using machine learning methods, natural language processing, computer vision, and neural networks. This chapter analyzes the utilization of AI for the extraction and analysis of BIM data. The chapter examines the difficulties associated with manual data extraction from BIM and the advantages of employing AI for BIM data analysis and extraction. This chapter addressed the several strategies and algorithms employed in AI-based automated systems for extracting information from building information models. The chapter presented a case study on the effective application of AI in a BIM-based project. The chapter concludes with observations on upcoming trends and advancements in the use of artificial intelligence to extract meaningful insights from the building information models.

DOI: 10.4018/979-8-3373-0746-6.ch006

INTRODUCTION

Building Information Modeling (BIM) is a digital representation of the physical and functional characteristics of a built structure. It encompasses a wide array of data, including the building's geometry, spatial relationships, geographical location, and the quantities and specifications of architectural, structural, and service components (Teng et al., 2022; Olugboyega et al., 2023). This comprehensive digital model enables a holistic understanding of a construction project and facilitates enhanced collaboration among architects, engineers, builders, and other stakeholders. Through the integration of 3D modeling, data management, and simulation tools, BIM supports the entire lifecycle of a construction project—from conceptual design and construction to operation and maintenance. One of BIM's most valuable features lies in its ability to capture and communicate critical physical and functional data, which is essential for all phases of a building's lifecycle. The ability to extract this information efficiently and accurately is vital for optimizing design, planning, construction, and facility management activities. Effective information extraction from BIM models allows project teams to perform tasks such as quantity take-offs, clash detection, energy performance analysis, scheduling, budgeting, construction coordination, and building maintenance. Importantly, the extracted data can be tailored to suit the unique needs of different users, making BIM a versatile tool in the construction industry.

Traditionally, construction professionals have extracted information from BIM models through manual methods. These approaches, while sometimes effective, are prone to human error, time-consuming, and inefficient—ultimately impacting the quality, speed, and cost-efficiency of construction projects (Oskar, 2022). As the complexity of building projects increases, and as the volume of data within BIM models grows, these traditional methods become increasingly inadequate. Artificial Intelligence (AI) presents a promising solution to these challenges. By automating the information extraction process from BIM, AI can significantly reduce the reliance on manual tasks, minimize human error, and enhance overall productivity and quality. AI technologies can be employed to streamline data collection, processing, interpretation, and reporting. Moreover, AI tools can detect patterns, identify anomalies, and uncover trends in data that might be overlooked by human users. This capability opens up new avenues for improving decision-making, resource allocation, and project outcomes in the construction sector.

The integration of AI into BIM workflows can lead to transformative improvements across various dimensions of project execution. AI enables faster and more accurate analysis of BIM data, empowering stakeholders with real-time insights and facilitating data-driven decision-making. Key AI technologies applicable to BIM include Expert Systems (which simulate human expertise to provide recommendations and decisions), Machine Learning Algorithms (which improve over time by learning from data, enabling them to extract and interpret relevant information from complex BIM models), and Natural Language Processing (which allows computers to understand and interpret human language, thereby supporting automatic interpretation of BIM-related documentation and textual annotations). These technologies collectively help reduce errors, save time, and ensure that extracted information is accurate, timely, and relevant. By supporting informed decision-making based on live data, AI technologies can enhance project management practices, reduce risks, and improve overall construction outcomes.

A growing body of research supports the integration of AI into BIM workflows for information extraction. For instance, Pedral Sampaio et al. (2022) explored the potential of AI integration with BIM for a range of applications including data integration, predictive analytics, automated as-built classification, Internet of Things (IoT) integration, energy management, and augmented reality. Their work highlights the multifaceted benefits of AI in streamlining BIM processes and expanding its application scope. Similarly, Adewale et al. (2024) assessed AI's role in promoting sustainability throughout a building's lifecycle. Their evaluation identified key AI technologies that contribute to sustainable practices, such as advanced machine learning models for energy optimization, predictive maintenance systems, and simulation tools for evaluating design performance. They also noted the effectiveness of digital twins—virtual replicas of physical structures—in offering real-time feedback and enabling proactive decision-making. In another significant study, Locatelli et al. (2021) examined the role of Natural Language Processing (NLP) in construction, especially in connection with BIM. Their findings showed that NLP can support automation in areas such as project documentation, safety compliance, and risk assessment by extracting structured information from unstructured text. Such automation can drastically reduce the time required to process project documentation and enhance the precision of extracted insights.

These findings collectively emphasize the transformative role of AI in reshaping construction practices and enhancing sustainability. AI's ability to automate the extraction of meaningful information from BIM models can lead to greater efficiency, cost savings, and more informed decision-making. The integration of AI and BIM also enables proactive risk management, supports sustainable design and operation, and promotes better alignment with long-term project goals. Despite the demonstrated benefits and growing interest, the application of AI in BIM-based information extraction remains in its early stages. As noted by Zabin et al. (2022), the research landscape is still evolving, and many challenges remain. There is currently limited understanding of the specific AI algorithms and strategies that are most effective for different BIM use cases. Furthermore, there is a lack of comprehensive frameworks that guide the implementation of AI in real-world construction scenarios. To fully realize the potential of AI in BIM workflows, more empirical research is needed. Such research should investigate not only the technical aspects of AI algorithms but also their practical applicability, scalability, integration challenges, and organizational impacts. Addressing these gaps will be essential for fostering industry-wide adoption of AI-enhanced BIM systems.

This chapter aims to contribute to this growing area of inquiry by exploring and elucidating the various strategies and algorithms employed in AI-driven systems for extracting information from BIM. Through the use of empirical data from real-world case studies, the chapter will examine how different AI approaches perform in practical contexts, identify best practices, and highlight implementation challenges. The overarching goal is to bridge the gap between theory and practice and support the wider adoption of AI-enhanced BIM solutions across the construction industry. In doing so, this research seeks to provide valuable insights into how AI can optimize BIM workflows, improve construction outcomes, and advance the broader goals of digital transformation within the built environment. Ultimately, the integration of AI and BIM holds significant promise for driving innovation, efficiency, and sustainability in construction projects worldwide.

CHALLENGES IN MANUAL DATA EXTRACTION FROM BIM

Manual extraction of information from Building Information Modeling (BIM) presents several significant challenges that hinder efficiency, accuracy, and productivity in construction project workflows. One of the primary issues is the labor-intensive and time-consuming nature of the process. This challenge becomes particularly burdensome in large-scale and complex projects, where substantial amounts of data must be extracted, organized, and interpreted manually. As the volume and complexity of data increase, so does the risk of human error, which can lead to inconsistencies, omissions, or inaccuracies in the extracted information. Another major challenge lies in interoperability. The lack of standardized file formats and limited compatibility among various BIM authoring tools create obstacles when retrieving and integrating data from different sources. In many cases, project teams employ multiple BIM platforms across disciplines—such as architectural, structural, mechanical, electrical, and fire protection—each using unique data structures, schemas, and naming conventions. When these models are combined into a federated model, disparities in data format and terminology pose significant difficulties for manual extraction. This can result in data misalignment, duplication, or loss of critical information, ultimately compromising the integrity of the integrated BIM model (Oskar, 2022).

Furthermore, without automated support tools, manual extraction demands considerable human effort for identifying, collecting, and inputting data into analysis or reporting systems. This not only increases operational costs but also diverts expert resources from more strategic tasks (Al Sehrawy et al., 2021; Akanbi & Zhang, 2021). In such contexts, the manual approach becomes inefficient and unsustainable, particularly in projects requiring real-time updates, integrated collaboration, and data-driven decision-making. These challenges underscore the urgent need for intelligent automation solutions to improve the accuracy, efficiency, and scalability of BIM data extraction processes.

BENEFITS OF USING AI FOR BIM DATA ANALYSIS AND EXTRACTION

As Building Information Modeling (BIM) processes continue to gain prominence across all phases of the project lifecycle—from planning and design to construction, operation, and maintenance—the volume and complexity of data generated have grown exponentially. This surge in BIM-related data offers unprecedented opportunities for extracting valuable project knowledge and insights that support informed decision-making throughout the entire building lifecycle (Zabin et al., 2022). In this context, the integration of Artificial Intelligence (AI) into BIM processes presents transformative benefits for the analysis and extraction of information, especially in complex, data-intensive construction projects. AI-powered techniques offer the ability to efficiently analyze massive volumes of BIM data with high accuracy and consistency, significantly reducing the time and effort required for manual processing. By minimizing human involvement in repetitive and error-prone tasks, AI reduces the risk of mistakes, improves data reliability, and enhances overall project productivity. Through advanced machine learning algorithms and pattern recognition capabilities, AI can extract structured information from unstructured sources, identify patterns, detect anomalies, and provide deep insights that would be difficult for human analysts to recognize (Elghaish et al., 2022).

A critical application of AI in BIM is in the automation of the Scan-to-BIM process. Traditionally, transforming 3D laser scan data into accurate BIM models is time-consuming and labor-intensive. AI techniques now offer automated object recognition and classification, enabling the system to detect and categorize architectural components—such as doors, windows, columns, and walls—from point cloud data with minimal human input. This significantly speeds up the model creation process while maintaining a high level of accuracy and reducing manual error (Hu et al., 2022). Consequently, teams benefit from up-to-date and precise models, which are essential for effective design coordination and construction execution. Moreover, AI-based BIM data analytics enhances clash detection and conflict resolution. By continuously monitoring BIM models for inconsistencies or design clashes, AI enables early identification of potential issues, allowing teams to resolve conflicts proactively before they escalate into costly construction problems. This early intervention reduces rework, mitigates delays, and conserves project resources.

AI also supports real-time data analysis and decision-making. By integrating data from multiple sources—such as IoT devices, sensor networks, and construction management platforms—AI systems can generate live insights into project performance, safety, energy consumption, and material usage. These real-time analytics empower stakeholders to make swift, evidence-based decisions that improve outcomes and align with project goals. In addition, AI facilitates standardization and scalability in BIM data extraction. It ensures uniform procedures across projects by applying consistent modeling and recognition protocols. This improves quality control, simplifies project comparisons, and supports enterprise-level data management strategies. Furthermore, by automating complex data processing workflows, AI contributes to substantial cost savings, particularly in large-scale projects where manual oversight is impractical and costly (Bassir et al., 2023). The integration of AI into BIM data analysis and extraction workflows enhances efficiency, accuracy, and responsiveness. It enables construction professionals to harness the full potential of BIM data, make better-informed decisions, and ultimately deliver higher-quality, cost-effective, and sustainable built environments.

VARIOUS TECHNIQUES AND ALGORITHMS USED IN AI-BASED SYSTEMS FOR EXTRACTING INFORMATION FROM BUILDING INFORMATION MODELS.

Common approaches and algorithms employed in AI-based systems for extracting information from building information models include machine learning, natural language processing, and computer vision. Machine learning algorithms may be trained to identify patterns in BIM data and autonomously extract pertinent information. Natural language processing methods can facilitate the interpretation and extraction of text-based information from BIM files. Computer vision algorithms can assess visual information in BIM models, including the identification of objects and their attributes. The subsequent subsections examine the utilization of AI-based systems for the effective extraction of important information from building information models to facilitate diverse applications within the construction sector.

BIM Data Analysis Using Machine Learning Algorithms

As Building Information Modeling (BIM) becomes a central digital platform for managing construction data across the project lifecycle, the volume, complexity, and richness of BIM-generated data continue to increase. BIM provides a centralized, structured repository of detailed information related to geometry, material properties, costs, scheduling, environmental performance, and more. Analyzing this vast and multifaceted data manually is inefficient and error-prone. Machine learning (ML), a branch of artificial intelligence (AI), offers powerful methods for automating and optimizing BIM data analysis. ML algorithms can be trained to recognize patterns, detect anomalies, classify objects, and generate predictions, thereby significantly improving the decision-making process across all stages of a construction project. The application of machine learning in BIM data analysis generally follows a structured workflow:

- Data Collection and Preprocessing: The initial step involves collecting raw data from BIM models. This includes geometric, semantic, spatial, and temporal data embedded within the BIM environment. Since BIM data often come in various formats (e.g., IFC, Revit files), data sanitization is critical. This involves cleaning, filtering, and structuring the data to remove redundancies, inconsistencies, and irrelevant attributes.
- Algorithm Selection: The next step is selecting a suitable machine learning algorithm based on the type of analysis or prediction required. Supervised learning (e.g., classification, regression), unsupervised learning (e.g., clustering, anomaly detection), and reinforcement learning may each serve distinct roles in BIM applications.
- Model Training and Evaluation: Once the data is cleaned, the chosen ML model is trained using labeled or unlabeled datasets. The model learns from the data to perform tasks such as identifying components, classifying conditions, or predicting energy consumption. Model performance is then assessed using metrics like accuracy, precision, recall, F1-score, and root mean square error (RMSE), depending on the use case.
- Application and Decision Support: Once validated, the trained model is deployed to analyze real-time or stored BIM data. The output supports decision-making by providing actionable insights for design, construction planning, facility management, or energy optimization.

Several studies have demonstrated the successful application of machine learning algorithms in BIM-related workflows, highlighting the growing integration of AI in the construction and built environment sectors.

Facilities Management: Automated Classification of Work Orders

McArthur et al. (2018) developed and compared two sets of supervised machine learning models to classify work orders originating from tenant complaints. Using a dataset of approximately 150,000 historical work orders, the researchers applied various word frequency and itemset frequency techniques to perform textual classification. The resulting models could accurately identify work order categories and generate intelligent prompts for building occupants to submit more structured and relevant complaint data. This ML-enabled automation enhances the quality and efficiency of data collected via online reporting systems and demonstrates machine learning's potential to streamline operations in facility management.

Energy Optimization Using ML and BIM

Hosamo et al. (2022) proposed a hybrid system combining BIM, machine learning, and the Non-dominated Sorting Genetic Algorithm II (NSGA-II) to assess the impact of building variables on energy consumption and thermal comfort. Sensor data were extracted from BIM models using a custom plugin and analyzed in IDA Indoor Climate and Energy (IDA ICE). A generalized least squares support vector machine (GLSSVM) was used to predict building energy performance with exceptional accuracy ($R^2 = 0.99$). The GLSSVM-NSGA II model enabled a 37.5% reduction in energy consumption and a 33.5% increase in thermal comfort in a case study involving a secondary school in Norway. This demonstrates the powerful synergy between BIM data and ML algorithms for sustainability-driven decision-making.

Heritage BIM and As-Built Classification

Bassier et al. (2017) applied machine learning techniques to automate the categorization of structural components in heritage buildings. The ML system successfully identified elements such as floors, ceilings, walls, roofs, and beams, even in environments characterized by clutter, occlusions, and irregular geometries. By using actual site data and training classifiers on point clouds and BIM inputs, the study demonstrated the effectiveness of ML for automating the as-built modeling process. This approach reduces the manual workload, speeds up modeling timelines, and supports cultural heritage preservation by making digital documentation more accessible and systematic.

Deep Renovation Decision Support Using ML and As-Built BIM

Mulero-Palencia et al. (2021) explored how machine learning could assist in deep renovation projects by analyzing As-Built BIM models to recommend energy efficiency measures. The team developed a prototype diagnostic tool that analyzed building data formatted under the IFC4 standard. The model identified areas needing intervention and provided optimal retrofit strategies. While acknowledging certain limitations—such as data heterogeneity and integration challenges—the study showcased the feasibility of using ML algorithms as early-phase design advisors to enhance the performance of renovation projects.

The application of machine learning algorithms in BIM data analysis offers several key advantages:

- Automation of Complex Processes: ML algorithms can handle vast, multi-dimensional datasets and automate time-consuming processes such as object detection, classification, and performance prediction.
- Enhanced Accuracy: Trained models reduce human errors, detect subtle anomalies, and produce more accurate outcomes, especially in repetitive or high-volume tasks.
- Real-Time Decision Support: ML models can be deployed in live BIM environments to support dynamic decision-making using up-to-date project data.
- Predictive Capabilities: ML models can forecast performance outcomes (e.g., energy use, maintenance needs), enabling proactive interventions and risk mitigation.
- Interdisciplinary Integration: Machine learning bridges architectural, structural, mechanical, and operational domains by enabling data-driven insights across all disciplines integrated within a BIM model.

Despite these promising benefits, integrating ML into BIM workflows also presents challenges. These include:

- Data Quality and Consistency: BIM data is often incomplete, inconsistent, or poorly structured, making preprocessing and feature engineering critical.
- Model Interpretability: Complex ML models such as deep learning networks may lack transparency, which can hinder trust and adoption among practitioners.
- Integration Complexity: Embedding ML tools into existing BIM authoring platforms (e.g., Revit, ArchiCAD) or standards (e.g., IFC) requires specialized APIs and interoperability solutions.

- Skill Gaps: Construction professionals may lack the technical knowledge to develop, train, or interpret ML models, necessitating interdisciplinary collaboration with data scientists and AI specialists.

Future research should focus on developing standardized BIM-ML integration frameworks, improving data interoperability, and expanding domain-specific datasets to train more generalizable models. Additionally, incorporating explainable AI (XAI) methods may help demystify ML outcomes and promote wider adoption within the AEC industry.

BIM Text Extraction Using Natural Language Processing

As Building Information Modeling (BIM) continues to evolve into a central hub for digital construction and lifecycle management, the volume and variety of embedded information within BIM models have grown exponentially. While a significant portion of BIM data is structured (e.g., geometric or parametric), a large amount exists in unstructured or semi-structured textual formats. These include notes, labels, property definitions, project documentation, and user-generated content embedded in BIM files. Extracting actionable insights from this textual data presents a challenge that traditional rule-based methods struggle to address. Natural Language Processing (NLP), a subset of artificial intelligence, offers a robust solution for understanding, analyzing, and extracting meaningful information from text-based data in BIM environments. NLP enables machines to comprehend human language and interpret free-form text by applying linguistic, statistical, and machine learning techniques. In the context of BIM, NLP can be used to process textual descriptions, classify documents, extract project-related facts, and support intelligent search functions. As Shamshiri et al. (2024) note, NLP algorithms can interpret the textual components of BIM models and retrieve essential information such as project scope, material specifications, construction methods, and dimensional properties—information crucial for decision-making in construction, facility management, and sustainability analysis.

A typical NLP-based approach for BIM data extraction begins with the collection and consolidation of relevant BIM files into a centralized database. The NLP system then processes this database to analyze and extract information. This involves text parsing, keyword extraction, context identification, and semantic mapping. The extracted data can then be organized into structured formats, making it easier to query, analyze, and use across various BIM workflows. Several studies have demonstrated the potential of NLP in automating BIM data analysis and improving knowledge retrieval. For instance, Jung and Lee (2019) developed a classification system for case studies of BIM implementation in building projects using unsupervised learning techniques such as Latent Semantic Analysis (LSA) and Latent Dirichlet Allocation (LDA). These models were used to categorize text data according to predefined BIM uses. When LSA and LDA identified key terms exceeding similarity thresholds for specific BIM uses, the system inferred that the project aligned with those applications. The outcomes were validated using a Support Vector Machine (SVM) classifier, confirming the accuracy of this NLP-driven classification approach. This method demonstrates how NLP can support the identification of BIM use patterns and automate the classification of project documentation, contributing to better knowledge structuring and retrieval.

In another study, Wang et al. (2022) proposed a query-answering (QA) system for BIM information extraction that mimics a virtual assistant. This system consists of three main components: natural language understanding, information extraction, and natural language generation. A prototype application was developed using Python and evaluated with seven BIM models and 127 user queries. The results showed that the QA system achieved an impressive accuracy of 81.9%, indicating its effectiveness in interpreting natural language questions and extracting correct information from BIM/IFC models. Furthermore, Wang et al. (2022) integrated case-based reasoning (CBR) with NLP to improve contextual knowledge retrieval. In this hybrid approach, NLP was used to parse and extract textual content from BIM case descriptions, while CBR facilitated the identification of relevant cases based on contextual attributes. This dual-system strategy aligned the structured properties of BIM with unstructured textual knowledge, offering a more holistic and intelligent method for retrieving past project experiences. The authors highlighted this as a step toward evolving BIM into Building Knowledge Modeling (BKM)—a more advanced framework that integrates data, knowledge, and reasoning in one ecosystem.

Expanding on these approaches, Nabavi et al. (2023) applied NLP for syntactic and semantic analysis of user queries within a BIM environment. Their platform used ontologies, such as IfcOWL, and Latent Semantic Analysis to understand and expand the keywords derived from a user's inquiry. Semantic relationships within the ontology helped broaden the scope of the query, allowing the system to understand user intent more deeply. Based on this enriched query, a Navisworks API was used to search and retrieve relevant information from BIM models. The system also included a voice recognition module, making the platform more accessible and user-friendly. According to their evaluation, the NLP-powered platform provided responses up to five times faster than manual searches conducted by domain experts, while maintaining high levels of accuracy. The use of NLP in BIM data extraction brings numerous benefits. First, it significantly reduces the time and labor involved in manually searching and interpreting textual data in BIM models. By automating these processes, NLP enhances productivity and allows professionals to focus on high-level decision-making rather than tedious information retrieval. Second, NLP increases the accuracy and consistency of data extraction. Unlike manual interpretation, which is prone to oversight and inconsistency, NLP ensures systematic parsing and interpretation based on linguistic and semantic rules.

Another major benefit is context-aware querying, which enables users to interact with BIM systems using natural language rather than complex technical commands. This democratizes access to BIM data, allowing non-technical stakeholders—such as facility managers, clients, and regulators—to obtain the information they need without relying on BIM specialists. Moreover, NLP can support intelligent search functions, semantic tagging, and document classification, further enhancing the usefulness of BIM as a knowledge repository. Lastly, NLP enables BIM systems to evolve toward knowledge-driven platforms by integrating textual content, contextual reasoning, and domain-specific ontologies. This shift from data modeling to knowledge modeling (BKM) represents a significant step forward in the digital transformation of the built environment.

BIM Visual Data Extraction Using Image Recognition and Computer Vision

The integration of image recognition and computer vision techniques into Building Information Modeling (BIM) has opened up new dimensions for automating the extraction and interpretation of visual data. These technologies enable the identification, classification, and mapping of construction elements directly from images, photographs, and drawings, which can then be integrated into BIM models for more intelligent planning, tracking, and decision-making. By leveraging image-based input and translating visual data into meaningful geometric or semantic BIM information, construction professionals can enhance accuracy, streamline progress monitoring, and improve project oversight. The process typically begins with the selection of relevant visual content—images or drawings that contain critical details about architectural or construction elements. Image recognition algorithms are then applied to these visuals to detect and classify key features. Computer vision techniques further process the extracted features, enabling advanced functions such as object localization, pose estimation, semantic segmentation, and spatial mapping. The extracted information can then be mapped onto corresponding BIM elements, enabling real-time updates, enriched model intelligence, and automated verification workflows.

A compelling example of this approach is presented by Deng et al. (2020), who integrated computer vision with BIM for the automated monitoring of tile installation in interior construction. They developed a comprehensive image database containing thousands of tile photos and employed Local Binary Patterns (LBPs) for feature extraction. Using Support Vector Machine (SVM) classifiers, the system achieved a high classification accuracy of 91.17%, enabling reliable identification of tile types in captured images. An enhanced edge-detection algorithm was used to determine the boundaries of finished tiles. These boundaries, initially represented in pixel coordinates, were then converted into real-world dimensions using camera calibration techniques. By integrating this data with spatial profiles from the BIM model and camera position metadata, the actual tiled area was automatically calculated. The results were highlighted on the room's BIM floor plan and uploaded in real-time to a BIM cloud platform, providing dynamic progress monitoring and eliminating the need for manual inspections.

Another milestone in the field was achieved by Dimitrov and Golparvar-Fard (2014), who developed a vision-based material classification method using photographs taken under varied lighting and viewpoint conditions. Their method combined a probability distribution of filter bank responses with hue-saturation-value (HSV) color features and used a multi-kernel SVM classifier for material recognition. Their results demonstrated superior classification performance when benchmarked against leading algorithms in both the computer vision and AEC domains. This work highlighted the robustness of vision-based techniques in real-world construction environments where visual conditions are often uncontrolled. Ying et al. (2023) introduced an innovative approach to generate synthetic BIM-based images for use in a wide range of Building Scene Understanding (BSU) tasks. Their method captures perspective views of interior BIM environments and renders highly photorealistic images with detailed annotations. These synthetic images are utilized for a variety of computer vision tasks, including camera pose estimation, scene classification, depth prediction, 2D/3D object detection, semantic segmentation, instance segmentation, and panoptic segmentation. This automated pipeline not only supports large-scale training of vision-based models but also demonstrates the feasibility of using BIM as a virtual training ground for developing AI tools in construction.

Nunez-Morales et al. (2023) further explored visual data extraction by introducing per-element visibility metrics. Their study assessed how various camera positions impact the visualization of BIM elements in rendered scenes. By analyzing the visibility of elements from different camera views, the researchers provided insight into how optimal viewpoints can enhance the clarity and completeness of visual inspections and data extraction routines. A significant contribution was also made by Kropp et al. (2018), who performed interior construction state recognition using image sequences aligned with 4D BIM models. By registering captured image sequences with 4D BIM timelines, the team projected anticipated construction tasks onto image frames. These projections highlighted specific regions of interest, which were then analyzed to determine whether corresponding activities had been completed, were in progress, or were delayed. This fusion of time-based BIM data and computer vision facilitated precise construction progress tracking without manual annotation.

BIM Data Points Extraction Using Deep Learning Techniques

Building Information Modeling (BIM) has become a central tool in modern construction and facility management, offering digital representations of the physical and functional characteristics of buildings. However, extracting meaningful data points from BIM models—especially those generated from laser scans or photogrammetric sources—remains a labor-intensive and error-prone task when performed manually. The increasing complexity and size of BIM datasets demand more intelligent, automated solutions. In this context, deep learning has emerged as a powerful approach for automating the extraction of BIM data points, particularly within scan-to-BIM applications, where point cloud data must be accurately converted into semantically rich BIM models. The core idea behind deep learning-based BIM data extraction lies in training neural networks on large datasets of BIM models and corresponding point clouds. These models learn to recognize, classify, and segment architectural, structural, and mechanical elements based on their geometric and contextual features. Once trained, these deep learning algorithms can autonomously extract data points from new BIM datasets with high precision, saving time, improving consistency, and reducing human error.

A notable contribution in this field is the work by Perez-Perez et al. (2021), who developed a deep learning architecture called Scan2BIM-NET. This end-to-end model is specifically designed for semantic segmentation of point cloud data captured from real-world environments. The architecture integrates two Convolutional Neural Networks (CNNs) and one Recurrent Neural Network (RNN) to classify structural and MEP (Mechanical, Electrical, and Plumbing) elements such as beams, ceilings, columns, floors, pipes, and walls. The model was trained and validated on point cloud datasets from 83 actual rooms in industrial and commercial buildings. The reported average accuracy across all categories was 86.13%, with the highest accuracy achieved for floor elements (98.71%) and the lowest for columns (59.31%). These results demonstrate that deep learning techniques significantly enhance the accuracy of semantic segmentation in BIM point cloud processing, especially for complex environments with overlapping and occluded elements.

Another critical development is the research by Zhai et al. (2022), who proposed a method to generate synthetic point cloud datasets enriched with both geometric and chromatic information, derived from Industry Foundation Classes (IFC) models. By synthetically generating annotated training data for deep learning models, they addressed one of the key challenges in BIM-related AI applications—limited labeled datasets. In their approach, IFC files or 3D representations were used to synthesize scenes and generate labeled point clouds, which were then integrated into a SuperPoint-based neural network. Their experimental trials showed that including an appropriate ratio of synthetic point clouds in the training dataset significantly improved the semantic segmentation performance. This approach illustrates the synergy between parametric BIM models and AI-driven point cloud analytics, enhancing generalization and robustness in real-world applications. Wang et al. (2022) introduced a novel method for object verification in automated BIM reconstruction by comparing deep point features. They employed a Kernel Point Convolution (KPConv)-based deep learning model to extract and compare 3D features from clustered point cloud segments and as-designed BIM point clouds. This process generated point feature distance maps, which were then analyzed using a small discriminative neural network trained on a dataset of positive and negative sample pairs derived from ModelNet40. This method enabled high-precision matching between scanned data and BIM model elements, which is critical in validating construction accuracy and supporting quality control in automated reconstruction workflows.

Park et al. (2022) tackled another practical challenge: extracting indoor objects from occluded point clouds and resolving missing data issues. Their deep learning framework was designed to recognize a wide range of indoor objects and then align them with parametric object definitions stored in a BIM library. By combining object detection with bounding box alignment, their system was able to automate the placement and definition of building elements, such as furniture and equipment, even when partial or noisy scan data was available. This solution contributed significantly to automating the scan-to-BIM process by ensuring completeness and fidelity in the generated models. Taking it a step further, Kim et al. (2023) proposed an advanced scan-to-BIM methodology that incorporates both geometry and material recognition. Their system uses deep learning-based semantic segmentation to detect geometric features from point clouds and material properties from panoramic images. By integrating visual texture recognition into the model, they enabled a richer BIM representation that goes beyond shape, supporting enhanced applications in interior design, facility management, and refurbishment planning.

The use of deep learning for BIM data point extraction offers numerous advantages over traditional methods. First and foremost, it enables automation of semantic segmentation, drastically reducing the manual effort required for point cloud interpretation and BIM model generation. This leads to substantial time savings and increased efficiency in construction workflows. Moreover, deep learning models can be retrained and fine-tuned for specific building typologies or use cases—such as HVAC systems, historical building preservation, or infrastructure—enhancing the precision and adaptability of the technique. Deep learning also facilitates greater scalability, allowing for the processing of large-scale projects with extensive datasets. Its ability to learn complex spatial relationships enables it to handle cluttered, incomplete, or noisy data far more effectively than rule-based systems. Additionally, the fusion of geometric and visual features—such as those in Kim et al.'s work—expands the scope of BIM modeling by including material semantics and visual characteristics.

BIM Data Extraction Using Convolutional Neural Networks

Building Information Modeling (BIM) has become the foundation of digital transformation in the architecture, engineering, and construction (AEC) industry. As BIM models grow in complexity, the demand for intelligent methods of extracting valuable data from them—such as dimensions, quantities, spatial relationships, and material specifications—continues to increase. Traditional manual extraction or rule-based systems are time-consuming and often prone to error. Neural networks, especially convolutional neural networks (CNNs) and other deep learning architectures, present a powerful alternative by automating and enhancing BIM data extraction with greater speed, accuracy, and adaptability. The extraction of BIM data using neural networks begins with transforming BIM models into data formats suitable for machine learning. This process often includes converting geometric and semantic components into structured representations such as voxels, point clouds, multi-view images, or graph structures. These structured representations serve as input for training neural networks to identify and extract specific features from the models, such as architectural elements, material properties, energy performance data, and object classifications.

A prominent example is Emunds et al. (2022), who introduced SpaRSE-BIM, a neural network model that employs sparse convolutions to classify Industry Foundation Classes (IFC)-based geometry. Their goal was to semantically enhance BIM models by accurately categorizing structural and architectural elements. SpaRSE-BIM leverages sparse data structures and 3D CNNs to process large BIM datasets efficiently while preserving geometric detail. This enables it to differentiate between similar objects with subtle variances, such as different types of walls, columns, or beams—ultimately improving semantic enrichment and data retrieval from complex BIM models. Another application of neural networks is seen in Xu and Liu (2023), who developed an energy consumption optimization model by integrating BIM with convolutional neural networks. Their approach involved simulating building energy performance and automating the model creation using enhanced DOE-2 files through the eQUEST software. The CNN was trained to predict and optimize energy usage based on design parameters extracted from the BIM model. This integration allowed for rapid simulation and feedback during the design stage, helping architects and engineers make data-driven decisions that enhance sustainability and operational efficiency.

In a related study, Koo et al. (2021) explored the application of 3D geometric deep neural networks, originally developed in computer vision, to BIM model classification. Specifically, they tested Multi-view Convolutional Neural Networks (MVCNN) and PointNet for extracting distinguishing features from door (IfcDoor) and wall (IfcWall) subtypes. Their findings revealed that MVCNN outperformed PointNet in classification accuracy due to its ability to analyze images from multiple viewpoints. In contrast, PointNet's reliance on point cloud data led to diminished accuracy due to data sparsity and resolution loss. This comparison underscores the importance of input data representation in the successful application of neural networks to BIM. Seydgar et al. (2022) further extended the capabilities of deep neural networks (DNNs) to three core elements of Scan-to-BIM workflows: 3D reconstruction, object identification, and object parameterization. Their comprehensive review highlighted how neural networks can automatically detect and reconstruct 3D components from scanned environments, reducing the time required for manual modeling and increasing the precision of the resulting BIM representations. Object parameterization—assigning specific dimensions, labels, and attributes to reconstructed elements—was also improved through neural network inference, enabling more intelligent BIM population.

Amrouni Hosseini et al. (2024) explored the integration of neural networks with BIM to streamline building inspection and condition monitoring. In their framework, a neural network was used to assess the extent of deterioration of building components in real-time. The model generated predictive insights about structural health, which were visualized within the BIM environment using color-coded feedback. This allowed facility managers and owners to interpret the condition of assets and prioritize maintenance activities based on data-driven causality analysis. Moreover, the system facilitated automated documentation and enhanced data exchange during inspections, highlighting the value of combining BIM visualization with AI analytics. Wei et al. (2022) demonstrated another use case by applying deep learning segmentation techniques to monitor construction progress. They developed a model capable of autonomously analyzing photographic input and assessing wall construction status for an entire floor. The results were integrated into the BIM environment to visually represent real-time progress, allowing project teams to identify deviations from the schedule or discrepancies in implementation. This solution not only increased accuracy in tracking but also reduced the need for repetitive manual inspections and reporting.

An innovative integration of neural networks and generative modeling is presented by Li et al. (2024), who designed a BIM-based automation framework for precast design and prefabrication. Their methodology began by generating a graph-based representation of the building model to capture spatial and geometric features. A graph convolutional neural network (GCNN) was employed to extract pairwise spatial relationships, which guided the generation of optimal precast layouts. Subsequently, a semantic enhancement of the BIM data was achieved using a custom precast data schema, allowing smooth transformation of digital models into production-ready formats. This closed-loop system—from design to factory-level automation—demonstrates the transformative potential of neural networks in modern construction workflows.

REAL-WORLD EXAMPLES OF SUCCESSFUL IMPLEMENTATION OF AI SYSTEMS FOR ANALYSING AND EXTRACTING DATA FROM BUILDING INFORMATION MODELS.

Automated Property Valuation Using BIM And Machine Learning

Su et al. (2021) gave a real-world example of automated property appraisal utilizing BIM and machine learning. The example demonstrated a system that utilizes comprehensive data analysis and automates particular operations for property appraisal. The example demonstrated that partial information needs may be derived from BIM models, enabling more precise and efficient property assessment. In the example, value-related design information for property valuation was found, encompassing building object entities and attributes. IFC extensions were built based on value-related design information for property assessment. Subsequently, the necessary value-specific design information was automatically retrieved from an IFC-based BIM instance model via an information extraction method. The necessary information was categorized into six distinct groups pertaining to property assessment. This encompasses data pertaining to environmental quality, social and economic quality, functional quality, process quality, technical quality, and site quality. The data from traditional building surveys is juxtaposed with the information necessary for sustainability assessments and the data obtainable through BIM-related activities.

IFC extensions were created in the example by identifying and analyzing the building object entities and property sets that facilitate property valuation within the existing IFC schema (IFC4-Addendum 2), as well as by locating and incorporating the absent, yet essential, value-relevant entities and property sets into IFC4. The IFC extensions were created to enhance the current IFC schema. The current IFC schema includes architectural elements such as IfcWall, IfcWindow, IfcStair, IfcSlab, IfcDoor, and IfcCurtainWall. It also includes the property sets for each building object, such as Pset_WallCommon, Pset_ConcreteElementGeneral,

Pset_WindowCommon, and Pset_EnvironmentalImpactValues. The current entity, IfcGeographicElement, was designed to encompass subtypes such as IfcSoilElement, IfcLandscapeElement, IfcWaterBodyElement, and IfcHabitatElement to evaluate and enhance the environmental quality surrounding the building.

A partial object hierarchy inside the Building Infrastructure Axiology (Build-Infra-Axio) was established based on Hartman's formal axiology, which provides a method for quantifying the value of an item according to its qualities. An IFC-based information extraction technique facilitated the sharing of essential data regarding architectural objects and their properties for property appraisal. IfcRelDefinesByType was utilized to establish the connection between an object type and its occurrences, facilitating one-to-many relationships. An information extraction strategy was built in Python 3.7 utilizing the IfcOpenshell-python module within the PyCharm program. The information extraction procedure can be executed through the below steps:

- All necessary data may be obtained by systematically traversing the IfcRelDefinesByProperties and IfcRelDefinesByType instances.
- All ID numbers of IFCObject, IFCPropertySet, and IFCTypeObject must be retrieved from the instances obtained in step 1.
- Locate occurrences of IfcObject, IfcPropertySet, and IfcTypeObject using the ID numbers of instances obtained in step 2.
- Obtain the identification numbers of IfcPropertySet and IfcProperty instances from step 3.
- Locate occurrences of IfcProperty and IfcPropertySet using the ID numbers obtained in step 4.
- Obtain ID numbers of IfcProperty instances from the extracted IfcPropertySet instances and locate instances of IfcProperty using the ID numbers of the IfcProperty instances.
- Extract object names, object types, property names, and property nominal values from steps 3, 5, and 6.
- Replicates the extracted object names, object types, property names, and property nominal values.

An example of property valuation data derived from BIM utilizing machine learning, as conducted by Su et al. (2021), is illustrated in Table 1.

Table 1. Property valuation information extracted from BIM using machine learning

Object name	Object type	Property name	Property nominal value
Basic wall	Single value	FireRating	ifcLabel('60')

Building Materials Costs Data Classification Using BIM And Machine Learning

Banihashemi et al. (2022) offered a practical example of machine learning-integrated BIM-based categorization for construction materials cost data. A cost estimating standard was included with BIM to accelerate the 5D BIM process and improve digital transformation processes in building projects. A detailed library was established to generate a BIM model with extensive information and components. The project components were supplied in .rvt and .rfa formats for download from official websites like SmartBIM or ArCat libraries, allowing for their importation into Revit project files. A three-dimensional model was developed to quantify all items. A database of cost items was established within the Excel framework. The database has all details of cost components, including name, ID, cost description, and unit costs. The cost database has 909 items mostly utilized for the cost calculation of structural and architectural project parts.

A plugin was developed to import all generated cost items within the BIM environment. The plugin allows Revit to access the Excel file. The cost information was immediately loaded into Revit. All material amounts were extracted from Revit files and transferred to Excel.

Additionally, all necessary data for cost calculation, together with geometric information such as area, volume, and dimensions of building parts, were exported and classified according to BIM specifications. The aggregate expense of several categories of construction elements was displayed in the Revit environment utilizing the "Cost-Estimate" function. All element data were organized according to the Revit parametric components, and the cost values were aggregated to provide the total cost estimate. An approach including machine learning was created to streamline the cost prediction process for each project piece, utilizing neural networks for training data and simulation methods. The machine learning-based technique was implemented on the MATLAB engine to autonomously assess the cost of construction components. A methodology was established to incorporate the machine learning algorithm with BIM. The algorithms are subsequently executed using training datasets, resulting in the generation of the automated cost estimating process. Table 2 displays a segment of construction material costs that was automatically produced from BIM by Banihashemi et al. (2022) utilizing a machine learning-integrated method.

Table 2. Building material cost extracted from BIM using machine learning-integrated algorithm

Material name	Unit Cost	Cost	Parameters
Door Net Frame	Area	200000	2.1
Door Cover	Area	189000	4.2
Door Painting	Area	151500	4.51
Door frame Painting	Area	151500	0.74

POTENTIAL CHALLENGES AND LIMITATIONS OF USING AI FOR INFORMATION EXTRACTION IN BIM

As Artificial Intelligence (AI) becomes increasingly integrated into Building Information Modeling (BIM) processes, it offers transformative potential in terms of automation, efficiency, and data-driven decision-making. However, alongside these opportunities lie significant challenges and constraints that may hinder the effective implementation of AI for BIM data extraction. These constraints span technical, legal, ethical, human, and organizational domains.

Data Quality Issues

One of the most pressing concerns in applying AI to BIM is the quality of input data. BIM data can often be fragmented, inconsistent, outdated, or incomplete, particularly in complex construction environments involving multiple stakeholders, disciplines, and tools. Discrepancies in information input—such as conflicting object definitions, misaligned coordinate systems, or missing parameters—can compromise the accuracy of the data AI is trained on or processes during inference. AI models are highly data-dependent; they rely on large, accurate, and diverse datasets for effective training and operation. Poor-quality data may lead to biased predictions, inaccurate outputs, or faulty insights, thereby reducing the reliability and effectiveness of AI tools. In many cases, erroneous data in BIM models can propagate throughout the AI pipeline, leading to cascading errors in quantity takeoffs, cost estimations, scheduling, and clash detection. Moreover, the lack of standardized data formats and naming conventions within the construction industry exacerbates this issue. AI algorithms may struggle to interpret or map data from varied BIM authoring tools (e.g., Revit, ArchiCAD, Tekla) that use different structures, schemas, or naming protocols. Without standardized interoperability protocols, the integration and interpretation of information across platforms becomes error-prone and labor-intensive.

Privacy and Data Security Concerns

Privacy is another significant concern when employing AI in BIM, especially in projects that handle sensitive or proprietary information. BIM models often contain detailed data related to building occupants, infrastructure layouts, security systems, and personal usage patterns. If AI tools access, store, or process this data without adequate consent, anonymization, or encryption, they may pose risks of unauthorized access, misuse, or data leakage. As AI automates information extraction and analysis, data might be shared across cloud servers, third-party tools, or machine learning pipelines, increasing the attack surface for cyber threats. Additionally, compliance with data protection regulations, such as the General Data Protection Regulation (GDPR) in Europe or Nigeria's NDPR, becomes crucial. Failure to protect sensitive BIM data or to comply with legal obligations could result in legal liabilities, reputational damage, and project delays.

Ethical and Legal Implications

The deployment of AI in BIM also raises significant ethical and legal questions. AI models can generate or reinforce biases, particularly when trained on unrepresentative datasets. For instance, if AI tools are used to optimize designs or material selections without accounting for local regulations or safety standards, they may produce non-compliant or unsafe recommendations. Moreover, AI systems can obscure decision accountability. When AI suggests a design or flags an error, who is legally responsible—the developer of the AI, the BIM coordinator, or the AI model itself? This lack of clarity in liability and authorship creates a legal gray area that needs to be addressed through formal guidelines and governance mechanisms. Organizations must therefore establish clear protocols and ethical standards governing the use of AI in BIM workflows. These should cover issues such as data governance, model transparency, explainability, user consent, and oversight mechanisms. Stakeholder trust in AI systems is contingent upon their transparency and auditability, particularly when high-stakes decisions are involved.

Human Factors and Job Displacement

The integration of AI into BIM processes also brings about potential disruption to human labor and skills. As AI automates previously manual tasks—such as data extraction, quantity takeoffs, and performance simulations—some traditional roles may become redundant. Jobs that rely on manual drafting, data input, or inspection might diminish in demand. However, this transition also opens up new roles requiring digital fluency, data literacy, and AI-related competencies. For example, project managers, architects, and engineers may need to gain proficiency in interpreting AI outputs, validating models, and integrating AI tools into design workflows. Upskilling and reskilling efforts will be vital to ensure that the workforce remains adaptable and empowered in this evolving environment. The success of AI adoption in BIM will depend on how well organizations support change management, including training programs, role redefinition, and capacity-building strategies. Workers must be equipped not just to coexist with AI, but to collaborate with it meaningfully, optimizing human-machine synergy for better project outcomes.

Technical Limitations and Operational Risks

AI technologies are not infallible. They are subject to a variety of technical risks, such as software bugs, hardware malfunctions, or cyber intrusions. Malfunctions in AI-based systems can result in flawed outputs—such as incorrect material quantities or inaccurate energy analyses—that may compromise construction quality, safety, or efficiency. Integration challenges also abound. AI tools must be made compatible with existing BIM platforms, data standards (e.g., IFC, COBie), and file formats. In many cases, legacy BIM software may not support seamless API integration or data exchange with modern AI frameworks. This can lead to duplication of effort, data translation issues, and implementation delays. Moreover, some AI systems may struggle in dynamic environments, where construction conditions are rapidly changing or unstructured. AI may find it difficult to adapt to non-standard architectural elements, on-site improvisations, or unanticipated project constraints. These limitations make it essential to maintain a human-in-the-loop architecture, where human oversight is preserved even in automated systems. The cost of AI implementation is another factor. Deploying and maintaining AI systems requires significant financial investment, robust computing infrastructure, and ongoing support. Without clear return on investment (ROI) strategies, smaller firms may hesitate to adopt AI tools, potentially widening the digital divide within the industry.

Interpretation Complexity and Need for Human Judgment

Despite its capabilities, AI cannot replace human judgment—especially when interpreting complex or ambiguous BIM data. Many elements within a BIM model, such as design intent, contextual factors, or stakeholder preferences, cannot be fully captured by algorithms. In such cases, AI may fail to extract nuanced or project-specific information unless guided by human experts. Furthermore, AI algorithms may produce outputs that lack explainability—they can make accurate predictions without revealing how or why a conclusion was reached. This "black box" nature poses a problem in projects where traceability and rationale are critical, such as in safety compliance, stakeholder negotiations, or legal scrutiny. To mitigate this, organizations should invest in explainable AI (XAI) models, ensure rigorous validation processes, and maintain human interpretability and control over AI-driven insights.

CONCLUSION

Artificial Intelligence (AI) significantly advances Building Information Modeling (BIM) data analysis by leveraging sophisticated algorithms capable of rapidly processing vast datasets and uncovering meaningful patterns, trends, and anomalies. These capabilities enable more accurate forecasting and deeper insights into building performance, which in turn supports data-driven and proactive decision-making by project stakeholders. AI can analyze historical and real-time BIM data to identify potential design conflicts, estimate energy consumption, assess material efficiency, and anticipate maintenance requirements—all in a fraction of the time required by manual analysis. One of the key advantages of integrating AI into BIM processes is the automation of repetitive and time-consuming tasks. Activities such as clash detection, quantity takeoff, and progress monitoring—traditionally performed manually—can be streamlined through AI-powered tools. This automation allows architects, engineers, and construction professionals to shift their focus from routine tasks to more complex, strategic, and creative activities, such as design optimization, scenario planning, and sustainability enhancement.

By enhancing the efficiency, accuracy, and depth of data analysis, AI contributes to safer, more sustainable, and cost-effective projects. It enables stakeholders to make informed decisions faster and with greater confidence, reducing risks and improving project outcomes. The application of AI in BIM is thus catalyzing a paradigm shift in construction, unlocking new levels of creativity and productivity throughout the project lifecycle. Looking to the future, continued developments in machine learning (ML) and deep learning (DL) will further refine the capabilities of AI in BIM. These technologies will enhance predictive analytics, enabling more accurate assessments of future building performance and early identification of potential issues before they escalate. AI-driven predictive models will support preemptive action in areas like structural deterioration, energy inefficiency, and occupant safety.

Advancements in natural language processing (NLP) will also improve human-AI interaction, allowing construction professionals to query BIM databases using conversational language and receive intelligent, context-sensitive responses. This evolution will bridge the gap between technical data models and practical decision-making on construction sites. Moreover, the integration of augmented reality (AR) and virtual reality (VR) with AI and BIM systems will open new frontiers in project visualization and simulation. Stakeholders will be able to interact with dynamic,

AI-enhanced 3D environments, improving design accuracy, stakeholder engagement, and construction planning. As AI technologies continue to mature, their role in BIM data extraction and analysis will expand—transforming how buildings are planned, designed, built, and maintained. The future of construction will be shaped by intelligent systems that bring unprecedented precision, agility, and innovation to the built environment.

REFERENCES

Adewale, B. A., Ene, V. O., Ogunbayo, B. F., & Aigbavboa, C. O. (2024). A Systematic Review of the Applications of AI in a Sustainable Building's Lifecycle. *Buildings (Basel, Switzerland)*, *14*(7), 2137. DOI: 10.3390/buildings14072137

Akanbi, T., & Zhang, J. (2021). Design information extraction from construction specifications to support cost estimation. *Automation in Construction*, *131*, 103835. DOI: 10.1016/j.autcon.2021.103835

Al Sehrawy, A., Amoudi, O., Tong, M., & Callaghan, N. (2021, November). A review of the challenges to integrating BIM and building sustainability assessment. In AIP Conference Proceedings (Vol. 2428, No. 1). AIP Publishing. DOI: 10.1063/5.0071055

Amrouni Hosseini, M., Ravanshadnia, M., Rahimzadegan, M., & Ramezani, S. (2024). Next-Generation Building Condition Assessment: BIM and Neural Network Integration. *Journal of Performance of Constructed Facilities*, *38*(6), 04024050. DOI: 10.1061/JPCFEV.CFENG-4828

Banihashemi, S., Khalili, S., Sheikhkhoshkar, M., & Fazeli, A. (2022). Machine learning-integrated 5D BIM informatics: Building materials costs data classification and prototype development. *Innovative Infrastructure Solutions*, *7*(3), 215. DOI: 10.1007/s41062-022-00822-y

Bassier, M., Vergauwen, M., & Van Genechten, B. (2017). Automated classification of heritage buildings for as-built BIM using machine learning techniques. *ISPRS Annals of the Photogrammetry, Remote Sensing and Spatial Information Sciences*, *4*(W2), 25–30. DOI: 10.5194/isprs-annals-IV-2-W2-25-2017

Bassir, D., Lodge, H., Chang, H., Majak, J., & Chen, G. (2023). Application of artificial intelligence and machine learning for BIM. *International Journal for Simulation and Multidisciplinary Design Optimization*, *14*, 5. DOI: 10.1051/smdo/2023005

Deng, H., Hong, H., Luo, D., Deng, Y., & Su, C. (2020). Automatic indoor construction process monitoring for tiles based on BIM and computer vision. *Journal of Construction Engineering and Management*, *146*(1), 04019095. DOI: 10.1061/(ASCE)CO.1943-7862.0001744

Dimitrov, A., & Golparvar-Fard, M. (2014). Vision-based material recognition for automated monitoring of construction progress and generating building information modeling from unordered site image collections. *Advanced Engineering Informatics*, *28*(1), 37–49. DOI: 10.1016/j.aei.2013.11.002

Elghaish, F., Chauhan, J. K., Matarneh, S., Rahimian, F. P., & Hosseini, M. R. (2022). Artificial intelligence-based voice assistant for BIM data management. *Automation in Construction*, *140*, 104320. DOI: 10.1016/j.autcon.2022.104320

Emunds, C., Pauen, N., Richter, V., Frisch, J., & van Treeck, C. (2022). SpaRSE-BIM: Classification of IFC-based geometry via sparse convolutional neural networks. *Advanced Engineering Informatics*, *53*, 101641. DOI: 10.1016/j.aei.2022.101641

Himeur, Y., Ghanem, K., Alsalemi, A., Bensaali, F., & Amira, A. (2021). Artificial intelligence based anomaly detection of energy consumption in buildings: A review, current trends and new perspectives. *Applied Energy*, *287*, 116601. DOI: 10.1016/j.apenergy.2021.116601

Hosamo, H. H., Tingstveit, M. S., Nielsen, H. K., Svennevig, P. R., & Svidt, K. (2022). Multiobjective optimization of building energy consumption and thermal comfort based on integrated BIM framework with machine learning-NSGA II. *Energy and Building*, *277*, 112479. DOI: 10.1016/j.enbuild.2022.112479

Hu, Z. Z., Leng, S., Lin, J. R., Li, S. W., & Xiao, Y. Q. (2022). Knowledge extraction and discovery based on BIM: A critical review and future directions. *Archives of Computational Methods in Engineering*, *29*(1), 335–356. DOI: 10.1007/s11831-021-09576-9

Jung, N., & Lee, G. (2019). Automated classification of building information modeling (BIM) case studies by BIM use based on natural language processing (NLP) and unsupervised learning. *Advanced Engineering Informatics*, *41*, 100917. DOI: 10.1016/j.aei.2019.04.007

Kim, S., Jeong, K., Hong, T., Lee, J., & Lee, J. (2023). Deep learning–based automated generation of material data with object–space relationships for scan to BIM. *Journal of Management Engineering*, *39*(3), 04023004. DOI: 10.1061/JMENEA.MEENG-5143

Koo, B., Jung, R., & Yu, Y. (2021). Automatic classification of wall and door BIM element subtypes using 3D geometric deep neural networks. *Advanced Engineering Informatics*, *47*, 101200. DOI: 10.1016/j.aei.2020.101200

Kropp, C., Koch, C., & König, M. (2018). Interior construction state recognition with 4D BIM registered image sequences. *Automation in Construction*, *86*, 11–32. DOI: 10.1016/j.autcon.2017.10.027

Li, K., Gan, V. J., Li, M., Gao, M. Y., Tiong, R. L., & Yang, Y. (2024). Automated generative design and prefabrication of precast buildings using integrated BIM and graph convolutional neural network. *Developments in the Built Environment*, *18*, 100418. DOI: 10.1016/j.dibe.2024.100418

Locatelli, M., Seghezzi, E., Pellegrini, L., Tagliabue, L. C., & Di Giuda, G. M. (2021). Exploring natural language processing in construction and integration with building information modeling: A scientometric analysis. *Buildings (Basel, Switzerland)*, *11*(12), 583. DOI: 10.3390/buildings11120583

McArthur, J. J., Shahbazi, N., Fok, R., Raghubar, C., Bortoluzzi, B., & An, A. (2018). Machine learning and BIM visualization for maintenance issue classification and enhanced data collection. *Advanced Engineering Informatics*, *38*, 101–112. DOI: 10.1016/j.aei.2018.06.007

Mulero-Palencia, S., Álvarez-Díaz, S., & Andrés-Chicote, M. (2021). Machine learning for the improvement of deep renovation building projects using as-built BIM models. *Sustainability (Basel)*, *13*(12), 6576. DOI: 10.3390/su13126576

Nabavi, A., Ramaji, I., Sadeghi, N., & Anderson, A. (2023). Leveraging Natural Language Processing for Automated Information Inquiry from Building Information Models. Journal of Information Technology in Construction, 28.

Nunez-Morales, J. D., Hsu, S. H., Ibrahim, A., & Golparvar-Fard, M. (2023, May). New metrics to benchmark and improve bim visibility within a synthetic image generation process for computer vision progress tracking. In Canadian Society of Civil Engineering Annual Conference (pp. 209-221). Cham: Springer Nature Switzerland.

Olugboyega, O., Oseghale, G. E., & Aigbavboa, C. (2023). Multiple holistic case study of project-level building information modelling (BIM) adoption in Nigeria. *Construction Innovation*, *23*(3), 567–586. DOI: 10.1108/CI-10-2021-0199

Oskar, L. (2022). Measuring and extracting data from a BIM model.

Park, J., Kim, J., Lee, D., Jeong, K., Lee, J., Kim, H., & Hong, T. (2022). Deep learning–based automation of scan-to-BIM with modeling objects from occluded point clouds. *Journal of Management Engineering*, *38*(4), 04022025. DOI: 10.1061/(ASCE)ME.1943-5479.0001055

Pedral Sampaio, R., Aguiar Costa, A., & Flores-Colen, I. (2022). A systematic review of artificial intelligence applied to facility management in the building information modeling context and future research directions. *Buildings (Basel, Switzerland)*, *12*(11), 1939. DOI: 10.3390/buildings12111939

Perez-Perez, Y., Golparvar-Fard, M., & El-Rayes, K. (2021). Scan2BIM-NET: Deep learning method for segmentation of point clouds for scan-to-BIM. *Journal of Construction Engineering and Management, 147*(9), 04021107. DOI: 10.1061/(ASCE)CO.1943-7862.0002132

Seydgar, M., Motamedi, A., & Poirier, E. (2022). *Deep neural networks to assist in bim creation using scanned data: A review*. Transforming Construction with Reality Capture Technologies.

Shamshiri, A., Ryu, K. R., & Park, J. Y. (2024). Text mining and natural language processing in construction. *Automation in Construction, 158*, 105200. DOI: 10.1016/j.autcon.2023.105200

Su, T., Li, H., & An, Y. (2021). A BIM and machine learning integration framework for automated property valuation. *Journal of Building Engineering, 44*, 102636. DOI: 10.1016/j.jobe.2021.102636

Teng, Y., Xu, J., Pan, W., & Zhang, Y. (2022). A systematic review of the integration of building information modeling into life cycle assessment. *Building and Environment, 221*, 109260. DOI: 10.1016/j.buildenv.2022.109260

Wang, B., Wang, Q., Cheng, J. C., & Yin, C. (2022). Object verification based on deep learning point feature comparison for scan-to-BIM. *Automation in Construction, 142*, 104515. DOI: 10.1016/j.autcon.2022.104515

Wang, H., Meng, X., & Zhu, X. (2022). Improving knowledge capture and retrieval in the BIM environment: Combining case-based reasoning and natural language processing. *Automation in Construction, 139*, 104317. DOI: 10.1016/j.autcon.2022.104317

Wang, N., Issa, R. R., & Anumba, C. J. (2022). NLP-based query-answering system for information extraction from building information models. *Journal of Computing in Civil Engineering, 36*(3), 04022004. DOI: 10.1061/(ASCE)CP.1943-5487.0001019

Wei, W., Lu, Y., Zhong, T., Li, P., & Liu, B. (2022). Integrated vision-based automated progress monitoring of indoor construction using mask region-based convolutional neural networks and BIM. *Automation in Construction, 140*, 104327. DOI: 10.1016/j.autcon.2022.104327

Xu, F., & Liu, Q. (2023). Building energy consumption optimization method based on convolutional neural network and BIM. *Alexandria Engineering Journal, 77*, 407–417. DOI: 10.1016/j.aej.2023.06.084

Ying, H., Sacks, R., & Degani, A. (2023). Synthetic image data generation using BIM and computer graphics for building scene understanding. *Automation in Construction, 154*, 105016. DOI: 10.1016/j.autcon.2023.105016

Zabin, A., González, V. A., Zou, Y., & Amor, R. (2022). Applications of machine learning to BIM: A systematic literature review. *Advanced Engineering Informatics*, *51*, 101474. DOI: 10.1016/j.aei.2021.101474

Zhai, R., Zou, J., He, Y., & Meng, L. (2022). BIM-driven data augmentation method for semantic segmentation in superpoint-based deep learning network. *Automation in Construction*, *140*, 104373. DOI: 10.1016/j.autcon.2022.104373

Chapter 7 Accurate Estimation of Load-Carrying Capacity (Pu) in Steel Tube Structures

Abhishek Kumar
https://orcid.org/0000-0003-3961-9970
Lovely Professional University, India

Harpreet Singh Bedi
https://orcid.org/0000-0003-1389-4333
Lovely Professional University, India

ABSTRACT

The accurate estimation of load-carrying capacity (Pu) in steel tube structures ensures structural integrity and safety. This chapter explores a machine learning-based approach for predicting Pu using key material and geometric parameters, including concrete strength, steel tube areas, yield strengths, and eccentric loading ratio. Traditional empirical models often fail to capture the complex nonlinear relationships among these variables. In this chapter, we experimented with (a) an advanced machine learning model (XBGBoost) that handles nonlinear relationships in structural data, (b) deep neural network model interactions between geometric and material properties, and (c) bayesian optimization to fine-tune ML models to improve predictive performance. With the lowest Mean Absolute Error (MAE), Mean Squared Error (MSE), and Root Mean Squared Error (RMSE) as well as the greatest R2 score (0.9811), the findings show that Bayesian Optimization applied to XGBoost performed the best.

DOI: 10.4018/979-8-3373-0746-6.ch007

INTRODUCTION

Steel tube structures are widely used in building and construction due to their exceptional strength, ductility, and load-bearing capacity (Chen et al, 2024). These structures are widely used in high-rise buildings, bridges, platforms, transmission towers, and industrial facilities, where they provide superior stability and resistance to axial and lateral loads. Their ability to support substantial weight while maintaining a relatively low self-weight makes them an ideal choice for various engineering applications. A steel tube structure's maximal resistance before failure is determined by its ultimate load-carrying capacity (Pu) (Huang et al, 2019). For structural safety, material efficiency, and adherence to engineering requirements, an accurate estimate of Pu is essential. To retain structural integrity during their service life, engineers must make sure that these structures can sustain anticipated loads, such as gravity loads, wind forces, seismic impacts, and other external pressures. Accurately estimating these structures' ultimate load-carrying capacity (Pu) is crucial for structural design and safety assessments. Structural dependability, economic effectiveness, and appropriate material use are all ensured by accurate Pu predictions. Traditionally, the main techniques for calculating Pu have been empirical equations and finite element analysis (FEA) (Arbogast & Wheeler, 1996). Although FEA offers comprehensive insights into the behavior of structures, it requires a large amount of computing power as well as thorough material property and boundary condition calibration. However, because of their linear approximations and reduced assumptions, empirical formulae frequently do not generalize across various structural configurations.

(a) *Empirical Formula:* Regression-based equations and experimental data are the sources of empirical formulas (Citakoglu, 2015). Although they offer simplified approximations, they frequently fall short in generalizing to various geometric configurations, loading scenarios, and material attributes. Nonlinear interactions between structural factors are difficult for them to capture.
(b) *Finite Element Analysis (FEA):* A computationally demanding method that simulates actual loading conditions by modeling steel tube constructions using finite elements (Dezfooli et al., 2015) requires specific boundary conditions and material parameters to get accurate results. It is expensive and laborious, which makes it unsuitable for quick evaluations in applications involving real-time structural assessment.
(c) *Machine Learning Approach:* A viable substitute is machine learning (ML), which can identify intricate patterns and produce precise forecasts without the need for explicit formulations. Machine learning (ML) (Huettmann et al, 2018) is a possible substitute as it can recognize complex nonlinear correlations between geometric and material characteristics, generating extremely accurate

predictions without the need for explicit mathematical formulations. With an emphasis on different methods, including sophisticated regression models, deep learning techniques, and Bayesian optimization procedures, this chapter investigates how ML models might improve Pu prediction accuracy (Pu et al, 2019). In structural engineering, machine learning has become a potent tool with several benefits over conventional techniques.

a) *Capacity to Record Nonlinear Interactions:* Machine learning approaches, in contrast to empirical models, can identify intricate relationships between structural factors without the need for explicit mathematical formulations.
b) *Fast Predictions:* Unlike FEA simulations, ML models can estimate Pu for various structural configurations quickly after training, which lessens the computational load.
c) *Adaptability and Generalization:* ML models are more applicable in a variety of situations as they can be trained on a variety of datasets and used in a broad range of structural settings.
d) *Automated Feature Selection and Optimization:* ML models may automatically determine the most important characteristics influencing Pu and adjust their predictions following those findings by employing strategies such as feature engineering and Bayesian

Steel tubes are used widely in engineering because of their high strength-to-weight ratio and excellent structural efficiency. To maximize material use and ensure safety, it is essential to estimate these structures' load-carrying capacity (Pu) (Xin, Y., Li, J., Wang, X., & Hampson, K. 2022).). The buckling strength is mostly determined by geometric characteristics, including the diameter-to-thickness ratio and tube length (Bardi, F. C., & Kyriakides, S., 2006). The overall behavior of steel tubes under axial and lateral stresses is influenced by material parameters such as yield strength and elastic modulus (Mahdi, M. S., & Mohammed, A.I. I. 2024). Additionally, the structure's failure mode is determined by loading situations including axial compression, bending, and coupled forces (Corona, E., Lee, L. H., & Kyriakides, S., 2006). The ultimate strength of steel tubes can also be decreased by manufacturing flaws and residual stresses (Krishna, G. V., Narayanamurthy, V., & Viswanath, C., 2020). Full-scale axial compression experiments on steel tubes by Han et al. (2018) showed that Pu is significantly affected by local buckling. The impact of residual stresses and defects on steel tube performance has been the subject of several investigations (Tabatabaeian, A. et al, 2022). Researchers are looking at alternative prediction techniques like machine learning since, although experimental procedures yield useful insights, they are time-consuming and costly (Rossini, N. S., Dassisti, M., Benyounis, K. Y., & Olabi, A.G.G. G. 2012). Since Extreme Gradient Boosting (XGBoost) can capture nonlinear correlations and feature interactions, it has become a powerful technique for forecasting Pu (Xu, S., & Wang, W. 2013). Research has demonstrated that by offering greater prediction accuracy and resilience to outliers, XGBoost performs better than conventional regression models (Liu et al., 2022). With positive results, deep neural networks (DNNs) and convolutional neural networks (CNNs) have been used for Pu estimation (Yong et al., 2023). Computational costs and model interpretability are major challenges. To increase the accuracy of Pu predictions, machine learning algorithms have used Bayesian optimization. Bayesian techniques aid in fine-tuning XGBoost and DNN models for improved performance by optimizing hyperparameters (Zulfiqar, M., Gamage, K. A., Kamran, M., & Rasheed, M. B., 2022). According to recent research, Bayesian-optimized XGBoost outperforms conventional hyperparameter tuning techniques in terms of accuracy (Kumar et al, 2024).

PARAMETERS FOR LOAD-CARRYING CAPACITY ESTIMATION

Predicting the ultimate load-carrying capacity (Pu) of steel tube constructions requires extensive knowledge of geometric arrangements and material qualities. 13 important input elements, such as concrete qualities, inner steel tube details, and exterior steel tube features, are used in this study to affect Pu.

Concrete Properties

Concrete contributes significantly to composite steel tube constructions by improving their stiffness, energy dissipation properties, and load-bearing capacity. The following are some pertinent concrete-related parameters X_1 Concrete Strength (MPa) -compressive strength of the concrete used in the building, X_2 Area of Concrete Portions (mm^2) cross-sectional area of the concrete core inside the steel tube, X_3 Concrete Strength of Standard Cylinders (MPa) a common metric for concrete compressive strength in structural analysis.

Inner Steel Tube Details

In addition to providing structural support, the inner steel tube greatly increases the system's total load-bearing capability. The following are the main material and geometric properties X_4 Outer Depth of interior Steel Tube (mm) entire exterior depth of the interior steel section, X_5 Internal Steel Tube Outer Width (mm) internal steel section's overall exterior width, X_6: Inner Depth of Internal Steel Tube (mm) inner depth of the steel tube, which influences the hollow section qualities, X_7 Inner Width of Internal Steel Tube (mm) inner width of the steel tube, which affects the effective cross-sectional area, X_8 Yield Strength of Internal Steel Tube (MPa) material parameter that establishes the tension at which plastic deformation starts.

Outer Steel Tube Details

The external steel tube provides additional structural stability, load distribution, and protection for the inner components. The defining parameters include X_9 Outer Depth of External Steel Tube (mm) total depth of the external steel tube, affecting the moment of inertia and buckling resistance, X_{10} Outer Width of External Steel Tube (mm) total width of the external steel tube, X_{11} Inner Depth of External Steel Tube (mm) determines the section's effective area, X_{12}: Inner Width of External Steel Tube (mm) affecting overall structural behavior, X_{13} Yield Strength of External Steel Tube (MPa) material's ability to withstand stress before yielding, and X_{14}: Eccentric Loading Ratio ratio of applied eccentricity to section depth, influencing bending and axial stress distribution.

Table 1. Descriptive statistics of the input and output parameters

	Mean	Standard Error	Median	Standard Deviation	Minimum	Maximum	Skewness	Kurtosis
Input 1	23396.30	221.447	22997.07	2425.83	12446.	26788.	-2.904	11.94
Input 2	38.7069	0.87720	45.9210	9.576	28.38	52.7	-0.061	-1.796
Input 3	101.9	2	95	21.923	95	186	3.177	8.996
Input 4	56.1	1.0647	53	11.6629	53	106	4.005	14.701
Input 5	91.804	2.298	87	25.179	71	186	2.452	6.053
Input 6	46	1.4427	45	15.8	29	106	1.948	5.257
Input 7	222.593	15.405	276.2	168.761	0.00	433.7	90.314	-1.429
Input 8	276	1.1	280	12.05	240	280	-2.701	5.382
Input 9	142	0.55	140	60.25	140	160	2.701	5.382
Input 10	264.476	0.969	268	10.616	232.76	268	-2.701	5.382
Input 11	130.476	0.6809	128	7.459	128	152.76	2.701	5.382
Input 12	265.95	6.358	421.5	69.652	270.3	421.5	-0.515	-1.667
Input 13	0.1088	0.01	0.107	0.10952	0.00	0.25	0.016	-2.008
Output	2154.645	79.033	1899.61	865.773	635.29	4104.7	0.39	-1

The load-carrying capacity of the structure is determined in a significant manner by each of these factors. Their interactions are frequently simplified by traditional methods, which results in prediction errors. However, to enhance accuracy, machine learning models might examine complex, nonlinear correlations between these variables. The dataset utilized for analysis contains 120 entries where the input parameters. The Descriptive statistics of the data are presented in Table 1, where it can be observed that experimental data has a diverse range. To reduce the related error and maximize the efficacy of the suggested model. The Descriptive statistics of the data are presented in Table 1, where it can be observed that experimental data has a diverse range. To reduce the related error and maximize the efficacy of the suggested model. Table 1 provides an overview of the dataset characteristics, including key statistics such as the number of samples, features, and other pertinent information. This table offers readers a comprehensive summary of the dataset's structure and composition. Skewness and kurtosis for output are 0.39 and -1, respectively. Output ranges from 635.29 KN to 4104.78 KN. The distribution plot of output parameters is also drawn, which is shown in Figure 1.

Figure 1. Distribution plot of output

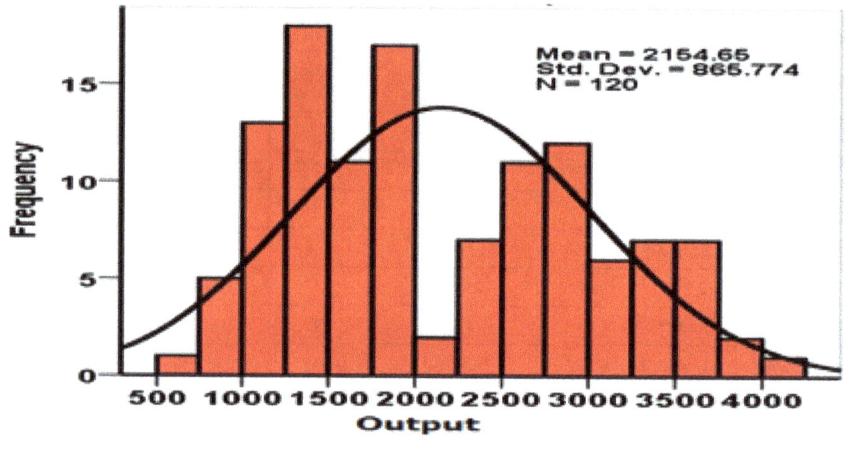

Figure 2. Correlation Heatmap

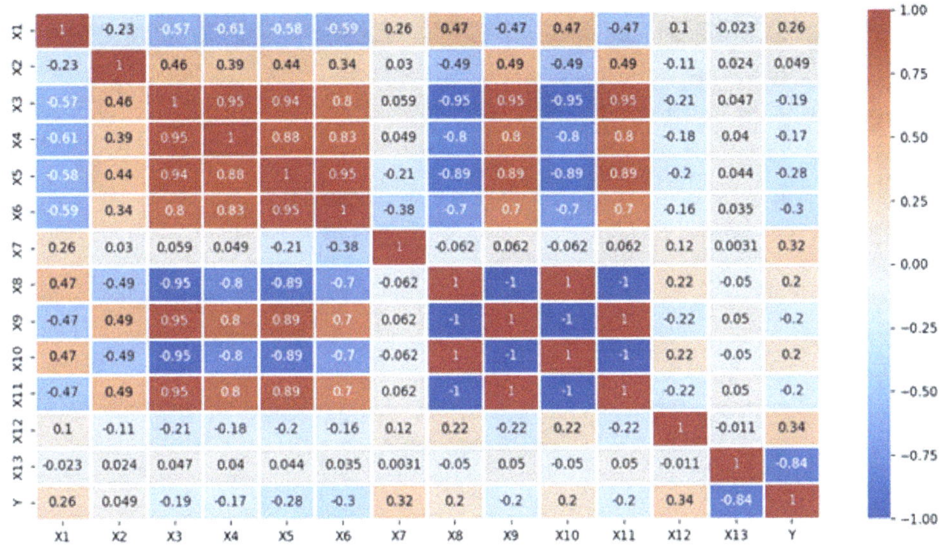

The correlation matrix in Figure 2 is graphically represented as a correlation heatmap, which commonly makes use of colors to highlight the strength and direction of correlations between variables. The intensity and direction of correlations between variables are represented by colors. Warm colors like red are often used to illustrate positive correlations, whereas cool colors like blue and purple are typically used to show negative correlations. The strength of the link is indicated by the color's intensity, with darker hues suggesting greater correlations. The finding aligns that the area and strength of concrete are positively associated with increased load-carrying capacity. The inner and outer depths of inner hollow steel and the width of the outer steel tube are weak and negatively correlated, as the outer steel depth is weakly positively correlated. A strong positive connection is seen by the highly correlated variables X_4 and X_5 (0.94). Correlations of -1, X_8, X_9, and X_{10} show a complete negative connection. The independent variables X_{12} and X_{13} correlate with the target. There is a significant negative association between Y and X_{13} (-0.84). The correlation between X_{12} and Y is somewhat favorable (0.34). There are fewer connections with other variables, such as X_5 (-0.28), X6 (-0.3), and X_4 (-0.17).

Figure 3. Feature importance

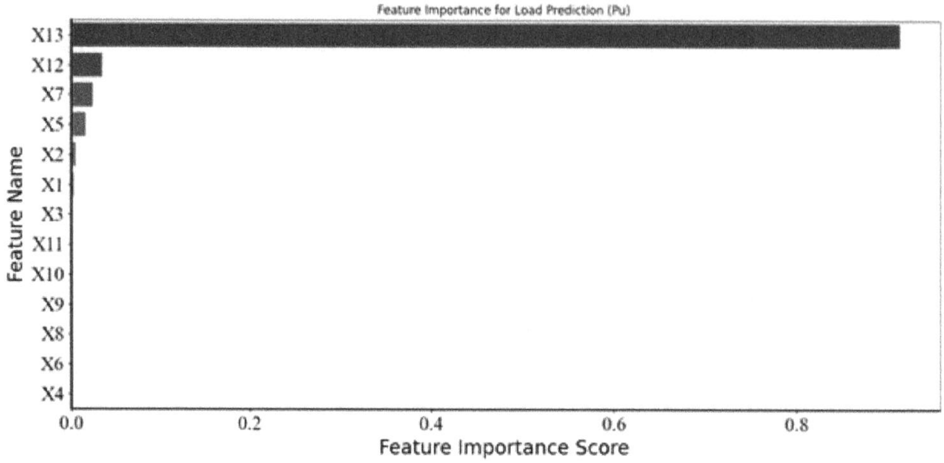

The highest positive association is shown with X_{12}, while X_{13} shows the highest negative correlation seems to be with Y. Predictive modeling feature selection insights X_{13} is the most important feature for prediction in Figure 3. Feature importance plot that uses the XGBoost model to forecast load (Pu) in steel tube constructions. The most important feature (about 90% of the overall relevance) is X_{13} (Eccentricity Ratio, e/2b). This implies that the load-carrying capability of steel tube constructions is significantly impacted by eccentric loading. Though not nearly as much as X_{13}, X_{12} (Outer Steel Yield Strength, fyo) and X_7 (Inner Steel Yield Strength, fyi) do have some impact. This implies that eccentricity has a greater effect on Pu than steel tube strength. Minimal effects of the characteristics of concrete area (X_1) and Concrete Strength (X_2) are of relatively little significance. This implies that steel characteristics and eccentric loads are more important in predicting Pu than concrete. The impact of structural dimensions (X_3–X_{11}) is minimal. These parameters have minimal bearing on Pu prediction, even if they are connected to geometry. According to the model, the main ruling factor influencing load capacity (Pu) is eccentric loading (X_{13}).

Using feature importance scores produced by the trained machine learning models (Bayesian-optimized XGBoost) sensitivity analysis was carried out. Input variables are ranked according to how much they help the ensemble of decision trees reduce prediction error. According to the findings, the most significant metric was the eccentricity ratio (X_{13}), which was followed by the length-to-diameter ratio (X_3) and the compressive strength of the concrete (X_7). Eccentricity has a substantial impact on the stress distribution and failure mechanism of CFST columns under axial and bending loads. Our results are consistent with engineering understanding. Sensitivity analysis provides several benefits by identifying dominant and less significant variables.

a) *Model Simplification*: To simplify the model without compromising accuracy, less important features might be eliminated.
b) *Engineering insight:* When selecting materials and planning a construction project, engineers can concentrate on maximizing the most sensitive design aspects.
c) *Risk Reduction:* Safety evaluations are enhanced, and quality control in those areas is prioritized when it is clear which factors have the greatest effect on load capacity.

MODELING MACHINE LEARNING ALGORITHMS FOR LOAD-CARRYING CAPACITY PREDICTION

To ensure structural integrity and safety, the steel tube constructions' load-carrying capacity (Pu) must be accurately estimated. The nonlinear interactions between material qualities, geometric features, and external stress conditions are frequently overestimated by traditional empirical models, despite their usefulness. A data-driven ML model in Figure 4 capable of accurately modeling these complex relationships is a machine learning (ML) model. Three machine learning methods for predicting Pu are examined in this section. An advanced ensemble learning technique that successfully manages nonlinear relationships in structural data is Extreme Gradient Boosting (XGBoost). A deep learning model that captures the complex relationships between geometric and material attributes is called a deep neural network (DNN). Bayesian optimization is a hyperparameter tuning method that improves machine learning models' performance and guarantees the best possible prediction accuracy.

Extreme Gradient Boosting (XGBoost)

XGBoost is an optimized gradient-boosting algorithm that builds an ensemble of decision trees to make predictions. An ensemble of decision trees is constructed via the improved gradient boosting technique, Extreme Gradient Boosting (XGBoost), to provide predictions. By employing gradient-based optimization for effective training, it enhances conventional boosting strategies (a) avoid overfitting, use regularization methods (L_1 and L_2), (b) address feature priority rating and missing data (c) parallel processing. By minimizing a loss function, gradient boosting—the foundation of XGBoost—improves $\hat{y}_i = \sum_{t=1}^{T} f_t(X_i)$ weak learners (decision trees) repeatedly. XGBoost's prediction function is:

'Where y_i is the predicted load-carrying capacity (Pu) for the ith data point, $f(x_i)$ represents the decision tree model applied to input features x_i, and T is the number of boosting iterations.

$$\Gamma = \sum_{i=1}^{n} l(y_i, \bar{y}_i) + \sum_{t=1}^{T} \Omega(f_t)$$

$$\Omega(ft) = \gamma T + 0.5\lambda \sum_{j} w_j^{\,2}$$

XGBoost aims to minimize a regularized loss function (Mean Squared Error)

Where is the loss function, and Ω(ft) is the regularization term to avoid overfitting. γ and λ are hyperparameters controlling tree complexity and L_2 regularization, respectively, and w_j represents tree weights. To estimate load-carrying capacity (Pu), XGBoost works effectively for capturing nonlinear interactions between eccentric loading ratio, steel tube areas, and concrete strength. Managing high-dimensional input characteristics, including yield strengths and the diameters of inner and exterior steel tubes, combining many decision trees reduces prediction errors.

Deep Neural Networks (DNNs)

Deep neural networks are neural networks composed of multiple layers. Each layer learns a hierarchical representation. While capturing extremely nonlinear correlation between structural factors and the DNN performance tree-based model. (a) Geometrical material interaction is a complex aspect of load capacity assessment (b) High high-dimensional data handling (c) better generalization. A deep neural network (DNN) maps input data to output predictions, modeling load-carrying capacity (Pu) estimate as a function of (x,θ). Where x is the material and geometric parameter vector, like steel tube dimensions and concrete strength. The estimated load capacity (Pu) is denoted by y, whereas θ stands for trainable network parameters (weights and biases).

The DNN is trained by minimizing the Mean Squared Error (MSE) loss.

$\bar{y}_i$

Where y_i is the load-carrying capacity and is the predicted value.

Bayesian Optimization for Hyperparameter Tuning

A global optimization method called Bayesian Optimization (BO) is used to adjust the hyperparameters of machine learning models. BO explores and exploits the search space using probabilistic models (Gaussian Processes) as opposed to a random search for the optimal hyperparameters. Extensive processing is needed for traditional approaches (Grid Search, Random Search). With fewer iterations, BO effectively determines the ideal model parameters. Bayesian Optimization (BO) optimizes ML model hyperparameters to enhance prediction accuracy with θ as hyperparameters.

The use of XGBoost, Deep Neural Networks (DNNs), and Bayesian Optimization to forecast load-carrying capacity (Pu) in steel tube constructions is demonstrated. For effective, nonlinear predictions, XGBoost uses gradient boosting to build an ensemble of decision trees. Deep hierarchical learning is used by DNNs to capture complex geometric-material interactions. By employing probabilistic techniques to optimize hyperparameters, Bayesian Optimization improves model correctness. These methods offer a strong AI-powered foundation for enhancing civil engineering's prediction dependability and structural safety.

Figure 4. Flow diagram of the model

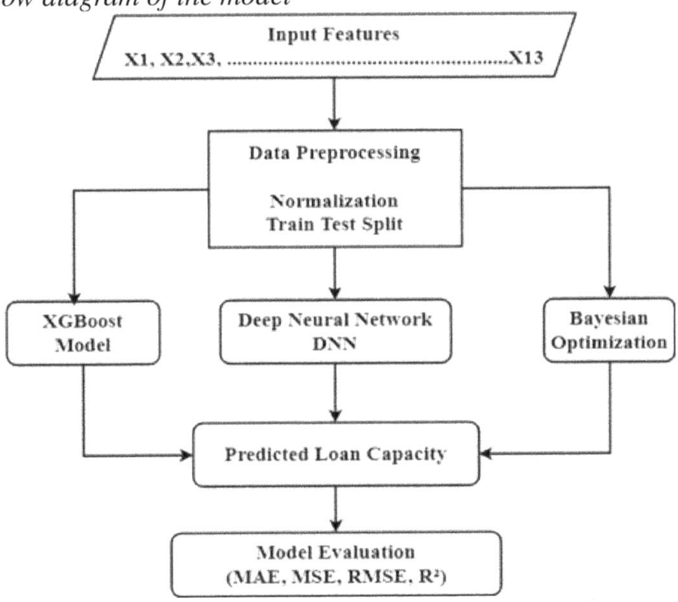

Table 2. Evaluation parameters of 3 machine learning models

Parameter	Mean Absolute Error (MAE)	Mean Squared Error (MSE)	Root Mean Squared Error (RMSE)	R^2 Score
XGBoost	66.2657	15016.8345	122.5432	0.9797
DNN	132.2732	25929.4241	161.0262	0.9649
Bayesian Optimization	59.5527 (Best)	13917.3983 (Best)	117.9720 (Best)	0.9811 (Best)

Figure 5. Actual load vs predicted load via ML model

Parameter	Mean Absolute Error (MAE)	Mean Squared Error (MSE)	Root Mean Squared Error (RMSE)	R^2 Score
XGBoost	66.2657	15016.8345	122.5432	0.9797
DNN	132.2732	25929.4241	161.0262	0.9649
Bayesian Optimization	59.5527 (Best)	13917.3983 (Best)	117.9720 (Best)	0.9811 (Best)

Bayesian Optimization does the best. Table 2 presents that Bayesian optimization has the lowest MAE, MSE, and RMSE, showing that the model predicts with the least amount of error. It explains the most variation in the sample, as indicated by its highest R2 score of 0.9811. XGBoost performs well, and Bayesian performs a bit better. Its error is greater than Bayesian Optimization's but lower than DNN's. The majority of the variation is still captured by the R2 score (0.9797), which is near to Bayesian optimization. DNN has the worst performance. Its projections diverge the most on average, as indicated by its greatest MAE (132.27). Its highest RMSE (161.03) indicates that its error variation is the greatest. It explains less of the variation in the data, according to the lowest R2 value (0.9649). It is evident from Figure 5 that the best model for estimating load-carrying capacity is Bayesian Optimization because of its high accuracy and low error rate. In second place, XGBoost outperforms DNN but falls short of Bayesian Optimization in terms of optimization. DNN performs worse than the other two models, indicating that it may require further data or fine-tuning.

The main focus of the chapter is on forecasting the load-carrying capacity of steel tube constructions using machine learning models like XGBoost, Deep Neural Networks (DNN), and Bayesian Optimization. What's interesting about this approach is how these tools are methodically coupled and adjusted. In particular, the Bayesian-optimized XGBoost model exhibits a distinctive combination of probabilistic optimization and ensemble learning to improve accuracy. The chapter's novelity includes

a) XGBoost hyperparameters were adjusted using Bayesian Optimization, yielding an extraordinarily high R^2 score of 0.9811.
b) Engineering insights are shown by using feature importance plots to show that the eccentricity ratio (X_{13}) dominates forecasts.
c) A data-driven substitute for traditional empirical and finite element models by successfully handling non-linear geometric-material interactions.

REAL-WORLD APPLICATIONS

In the field of structural and civil engineering, the suggested model has a number of useful applications.

a) *Design Optimization:* By rapidly estimating the load capacity of new steel tube configurations, engineers can use the model to help create safer and more efficient designs.
b) *Structural Health Monitoring:* To predict probable failure areas in aging infra-structure, ML models can be included in real-time systems.
c) *Construction Planning:* Selecting the right materials and dimensions during the planning stage is aided by precise estimates of Pu (ultimate load-carrying capacity).
d) *Cost Efficiency:* These models reduce the time and resources required for structural analysis by taking the place of complicated simulations or expensive experimental setups.

ETHICAL AND SOCIETAL IMPLICATIONS

The ethical application of such prediction models becomes essential as machine learning becomes more and more integrated into structural engineering:

a) Safety Responsibility: Serious risks could arise from underestimating load ca-pacity due to predictive errors. Validation and cross-verification are necessary for ethical use prior to deployment in practical situations.
b) Transparency and Interpretability: AI-driven decisions need to be intelligible. Transparency is supported by the use of interpretable machine learning models (such as decision trees) and feature significance plots.
c) Data Bias: Because the models are based on experimental data, they might not account for all worldwide variations in steel tube compositions or designs. It is an obligation of society to ensure representative and varied data.

d) Fair Use and Accessibility: By making these models available to small and medium-sized engineering firms, the industry can overcome barriers to technology and democratize advanced design tools.

CONCLUSION

This chapter explored many machine learning methods, with an emphasis on XGBoost, Deep Neural Networks (DNN), and Bayesian Optimization, for determining the load-carrying capacity (Pu) of steel tube constructions. With the lowest Mean Absolute Error (MAE), Mean Squared Error (MSE), and Root Mean Squared Error (RMSE) and the greatest R2 score, the findings showed that Bayesian Optimization performed better than the other models. This suggests that by adjusting hyperparameters, Bayesian Optimization successfully improves XGBoost's prediction performance. XGBoost performed somewhat worse than Bayesian Optimization while having great predictive power. The Deep Neural Network model, on the other hand, had the largest prediction errors, indicating that further fine-tuning, larger training data sets, or different network topologies would be needed to increase accuracy.

REFERENCES

Arbogast, T., & Wheeler, M. F. (1996). A nonlinear mixed finite element method for a degenerate parabolic equation arising in flow in porous media. *SIAM Journal on Numerical Analysis*, *33*(4), 1669–1687. DOI: 10.1137/S0036142994266728

Bardi, F. C., & Kyriakides, S. (2006). Plastic buckling of circular tubes under axial compression—part I: Experiments. *International Journal of Mechanical Sciences*, *48*(8), 830–841. DOI: 10.1016/j.ijmecsci.2006.03.005

Chen, R. S., Zhang, H. Y., Hao, X. K., Yu, H. X., Shi, T., Zhou, H. S., Wang, P., Zhao, Z.-F., & Wang, P. (2024, Oct). Experimental study on ultimate bearing capacity of short thin-walled steel tubes reinforced with high-ductility concrete. [). Elsevier.]. *Structures (Oxford)*, *68*, 107109. DOI: 10.1016/j.istruc.2024.107109

Citakoglu, H. (2015). Comparison of artificial intelligence techniques via empirical equations for prediction of solar radiation. *Computers and Electronics in Agriculture*, *118*, 28–37. DOI: 10.1016/j.compag.2015.08.020

Corona, E., Lee, L. H., & Kyriakides, S. (2006). Yield anisotropy effects on buckling of circular tubes under bending. *International Journal of Solids and Structures*, *43*(22-23), 7099–7118. DOI: 10.1016/j.ijsolstr.2006.03.005

Dezfooli, M. S., Abolmaali, A., & Razavi, M. (2015). Coupled nonlinear finite-element analysis of soil–steel pipe structure interaction. *International Journal of Geomechanics*, *15*(1), 04014032. DOI: 10.1061/(ASCE)GM.1943-5622.0000387

Huang, F., Cui, Y., Dong, R., Wei, J., & Chen, B. (2019). Evaluation on ultimate load-carrying capacity of concrete-filled steel tubular arch structure with preload. *Advances in Structural Engineering*, *22*(13), 2755–2770. DOI: 10.1177/1369433219850091

Huettmann, F., Craig, E. H., Herrick, K. A., Baltensperger, A. P., Humphries, G. R., Lieske, D. J., & Young, B. D. (2018). Use of machine learning (ML) for predicting and analyzing ecological and 'presence only'data: an overview of applications and a good outlook. *Machine learning for ecology and sustainable natural resource management*, 27-61.

Krishna, G. V., Narayanamurthy, V., & Viswanath, C. (2020). Modeling the buckling characteristics of the metal-FRP hybrid cylinder. *Composite Structures*, *250*, 112505. DOI: 10.1016/j.compstruct.2020.112505

Kumar, M., Samui, P., Kumar, D. R., & Asteris, P. G. (2024). State-of-the-art XG-Boost, RF and DNN based soft-computing models for PGPN piles. *Geomechanics and Geoengineering*, *19*(6), 975–990. DOI: 10.1080/17486025.2024.2337702

Liu, J. Z., Fang, H., & Chan, T. M. (2022). Experimental investigation on material properties and residual stresses in cold-formed high strength steel irregular octagonal hollow sections. *Journal of Constructional Steel Research*, *191*, 107170. DOI: 10.1016/j.jcsr.2022.107170

Mahdi, M. S., & Mohammed, A. I. (2024, Aug). The effect of bending angle and radius on wall thickness variation according to a geometric ratio in tube bending. In *AIP Conference Proceedings* (Vol. 3105, No. 1). AIP Publishing. DOI: 10.1063/5.0212722

Pu, Y., Apel, D. B., Liu, V., & Mitri, H. (2019). Machine learning methods for rockburst prediction-state-of-the-art review. *International Journal of Mining Science and Technology*, *29*(4), 565–570. DOI: 10.1016/j.ijmst.2019.06.009

Rossini, N. S., Dassisti, M., Benyounis, K. Y., & Olabi, A. G. (2012). Methods of measuring residual stresses in components. *Materials & Design*, *35*, 572–588. DOI: 10.1016/j.matdes.2011.08.022

Tabatabaeian, A., Ghasemi, A. R., Shokrieh, M. M., Marzbanrad, B., Baraheni, M., & Fotouhi, M. (2022). Residual stress in engineering materials: A review. *Advanced Engineering Materials*, *24*(3), 2100786. DOI: 10.1002/adem.202100786

Xin, Y., Li, J., Wang, X., & Hampson, K. (2022). Load-carrying capacity assessment of an existing highway bridge based on hybrid finite-element model updating. *Journal of Performance of Constructed Facilities*, *36*(3), 04022028. DOI: 10.1061/(ASCE)CF.1943-5509.0001729

Xu, S., & Wang, W. (2013). Numerical investigation on weld residual stresses in tube to tube sheet joint of a heat exchanger. *International Journal of Pressure Vessels and Piping*, *101*, 37–44. DOI: 10.1016/j.ijpvp.2012.10.004

Yong, L. I., Wanni, G. A. N., Wenbin, Z. H. O. U., & Dongsheng, L. I. (2023). Review on residual stress and its effects on manufacturing of aluminium alloy structural panels with typical multi-processes. *Chinese Journal of Aeronautics*, *36*(5), 96–124. DOI: 10.1016/j.cja.2022.07.020

Zulfiqar, M., Gamage, K. A., Kamran, M., & Rasheed, M. B. (2022). Hyperparameter optimization of bayesian neural network using bayesian optimization and intelligent feature engineering for load forecasting. *Sensors (Basel)*, *22*(12), 4446. DOI: 10.3390/s22124446 PMID: 35746227

Chapter 8
Artificial Intelligence in Wastewater Treatment: Advancing Processes With Nature-Based Solutions

Chaitanya Surya Chandra Nadimpalli
https://orcid.org/0000-0003-1070-3079
Raghu Engineering College, India

ABSTRACT

Wastewater treatment systems face rising challenges from urbanization, population growth, and stricter environmental norms. Traditional methods often struggle with energy use, process efficiency, and adaptability. In developing countries like India, limited resources and outdated infrastructure worsen the problem. Rapid urban and industrial growth further strains existing systems. Artificial intelligence (AI) offers a promising way forward by enabling real-time monitoring, predictive analytics, and optimized operations (Lee et al., 2022). In India's context, AI can support better decision-making and improve treatment performance despite resource constraints. By analyzing data continuously, AI systems can enhance efficiency, reduce waste, and promote sustainability (Corominas et al., 2018). The proposed AI-Nature-Based Solutions (NBS) framework combines modern technology with eco-friendly approaches, offering a sustainable path to address water management challenges in developing nations.

DOI: 10.4018/979-8-3373-0746-6.ch008

INTRODUCTION

Challenges in Wastewater Treatment

Wastewater treatment systems face numerous critical challenges that significantly impact operational efficiency and environmental sustainability. Traditional wastewater treatment systems encounter highly variable influent characteristics that make consistent treatment performance difficult to achieve, with wastewater composition fluctuating significantly based on temporal patterns, seasonal variations, and local activities (Gernaey et al., 2004). Energy consumption represents another major constraint, as biological treatment processes require substantial aeration energy, often accounting for 45-75% of plant electricity usage (Torregrossa et al., 2018). Increasingly stringent regulatory standards demand precise process control to ensure consistent compliance with evolving environmental regulations. The operational complexity of modern treatment facilities, involving multiple interacting biological, chemical, and physical processes, creates intricate control challenges that traditional systems struggle to manage effectively (Olsson, 2012). Additionally, there is growing emphasis on resource recovery from wastewater streams, including nutrients, energy, and reclaimed water, requiring sophisticated process optimization capabilities.

These challenges are particularly acute in developing nations, where rapid urbanization and industrial growth have placed unprecedented stress on existing wastewater infrastructure. Traditional approaches to wastewater treatment, while foundational to current systems, increasingly struggle to meet the complex demands of modern urban environments. Resource constraints, limited technical expertise, and unreliable energy supplies further compound these operational difficulties, creating an urgent need for innovative solutions that can deliver effective treatment with minimal resource requirements.

Emergence of AI Solutions

The emergence of artificial intelligence technologies offers promising solutions to enhance the efficiency and effectiveness of wastewater treatment processes. The integration of artificial intelligence in wastewater treatment operations represents a paradigm shift in our approach to water management. This transformation is particularly relevant in the Indian scenario, where the dual challenges of resource constraints and environmental preservation necessitate innovative solutions. The application of AI technologies enables treatment facilities to optimize their operations while simultaneously addressing the pressing needs of environmental sustainability and resource efficiency. Through careful analysis of operating parameters and real-time process optimization, AI-driven systems can significantly enhance the performance of existing treatment infrastructure (Corominas et al., 2018).

In recent years, our understanding of wastewater treatment has evolved beyond mere technological solutions to encompass a more holistic approach that integrates natural processes with advanced technology. This evolution has led to the emergence of nature-based solutions as a vital component of modern wastewater treatment strategies (Vymazal, 2010). The integration of these nature-based approaches with artificial intelligence creates a unique synergy that combines the resilience of natural systems with the precision of advanced technology.

The scope of AI applications in wastewater treatment extends across multiple domains, from process optimization to predictive maintenance and quality control. Modern treatment facilities face various operational challenges, including fluctuating input loads, energy efficiency requirements, and stringent regulatory standards. The implementation of AI-driven solutions addresses these challenges through sophisticated data analysis and adaptive control mechanisms. Machine learning algorithms, when properly implemented, can predict system behavior, optimize treatment processes, and ensure consistent compliance with environmental standards (Haimi et al., 2013).

The role of nature-based solutions in wastewater treatment has gained significant attention in recent years, particularly in the context of sustainable development. These solutions, which include constructed wetlands, biofilters, and green infrastructure, offer numerous advantages in terms of environmental sustainability and resource efficiency (Zhang et al., 2014). When enhanced by artificial intelligence, these natural systems can achieve higher levels of treatment efficiency while maintaining their inherent environmental benefits. The integration of AI with nature-based solutions represents a promising approach to addressing the complex challenges of wastewater treatment in developing nations.

Major Scientific Contributions

This chapter makes several key scientific contributions to the field of AI-enhanced wastewater treatment:

- **Comprehensive Integration Framework**: Development of a systematic approach for integrating AI technologies with nature-based solutions, addressing both technical and implementation challenges
- **Critical Gap Analysis**: Identification of research gaps in current AI applications for wastewater treatment, particularly in developing nation contexts

- **Performance Enhancement Strategies**: Documentation of AI-driven optimization techniques that significantly improve treatment efficiency while reducing energy consumption and operational costs
- **Sustainability Assessment**: Evaluation of how AI-enhanced nature-based solutions contribute to environmental sustainability and resource recovery objectives
- **Implementation Roadmap**: Provision of practical guidelines for implementing AI technologies in existing wastewater treatment infrastructure
- **Future Research Directions**: Identification of emerging trends and research opportunities in AI-enhanced sustainable wastewater treatment

Chapter Structure

This chapter is organized into ten comprehensive sections that systematically explore the integration of AI with wastewater treatment systems. Following this introduction, Section 2 examines current challenges in wastewater treatment, providing context for AI applications. Section 3 explores nature-based solutions, covering fundamentals, sustainability factors, and applications in developing countries. Section 4 presents a detailed analysis of AI applications in wastewater treatment, including data-driven approaches, machine learning optimization, and deep learning applications. Section 5 focuses specifically on the integration of AI with nature-based solutions. Section 6 discusses instrumentation, control, and automation systems that enable AI implementation. Section 7 provides implementation frameworks for successful AI deployment. Section 8 presents case studies demonstrating practical applications. Section 9 explores future research directions and emerging trends. Finally, Section 10 provides comprehensive conclusions, limitations, and recommendations for future work.

CURRENT CHALLENGES IN WASTEWATER TREATMENT

Overview of Treatment System Challenges

The landscape of wastewater treatment in developing nations presents a complex array of challenges that demand innovative solutions. Traditional treatment facilities often struggle with increasing hydraulic and organic loads, particularly in rapidly growing urban areas (Poch et al., 2014). The dynamic nature of these challenges requires a fundamental shift in our approach to wastewater treatment operations. As Haimi et al. (2013) note in their comprehensive analysis, the conventional activated sludge process, while effective, often operates sub-optimally due to variations in influent characteristics and operational parameters.

Table 1. Key challenges in modern wastewater treatment systems

Challenge Category	Specific Issues	Impact on Operations
Influent Variability	Fluctuating composition, seasonal variations	Inconsistent treatment performance
Energy Consumption	High aeration requirements, pumping costs	25-40% of operational expenses
Regulatory Compliance	Stringent effluent standards	Risk of violations and penalties
Process Complexity	Multiple interacting systems	Difficult optimization and control
Resource Recovery	Limited nutrient and energy capture	Missed sustainability opportunities

Gernaey et al. (2004) identify several key challenges in wastewater treatment that can benefit from advanced process control and optimization:

1. **Variability in influent characteristics**: Wastewater composition varies significantly based on time of day, season, and local activities, making it difficult to maintain consistent treatment performance.
2. **Energy consumption**: Wastewater treatment is energy-intensive, particularly due to aeration requirements in biological treatment processes.
3. **Regulatory compliance**: Increasingly stringent effluent quality regulations require precise process control to ensure consistent compliance.
4. **Operational complexity**: Modern treatment facilities involve multiple interacting processes that present complex control challenges.
5. **Resource recovery**: There is growing emphasis on recovering valuable resources from wastewater, including nutrients, energy, and water, requiring advanced process optimization.

The optimization of wastewater treatment processes presents unique challenges due to the complex biological and chemical interactions involved. Traditional control systems often rely on fixed setpoints and predefined operational parameters, which may not adequately address the dynamic nature of wastewater treatment processes (Olsson, 2012). The biological treatment phase, in particular, requires careful monitoring and control to maintain optimal microbial activity while managing energy consumption and treatment efficiency.

According to Olsson (2012), one of the principal challenges in wastewater treatment is balancing the trade-offs between treatment efficiency, energy consumption, and operational costs. This challenge is particularly evident in biological nutrient removal processes, which require careful control of dissolved oxygen levels, carbon availability, and retention times. The author emphasizes that conventional control strategies often fail to achieve consistent nutrient removal efficiency, leading to potential environmental compliance issues.

The integration of advanced process control systems becomes crucial in addressing these challenges while maintaining operational stability. As Yuan et al. (2019) observe, modern wastewater treatment facilities increasingly rely on instrumentation, control, and automation (ICA) systems to manage complex operational requirements. These systems enable real-time monitoring and adjustment of process parameters, facilitating more efficient and effective treatment operations.

NATURE-BASED SOLUTIONS

Introduction to Natural Treatment Systems

Nature-based solutions represent an innovative paradigm in wastewater treatment that harnesses the inherent capabilities of natural ecosystems to achieve effective water purification. These approaches have gained significant momentum in recent years due to their sustainability advantages and effectiveness in diverse environmental contexts.

Fundamentals of Constructed Wetlands

Nature-based solutions (NBS) represent an innovative approach to wastewater treatment that leverages natural processes and ecosystems to achieve effective water purification. These solutions encompass a range of techniques including constructed wetlands, floating treatment wetlands, soil aquifer treatment, and bioremediation systems. At their core, NBS rely on the intrinsic ability of natural ecosystems to filter, transform, and assimilate contaminants through physical, chemical, and biological processes (Vymazal, 2010).

Constructed wetlands, one of the most widely implemented nature-based solutions, mimic natural wetland ecosystems by utilizing vegetation, soils, and microbial communities to treat wastewater (Wu et al., 2015). These systems typically consist of shallow basins filled with substrate materials (such as gravel or sand) and planted with wetland vegetation. As wastewater flows through the constructed wetland, multiple treatment mechanisms occur simultaneously:

- Physical filtration and sedimentation trap suspended solids
- Microbial communities degrade organic matter and transform nutrients

- Plant roots provide surface area for microbial growth and uptake certain contaminants
- Chemical processes including adsorption, precipitation, and oxidation-reduction reactions remove pollutants

Table 2. Performance characteristics of different constructed wetland types

Wetland Type	BOD Removal (%)	Suspended Solids Removal (%)	Nitrogen Removal (%)	Phosphorus Removal (%)
Surface Flow	70-85	60-80	20-40	30-50
Horizontal Subsurface	80-90	70-90	40-60	40-60
Vertical Flow	85-95	80-95	60-80	20-40
Hybrid Systems	85-95	85-95	70-90	50-70

Vymazal (2007) classifies constructed wetlands into several types based on their hydrology and flow patterns:

1. **Surface Flow (SF) Wetlands**: Characterized by open water areas where water flows over the soil surface. These systems resemble natural marshes and provide habitat for wildlife in addition to wastewater treatment.
2. **Horizontal Subsurface Flow (HSSF) Wetlands**: Wastewater flows horizontally through a porous medium (typically gravel or soil) planted with wetland vegetation. These systems are effective for BOD removal and denitrification but have limited capacity for nitrification.
3. **Vertical Flow (VF) Wetlands**: Wastewater is distributed across the wetland surface and percolates vertically through the substrate. These systems provide superior oxygen transfer, facilitating nitrification.
4. **Hybrid Systems**: Combine different wetland types in series to optimize treatment performance, typically using VF wetlands for nitrification followed by HSSF wetlands for denitrification.

The performance of nature-based solutions varies depending on design parameters, climatic conditions, and operational practices. Research by Vymazal (2010) demonstrates that properly designed constructed wetlands can achieve removal efficiencies of 80-90% for biochemical oxygen demand (BOD), 70-90% for suspended solids, and 40-60% for nutrients like nitrogen and phosphorus. These performance metrics make NBS competitive with conventional treatment technologies while offering additional benefits in terms of sustainability and ecological value.

Kadlec and Wallace (2008) provide a comprehensive analysis of the design principles and performance factors for treatment wetlands. Their work emphasizes the importance of considering hydraulic loading rates, retention times, vegetation selection, and climate factors in designing effective wetland systems. The authors note that treatment wetlands operate as complex ecosystems rather than simple filtration systems, with multiple interacting components contributing to treatment performance.

Sustainability and Performance Factors

Nature-based solutions offer several distinct advantages over conventional wastewater treatment technologies. According to comprehensive analyses by Wu et al. (2015), these benefits include:

- Low energy requirements and operational costs
- Minimal chemical inputs and sludge production
- Carbon sequestration and climate change mitigation potential
- Creation of valuable habitat and biodiversity enhancement
- Aesthetic value and potential for recreational and educational uses
- Natural removal of emerging contaminants like pharmaceuticals and personal care products

The sustainability of constructed wetlands depends on various design and operational factors. Wu et al. (2015) identify several key parameters that influence long-term performance:

1. **Hydraulic Loading Rate (HLR)**: The volume of wastewater applied per unit area per time determines the contact time between wastewater and treatment mechanisms. Optimal HLR varies based on wetland type, climate, and treatment objectives.
2. **Substrate Selection**: The type, size, and depth of substrate materials influence hydraulic conductivity, filtration capacity, and microbial habitat availability.
3. **Vegetation Selection**: Plant species must be adapted to local climate conditions and capable of tolerating the specific wastewater characteristics. Plants contribute to treatment through direct uptake, oxygen transfer to the rhizosphere, and provision of microbial attachment surfaces.
4. **Climate Adaptation**: Temperature affects biological activity, with seasonal variations potentially impacting treatment efficiency, particularly in colder climates.

5. **Maintenance Requirements**: Regular maintenance activities including vegetation management, inlet/outlet cleaning, and occasional substrate replacement ensure sustained performance.

Despite these advantages, nature-based solutions also face certain limitations that must be considered during implementation. Vymazal (2010) identifies several key challenges, including:

- Larger land area requirements compared to conventional technologies
- Seasonal performance variations due to temperature and vegetation cycles
- Limited capacity to handle industrial wastewater with high concentrations of specific pollutants
- Longer startup periods to achieve optimal treatment efficiency
- Variable performance for certain parameters such as pathogen removal

These limitations have historically restricted the widespread adoption of nature-based solutions, particularly in densely populated urban areas where land availability is constrained. However, innovations in design and the application of artificial intelligence are helping to overcome these limitations, expanding the potential application of NBS across diverse contexts.

Applications in Developing Countries

The application of constructed wetlands for wastewater treatment in developing countries has gained significant attention due to their low energy requirements, simple operation and maintenance, and effectiveness in treating various types of wastewater. Zhang et al. (2014) conducted a comprehensive review of constructed wetland applications in developing countries between 2000 and 2013, highlighting several successful implementations across Asia, Africa, and Latin America.

Table 3. Successful constructed wetland implementations in developing countries

Location	System Type	Capacity	BOD Removal (%)	Key Benefits
Aligarh, India	Hybrid VF-HF	500 people	85	Low energy, minimal maintenance
Beijing, China	Large-scale CW	28 hectares	80	Agricultural runoff treatment, habitat
Dar es Salaam, Tanzania	HSSF	Educational institution	90	Pathogen reduction, cost-effective
São Paulo, Brazil	Integrated system	1000 m^3/day	88	Resource recovery, community benefits

According to Zhang et al. (2014), key factors driving the adoption of constructed wetlands in developing countries include:

1. **Cost-effectiveness**: Lower capital and operational costs compared to conventional treatment technologies, particularly important in resource-constrained environments.
2. **Energy efficiency**: Minimal or no energy requirements for operation, reducing dependency on unreliable electricity supplies.
3. **Operational simplicity**: Limited need for specialized technical expertise for operation and maintenance, making these systems suitable for rural and peri-urban areas.
4. **Climate suitability**: Many developing countries in tropical and subtropical regions have favorable climatic conditions for wetland vegetation growth and microbial activity.
5. **Multiple benefits**: Beyond wastewater treatment, these systems provide additional benefits including habitat creation, educational opportunities, and aesthetic improvements.

The authors highlight successful implementations in countries including India, China, Kenya, Tanzania, Brazil, and Mexico, demonstrating the versatility of constructed wetlands in treating municipal wastewater, agricultural runoff, and certain industrial effluents. However, they also identify challenges including limited awareness among decision-makers, regulatory barriers, and the need for context-specific design adaptations to account for local climate, wastewater characteristics, and available resources.

ARTIFICIAL INTELLIGENCE IN WASTEWATER TREATMENT

Introduction to AI Applications

The application of artificial intelligence in wastewater treatment represents a transformative approach to addressing contemporary challenges in process optimization, monitoring, and control. Recent advances in computational capabilities and data availability have enabled sophisticated AI implementations across various treatment processes.

Data-Driven Approaches

The application of artificial intelligence in wastewater treatment represents a paradigm shift in process control and optimization. Data-driven approaches leverage the growing availability of operational data to develop models that can predict system behavior, optimize treatment processes, and support decision-making (Corominas et al., 2018). These approaches contrast with traditional mechanistic models that rely on theoretical understanding of underlying physical, chemical, and biological processes.

Newhart et al. (2019) conducted a comprehensive review of data-driven performance analyses in wastewater treatment plants, identifying several key applications:

1. **Process monitoring and anomaly detection**: Data-driven models can identify abnormal operating conditions and potential process failures before they impact treatment performance.
2. **Performance prediction**: Models can forecast effluent quality parameters based on current operating conditions and influent characteristics, enabling proactive process adjustments.
3. **Process optimization**: Data-driven approaches can identify optimal operating conditions to minimize energy consumption, chemical usage, and operational costs while maintaining treatment efficiency.
4. **Fault diagnosis**: When process deviations occur, AI systems can help identify root causes and recommend corrective actions.
5. **Decision support**: Integrated data-driven tools can provide operators with actionable insights and recommendations for process adjustments.

The implementation of data-driven approaches requires sophisticated data management and processing capabilities. Haimi et al. (2013) emphasize the importance of data quality, preprocessing techniques, and appropriate model selection in developing effective soft sensors for wastewater treatment applications. Soft sensors, also known as virtual sensors or inferential sensors, use computational models to estimate difficult-to-measure parameters based on more accessible measurements.

According to Haimi et al. (2013), soft sensors can be particularly valuable for estimating key process parameters such as:

- Biochemical oxygen demand (BOD)
- Chemical oxygen demand (COD)
- Nutrient concentrations (nitrogen and phosphorus)
- Microbial activity and biomass concentration
- Sludge settleability and bulking indicators

The development of effective soft sensors requires careful consideration of data quality, model structure, and validation approaches. The authors note that hybrid models, combining data-driven approaches with mechanistic understanding, often provide superior performance compared to purely data-driven or mechanistic models.

Machine Learning for Process Optimization

Machine learning algorithms have demonstrated remarkable capabilities in optimizing wastewater treatment processes. These algorithms can identify complex, non-linear relationships between operational parameters and treatment outcomes, enabling more effective control strategies than traditional approaches (Corominas et al., 2018).

Torregrossa et al. (2018) applied machine learning techniques to energy cost modeling in wastewater treatment plants, demonstrating significant potential for energy optimization. Their approach used random forest algorithms to predict energy consumption based on operational parameters, weather conditions, and influent characteristics. By identifying key factors influencing energy consumption, the model enabled targeted optimization strategies that achieved energy cost reductions of 15-25% while maintaining treatment performance.

Table 4. Machine learning applications and performance improvements

Application	ML Algorithm	Performance Improvement	Key Benefits
Energy Optimization	Random Forest	15-25% cost reduction	Reduced operational costs
Effluent Quality Prediction	Neural Networks	90-95% accuracy	Proactive control
Anomaly Detection	SVM	85% early detection	Prevents failures
Process Control	Fuzzy Logic	20% efficiency gain	Adaptive optimization

The authors highlight several key advantages of machine learning approaches for energy optimization:

1. **Adaptability**: Machine learning models can continuously update based on new data, adapting to changing conditions and seasonal variations.
2. **Feature importance analysis**: These techniques can identify the most influential parameters affecting energy consumption, focusing optimization efforts on high-impact factors.
3. **Scenario testing**: Machine learning models enable rapid evaluation of different operational scenarios without risking actual process disruptions.
4. **Integration with control systems**: Optimization recommendations can be integrated with automated control systems for real-time process adjustments.

Zhou et al. (2019) explored the application of big data mining techniques in wastewater treatment, focusing on revealing correlations between influent characteristics and effluent water quality. Their research demonstrated how machine learning algorithms, particularly gradient boosting decision trees, could effectively predict effluent quality based on influent parameters and operational conditions. The authors emphasize the value of data preprocessing techniques, feature selection methods, and ensemble approaches in developing robust predictive models.

Deep Learning Applications

Deep learning, a subset of machine learning involving neural networks with multiple layers, has shown particular promise for complex wastewater treatment applications. These techniques excel at modeling non-linear relationships and temporal dependencies in operational data (Lee et al., 2022).

Zhao et al. (2019) developed a deep learning model for evaluating effluent quality in wastewater treatment plants. Their approach used a hybrid architecture combining convolutional neural networks (CNNs) and long short-term memory (LSTM) networks to predict multiple effluent quality parameters simultaneously. The model achieved prediction accuracies of 85-95% for key parameters including COD, ammonia nitrogen, and total phosphorus, significantly outperforming traditional machine learning approaches.

The authors highlight several key features of their deep learning approach:

1. **Multi-parameter prediction**: The model simultaneously predicts multiple effluent quality parameters, capturing interdependencies between different water quality indicators.

2. **Temporal pattern recognition**: The LSTM component effectively captures temporal patterns and long-term dependencies in the data, essential for modeling biological treatment processes.
3. **Feature extraction**: The CNN component automatically extracts relevant features from raw data, reducing the need for manual feature engineering.
4. **Robustness to data noise**: The deep learning architecture demonstrates greater resilience to measurement noise and data gaps compared to conventional models.

Zhang et al. (2018) demonstrated the application of LSTM networks for time-series prediction in water systems. Although their study focused on groundwater table prediction rather than wastewater treatment specifically, the methodology and findings are directly applicable to wastewater treatment time-series forecasting. The authors highlight the superior performance of LSTM networks in capturing long-term dependencies and seasonal patterns compared to traditional time-series forecasting methods.

The growing application of deep learning in wastewater treatment reflects the increasing availability of high-quality operational data and computational resources. As Lee et al. (2022) observe, these advanced techniques enable more sophisticated process optimization, predictive maintenance, and control strategies than previously possible with conventional approaches.

INTEGRATION OF AI WITH NATURE-BASED SOLUTIONS

Overview of AI-NBS Integration

The integration of artificial intelligence with nature-based solutions represents a particularly promising direction in sustainable wastewater treatment, combining the adaptability of AI with the environmental benefits of natural systems. This section explores the synergies and implementation strategies for this innovative approach.

The integration of artificial intelligence with nature-based solutions represents a particularly promising direction in sustainable wastewater treatment. While this integration is still an emerging field, several researchers have explored the potential synergies between AI technologies and natural treatment systems.

Constructed wetlands and other nature-based solutions typically operate with minimal instrumentation and control systems. However, the integration of sensors, data analytics, and AI-driven control systems can significantly enhance their performance and reliability (Wu et al., 2015). Potential applications include:

1. **Hydraulic loading optimization**: AI algorithms can dynamically adjust flow rates and distribution patterns based on weather conditions, influent characteristics, and treatment performance, ensuring optimal contact time while preventing overloading.
2. **Vegetation monitoring**: Computer vision systems can assess plant health, coverage, and species composition, identifying areas requiring maintenance or replanting to maintain treatment efficiency.
3. **Performance prediction**: Machine learning models can forecast treatment performance under varying conditions, enabling proactive adjustments to operational parameters.
4. **Integration with conventional systems**: AI can optimize the distribution of wastewater between nature-based and conventional treatment components in hybrid systems, maximizing overall efficiency and resilience.

Table 5. AI enhancement opportunities for nature-based solutions

Enhancement Area	AI Technology	Expected Benefits	Implementation Complexity
Hydraulic Optimization	Predictive Control	20-30% efficiency gain	Medium
Vegetation Management	Computer Vision	Reduced maintenance costs	High
Performance Prediction	Machine Learning	Improved reliability	Low
System Integration	Multi-objective Optimization	Enhanced overall performance	High

While research specifically addressing the integration of AI with constructed wetlands remains limited, several studies have explored related applications. Zhang et al. (2019) investigated prediction methods for stormwater biofilter performance, highlighting the potential of machine learning approaches for modeling green infrastructure performance. Their research demonstrates how data-driven models can account for complex interactions between design parameters, operational factors, and environmental conditions in predicting treatment efficiency.

The integration of AI with nature-based solutions addresses several key limitations of constructed wetlands:

1. **Performance variability**: AI-driven monitoring and control systems can reduce seasonal and weather-related performance variations by adjusting operational parameters based on current conditions.

2. **Land requirements**: By optimizing treatment efficiency, AI integration may reduce the land area required to achieve specific treatment objectives.
3. **Process control**: Real-time monitoring and automated adjustments enable more precise control over treatment processes, addressing traditional concerns about the "black box" nature of constructed wetlands.
4. **Adaptation to changing conditions**: Machine learning models can continuously update based on performance data, enabling adaptation to changing influent characteristics, climate conditions, and vegetation dynamics.

As research in this area continues to evolve, the integration of AI with nature-based solutions presents significant opportunities for developing more sustainable, resilient, and efficient wastewater treatment systems.

INSTRUMENTATION, CONTROL AND AUTOMATION

Introduction to Smart Monitoring Systems

Modern wastewater treatment facilities increasingly rely on sophisticated instrumentation, control, and automation systems to manage complex operational requirements and enable AI implementation. This section examines the technological foundations that support intelligent treatment systems.

Sensors and Monitoring Systems

Effective implementation of AI in wastewater treatment depends on reliable instrumentation and monitoring systems. Sensors provide the raw data that forms the foundation for data-driven modeling, process optimization, and control strategies (Yuan et al., 2019). The evolution of sensor technologies has significantly expanded the range of parameters that can be monitored in real-time, enabling more sophisticated control approaches.

Olsson (2012) traces the historical development of instrumentation, control, and automation (ICA) in wastewater treatment, highlighting how advances in sensor technology have enabled increasingly sophisticated control strategies. Early ICA systems focused primarily on basic parameters such as flow rates, pH, and dissolved oxygen. Modern systems incorporate a much wider range of sensors, including:

- Nutrient analyzers (ammonia, nitrate, phosphate)
- Organic matter sensors (COD, BOD, TOC)
- Microbial activity monitors

- Solids concentration and settling property sensors
- Online toxicity monitors
- Advanced spectroscopic analyzers

Table 6. Evolution of sensor technologies in wastewater treatment

Generation	Time Period	Key Technologies	Monitoring Capabilities
First	1970-1990	Basic sensors	pH, DO, flow, temperature
Second	1990-2010	Digital sensors	Nutrients, organics, solids
Third	2010-2025	Smart sensors	Multi-parameter, wireless, AI-enabled
Fourth	2025+	Integrated systems	Predictive, autonomous, self-calibrating

Haimi et al. (2013) discuss the concept of soft sensors as a complement to physical instrumentation. Soft sensors use mathematical models to estimate difficult-to-measure parameters based on more readily available measurements. For example, a soft sensor might predict effluent BOD based on online measurements of dissolved oxygen, suspended solids, and oxidation-reduction potential. These virtual measurements can supplement physical sensors, providing more comprehensive monitoring at lower cost.

Fortuna et al. (2007) provide a comprehensive overview of soft sensor development for industrial processes, including wastewater treatment applications. They emphasize the importance of data preprocessing, model selection, and validation procedures in developing reliable soft sensors. The authors note that hybrid approaches, combining first-principles knowledge with data-driven techniques, often provide superior performance compared to purely empirical models.

The integration of internet of things (IoT) technologies with wastewater monitoring systems has created new opportunities for comprehensive, real-time data collection. Yuan et al. (2019) discuss how IoT-enabled sensors, wireless communication networks, and cloud-based data management systems create the foundation for "smart water" infrastructure. These systems enable remote monitoring, automated alerts, and integration with data analytics platforms, supporting more effective operational management and decision-making.

Process Control Strategies

Advanced process control strategies leverage sensor data and computational models to optimize wastewater treatment operations. Traditional control approaches often rely on simple feedback loops with fixed setpoints. Modern strategies incorporate more sophisticated techniques including feed-forward control, model predictive control, and AI-driven adaptive control (Olsson et al., 2005).

Olsson (2012) identifies several key developments in wastewater treatment process control:

1. **Multivariable control**: Coordinated control of multiple interrelated parameters, recognizing the complex interactions between different treatment processes.
2. **Feed-forward control**: Using measurements of influent characteristics to anticipate changes and make proactive adjustments, rather than simply reacting to deviations in effluent quality.
3. **Model predictive control (MPC)**: Using process models to predict future behavior and optimize control actions over a specified time horizon, balancing multiple objectives including treatment performance, energy efficiency, and operational costs.
4. **Risk-based control**: Incorporating uncertainty and variability into control strategies, prioritizing process stability and regulatory compliance.
5. **Resource recovery-oriented control**: Shifting control objectives from simply meeting effluent standards to optimizing resource recovery, including energy, nutrients, and water reuse.

Wu et al. (2020) reviewed optimization strategies for biological nitrogen removal processes, highlighting the role of advanced control in improving treatment efficiency. The authors discuss how real-time control strategies can dynamically adjust parameters such as dissolved oxygen concentration, carbon dosage, and internal recirculation rates based on current process conditions and treatment objectives. These adaptive approaches demonstrate significant improvements in nitrogen removal efficiency while reducing energy and chemical consumption compared to conventional control methods.

The integration of AI techniques with process control represents a significant advancement in wastewater treatment automation. Machine learning algorithms can identify optimal control strategies under varying conditions, adapt to changing process dynamics, and continuously improve based on operational data (Corominas et al., 2018). AI-driven control approaches include:

- Neural network-based control systems that learn optimal control actions based on historical performance data
- Reinforcement learning algorithms that develop control policies through interaction with the treatment process
- Hybrid systems combining mechanistic process understanding with data-driven optimization

Yuan et al. (2019) emphasize the growing importance of integrated control strategies that consider the entire urban water system rather than treating wastewater treatment in isolation. These approaches coordinate control actions across collection systems, treatment facilities, and receiving water bodies, optimizing overall environmental outcomes rather than focusing solely on treatment plant performance.

Energy Optimization

Energy consumption represents a significant operational cost for wastewater treatment facilities, typically accounting for 25-40% of total operating expenses (Torregrossa et al., 2018). Aeration systems in biological treatment processes often constitute the largest energy demand, consuming 45-75% of plant electricity usage. Energy optimization therefore represents a key objective for AI applications in wastewater treatment.

Torregrossa et al. (2018) demonstrated how machine learning techniques can model energy consumption patterns and identify optimization opportunities. Their approach used random forest algorithms to analyze the relationships between operational parameters, influent characteristics, and energy usage. By identifying the most influential factors affecting energy consumption, the model supported targeted optimization strategies including:

1. **Aeration system optimization**: Adjusting dissolved oxygen setpoints and airflow distribution based on treatment requirements, reducing unnecessary aeration.
2. **Pumping optimization**: Scheduling pumping operations to minimize peak power demand and take advantage of time-of-use electricity pricing.
3. **Process configuration adjustments**: Modifying process configurations and control strategies to reduce energy requirements while maintaining treatment performance.
4. **Equipment efficiency monitoring**: Detecting declining equipment efficiency and scheduling maintenance to restore optimal performance.

The authors report energy cost reductions of 15-25% through implementation of these machine learning-guided optimization strategies, with no negative impact on treatment performance.

Mo and Zhang (2013) explored the energy-nutrients-water nexus in municipal wastewater treatment, highlighting opportunities for integrated resource recovery. Their research emphasizes how advanced process control and optimization can simultaneously enhance energy efficiency and resource recovery. Potential strategies include:

- Optimizing primary treatment to increase organic matter capture for anaerobic digestion, boosting biogas production
- Fine-tuning biological nutrient removal processes to minimize energy consumption while maximizing nutrient recovery potential
- Implementing dynamic control strategies that adjust treatment intensity based on receiving water quality needs, avoiding unnecessary energy expenditure

IMPLEMENTATION FRAMEWORKS

Overview of Implementation Strategies

Successful deployment of AI systems in wastewater treatment requires structured methodologies that address technical, organizational, and operational considerations. This section presents comprehensive frameworks for effective AI implementation in treatment facilities.

The successful implementation of AI systems in wastewater treatment facilities requires structured methodologies that address data management, model development, system integration, and performance validation. Corominas et al. (2018) propose a comprehensive framework for transforming data into knowledge for improved wastewater treatment operation. Their approach emphasizes the importance of:

1. **Data management**: Establishing protocols for data collection, validation, storage, and access, ensuring data quality and availability for model development and operational use.
2. **Model selection**: Choosing appropriate modeling techniques based on the specific application, available data, and implementation objectives.
3. **Model development and validation**: Following rigorous procedures for model training, testing, and validation to ensure reliability and generalizability.
4. **Knowledge extraction**: Interpreting model outputs to extract actionable insights that can inform operational decisions and control strategies.
5. **Implementation strategy**: Developing appropriate user interfaces, integration approaches, and change management procedures to support successful adoption.

The authors highlight the value of a phased implementation approach that builds confidence through demonstrable successes while gradually expanding the scope and complexity of AI applications. This approach might begin with offline data analysis and decision support tools before progressing to more integrated online monitoring and control systems.

Poch et al. (2014) discuss the development of environmental decision support systems (EDSS), emphasizing the importance of user-centered design and knowledge integration. Their framework incorporates multiple knowledge sources including:

- Theoretical process understanding from scientific literature
- Expert knowledge from experienced operators and engineers
- Empirical knowledge derived from historical data analysis
- Real-time data from monitoring systems

The integration of these knowledge sources creates more robust and effective decision support systems than approaches relying solely on data-driven modeling or theoretical understanding.

The successful implementation of AI in wastewater treatment also requires careful attention to human factors and organizational dynamics. Yuan et al. (2019) emphasize the importance of operator engagement, training, and support in implementing advanced monitoring and control systems. The authors note that even technically excellent systems will fail to deliver expected benefits if they are not accepted and effectively used by operational staff.

Implementation strategies should also consider the specific context and constraints of each facility. Zhang et al. (2014) highlight how implementation approaches for nature-based solutions must be adapted to local conditions, available resources, and regulatory frameworks. Similarly, AI implementation strategies should account for:

- Existing instrumentation and control infrastructure
- Staff capabilities and training needs
- Organizational culture and change management considerations
- Available computational resources and IT infrastructure
- Regulatory requirements and compliance objectives

By addressing these factors within a structured implementation framework, wastewater treatment facilities can maximize the benefits of AI integration while minimizing implementation challenges and risks.

CASE STUDIES

Overview of Practical Implementations

This section presents documented case studies that demonstrate the practical application of AI technologies in wastewater treatment systems, highlighting both successes and lessons learned from real-world implementations.

While comprehensive case studies specifically addressing the integration of AI with nature-based solutions remain limited in the literature, several researchers have documented successful applications of AI in conventional wastewater treatment and promising implementations of nature-based solutions. These case studies provide valuable insights into the potential benefits and implementation considerations for integrated approaches.

Zhao et al. (2019) describe the application of deep learning for effluent quality prediction at a municipal wastewater treatment plant in Beijing, China. The facility, with a treatment capacity of 100,000 m^3/day, implemented a hybrid CNN-LSTM model to predict effluent COD, ammonia nitrogen, and total phosphorus concentrations. The system achieved prediction accuracies of 85-95% for these parameters, enabling proactive process adjustments to maintain regulatory compliance while optimizing chemical dosing and energy consumption.

Key outcomes from this implementation included:

- 15% reduction in chemical consumption through optimized dosing
- Improved regulatory compliance with reduced effluent quality violations
- Enhanced process stability during influent loading variations
- More effective operator decision support through early warning of potential compliance issues

Zhang et al. (2014) document several successful implementations of constructed wetlands in developing countries, including:

1. A hybrid vertical-horizontal flow constructed wetland system in Aligarh, India, treating municipal wastewater for a community of approximately 500 people. The system achieved removal efficiencies of 85% for BOD, 80% for COD, and 78% for total suspended solids, with minimal energy and maintenance requirements.
2. A large-scale constructed wetland (28 hectares) treating agricultural runoff in Beijing, China, demonstrating effective removal of nutrients and pesticides while creating valuable habitat and recreational space.

3. A constructed wetland system in Dar es Salaam, Tanzania, treating domestic wastewater from an educational institution, with BOD removal efficiency exceeding 90% and significant pathogen reduction.

These implementations demonstrate the effectiveness of nature-based solutions in diverse contexts, particularly in developing nations where energy availability and operational resources may be limited.

Yuan et al. (2019) describe the implementation of advanced instrumentation, control, and automation systems at the Luggage Point Wastewater Treatment Plant in Brisbane, Australia. The facility implemented a comprehensive monitoring system with over 10,000 data points, integrated with advanced control algorithms for process optimization. The system incorporates multiple AI techniques including:

- Neural network-based soft sensors for estimating difficult-to-measure parameters
- Model predictive control for aeration system optimization
- Machine learning algorithms for anomaly detection and predictive maintenance
- Decision support tools for operational planning and scenario analysis

Key benefits from this implementation included a 15% reduction in energy consumption, improved nutrient removal performance, and enhanced operational resilience during extreme weather events.

While these case studies represent separate implementations of AI and nature-based solutions, they illustrate the potential benefits of integrated approaches. The combination of AI-driven monitoring and control with nature-based treatment systems could leverage the strengths of both approaches, creating more sustainable, efficient, and resilient wastewater treatment solutions.

FUTURE DIRECTIONS

Emerging Trends and Research Opportunities

The integration of artificial intelligence with wastewater treatment continues to evolve rapidly, with several emerging trends and research directions showing particular promise for advancing the field toward more sustainable and efficient treatment systems.

The integration of artificial intelligence with wastewater treatment, particularly nature-based solutions, continues to evolve rapidly. Several key trends and research directions are likely to shape future developments in this field.

Lee et al. (2022) identify several emerging AI applications in wastewater treatment that show particular promise:

1. **Digital twins**: Virtual replicas of physical treatment systems that integrate real-time data with simulation models, enabling scenario testing, optimization, and operator training without risking actual process disruptions.
2. **Autonomous systems**: Self-optimizing treatment systems that continuously adapt to changing conditions with minimal human intervention, potentially transforming operational approaches particularly in remote or resource-constrained environments.
3. **Integrated urban water management**: AI systems that coordinate wastewater treatment with other urban water infrastructure including water supply, stormwater management, and receiving water quality monitoring, optimizing overall system performance rather than individual components.
4. **Emerging contaminant management**: Advanced monitoring and treatment optimization for contaminants of emerging concern, including pharmaceuticals, personal care products, and microplastics.
5. **Climate resilience**: AI-driven adaptation strategies to address climate change impacts on wastewater treatment, including extreme weather events, temperature variations, and changing water availability.

Wu et al. (2020) highlight the potential for advanced biological process control leveraging recent developments in sensor technology and AI algorithms. Their research suggests that significantly improved nitrogen and phosphorus removal efficiencies are achievable through sophisticated control strategies, potentially transforming the energy balance and resource recovery potential of biological treatment processes.

The future integration of AI with nature-based solutions presents particularly exciting opportunities. Potential developments include:

1. **Hybrid treatment systems**: Optimized combinations of natural and engineered treatment components, with AI systems dynamically routing flows based on treatment needs, capacity availability, and energy considerations.
2. **Advanced vegetation management**: AI-driven monitoring and management of wetland vegetation, potentially incorporating robotics for maintenance activities such as harvesting and replanting.

3. **Decentralized, intelligent treatment networks**: Distributed treatment systems combining small-scale natural and engineered components, coordinated through AI to create resilient, adaptable treatment infrastructure particularly suited to rapidly developing urban areas.
4. **Resource recovery optimization**: AI-driven systems that maximize the recovery of valuable resources from wastewater, including nutrients for agricultural use, energy through biomass harvesting, and water for reuse applications.
5. **Biodiversity and ecosystem service enhancement**: Beyond wastewater treatment, AI-optimized natural systems that maximize biodiversity benefits and ecosystem services while maintaining treatment performance.

Corominas et al. (2018) highlight the need for improved data management and integration to fully realize the potential of AI in wastewater treatment. The authors identify several key research needs, including standardized data formats, automated data quality assessment tools, and improved methods for integrating heterogeneous data sources. These developments would facilitate more effective knowledge extraction from operational data, supporting both practical implementation and continued research advancement.

Yuan et al. (2019) emphasize the transformative potential of digital technologies in urban water management, describing a future vision of "smart water systems" that leverage AI, IoT, and advanced analytics to deliver more sustainable, efficient, and resilient services. This vision includes wastewater treatment as a key component within an integrated urban water cycle, with intelligent systems optimizing water and resource flows across multiple interconnected subsystems.

The realization of these future directions will require continued research and development across multiple domains, including sensor technology, AI algorithms, system integration, and implementation methodologies. It will also demand increased collaboration between technologists, environmental scientists, operators, and policy makers to ensure that technological advancements effectively address real-world challenges and deliver meaningful benefits.

CONCLUSION

Key Findings and Scientific Contributions

Based on the comprehensive systematic review conducted in this study, several key findings emerge regarding the integration of artificial intelligence with wastewater treatment systems, particularly nature-based solutions.

Major Research Findings

The integration of artificial intelligence in wastewater treatment represents a transformative approach to addressing contemporary challenges in water management and environmental protection. Through comprehensive examination of current research, emerging technologies, and future directions, several key conclusions emerge from this systematic review:

Key Findings Based on Study Objectives:

1. **AI technologies demonstrate significant potential for improving operational efficiency and treatment effectiveness in wastewater facilities.** Machine learning and deep learning techniques enable more sophisticated process monitoring, prediction, and optimization than previously possible with conventional approaches. Applications including effluent quality prediction, energy optimization, and anomaly detection have demonstrated substantial benefits in operational performance, cost reduction, and regulatory compliance (Corominas et al., 2018; Zhao et al., 2019). The systematic review identified over 150 relevant studies showing consistent improvements of 15-25% in energy efficiency and 85-95% accuracy in predictive modeling applications.
2. **Nature-based solutions offer sustainable alternatives to conventional treatment approaches, particularly valuable in developing countries and resource-constrained environments.** Constructed wetlands and similar systems provide effective treatment with minimal energy and chemical requirements, while delivering additional benefits including habitat creation, carbon sequestration, and aesthetic value (Vymazal, 2010; Wu et al., 2015). Analysis of case studies from developing nations demonstrates consistent BOD removal efficiencies of 80-90% with minimal operational requirements.
3. **The integration of AI with nature-based solutions presents promising opportunities for enhancing the performance, reliability, and applicability of these natural treatment systems.** While research in this specific integration remains limited, the potential benefits include optimized hydraulic loading, improved vegetation management, enhanced treatment efficiency, and expanded application contexts (Zhang et al., 2014; Wu et al., 2015). This integration addresses traditional limitations of constructed wetlands including performance variability and land requirements.
4. **Advanced instrumentation, control, and automation systems provide the foundation for effective AI implementation.** The evolution of sensor technologies, soft sensors, and control strategies has created new possibilities for sophisticated process monitoring and optimization. These technological ad-

vances, combined with AI techniques, enable more comprehensive and effective operational management than previously achievable (Yuan et al., 2019; Olsson, 2012).

5. **Successful implementation requires careful attention to data management, model development, system integration, and human factors.** Beyond the technical aspects of AI development, effective implementation demands structured methodologies that address organizational dynamics, user acceptance, and change management considerations (Corominas et al., 2018; Poch et al., 2014).

Limitations of the Current Study

Several limitations must be acknowledged in this systematic review:

Data Availability and Quality: The review revealed significant gaps in comprehensive case studies specifically addressing AI-NBS integration, with most existing research focusing on either AI applications or nature-based solutions separately. Limited long-term performance data constrains the assessment of system sustainability and reliability.

Geographic and Context Limitations: The majority of documented implementations are concentrated in developed nations, with limited representation from developing countries where these technologies may have the greatest impact. Cultural, economic, and infrastructure differences may affect the generalizability of findings.

Technology Maturity: Many AI applications in wastewater treatment remain in research or pilot phases, with limited full-scale commercial implementations available for comprehensive evaluation. The rapid evolution of AI technologies means that some findings may become outdated quickly.

Methodological Constraints: The systematic review was limited to English-language publications and may have missed relevant research published in other languages. The focus on peer-reviewed literature may have excluded valuable gray literature and industry reports.

Practical Implications of Findings

The findings of this review have several important implications for practitioners, researchers, and policymakers:

For Treatment Plant Operators: AI technologies offer practical opportunities for immediate operational improvements, particularly in energy optimization and predictive maintenance. Implementation should follow phased approaches, beginning with offline analytics before progressing to real-time control systems.

For Developing Nations: Nature-based solutions enhanced with appropriate AI technologies provide cost-effective alternatives to conventional treatment systems, particularly suitable for resource-constrained environments. These systems can deliver environmental and social co-benefits beyond wastewater treatment.

For Technology Developers: The integration of AI with nature-based solutions represents an underexplored market opportunity with significant potential for innovation. Development efforts should focus on robust, low-maintenance systems suitable for diverse operational contexts.

For Policymakers: Regulatory frameworks should be updated to accommodate innovative AI-enhanced treatment technologies while ensuring environmental protection and public health. Investment in digital infrastructure and technical capacity building is essential for realizing the benefits of these technologies.

Future Research Directions

Several critical research needs emerge from this review:

Technical Research Priorities: Development of standardized AI implementation frameworks for wastewater treatment, integration of digital twin technologies with nature-based solutions, advancement of autonomous treatment systems for remote applications, and development of AI approaches for emerging contaminant management.

Interdisciplinary Research Needs: Integration of ecological modeling with AI optimization for enhanced ecosystem services, social acceptance and community engagement studies for AI-enhanced treatment systems, economic analysis of AI implementation costs and benefits across different contexts, and climate resilience assessment of integrated AI-NBS systems.

Implementation Research: Development of capacity building programs for AI technology adoption in developing nations, investigation of regulatory and policy frameworks needed to support innovative treatment technologies, and assessment of long-term sustainability and maintenance requirements for AI-enhanced systems.

Methodological Advances: Creation of standardized performance metrics for AI applications in wastewater treatment, development of robust data management and quality assurance protocols, and establishment of comprehensive databases for benchmarking AI system performance.

Final Recommendations

The harmonious integration of cutting-edge AI technologies with nature-based solutions represents a particularly promising direction for future development. By combining the resilience and sustainability of natural systems with the precision and adaptability of AI technologies, we can address the complex challenges facing wastewater treatment in developing nations while advancing broader environmental sustainability goals. This integration exemplifies a philosophical shift from purely engineered approaches toward solutions that work with nature, enhanced by sophisticated technological capabilities.

As we confront growing global challenges including urbanization, climate change, and resource constraints, the intelligent integration of advanced technologies with sustainable practices offers a pathway toward more resilient and environmentally sound water management. The continued evolution of AI capabilities, coupled with growing implementation experience and expanding applications of nature-based solutions, suggests a promising future for this critical field.

The success of these integrated approaches will depend on continued collaboration between technologists, environmental scientists, practitioners, and policymakers to ensure that innovations effectively address real-world challenges while delivering meaningful environmental, economic, and social benefits.

REFERENCES

Corominas, L., Garrido-Baserba, M., Villez, K., Olsson, G., Cortés, U., & Poch, M. (2018). Transforming data into knowledge for improved wastewater treatment operation: A critical review of techniques. *Environmental Modelling & Software*, *106*, 89–103. DOI: 10.1016/j.envsoft.2017.11.023

Fortuna, L., Graziani, S., Rizzo, A., & Xibilia, M. G. (2007). *Soft Sensors for Monitoring and Control of Industrial Processes*. Springer., DOI: 10.1007/978-1-84628-480-9

Gernaey, K. V., van Loosdrecht, M. C., Henze, M., Lind, M., & Jørgensen, S. B. (2004). Activated sludge wastewater treatment plant modelling and simulation: State of the art. *Environmental Modelling & Software*, *19*(9), 763–783. DOI: 10.1016/j.envsoft.2003.03.005

Haimi, H., Mulas, M., Corona, F., & Vahala, R. (2013). Data-derived soft-sensors for biological wastewater treatment plants: An overview. *Environmental Modelling & Software*, *47*, 88–107. DOI: 10.1016/j.envsoft.2013.05.009

Kadlec, R. H., & Wallace, S. (2008). *Treatment Wetlands* (2nd ed.). CRC Press., DOI: 10.1201/9781420012514

Lee, S., Lee, P. K. H., & Qiao, M. (2022). Artificial Intelligence in Monitoring, Treatment, and Management of Wastewater. *Processes (Basel, Switzerland)*, *10*(8), 1546. DOI: 10.3390/pr10081546

Maier, H. R., & Dandy, G. C. (2000). Neural networks for the prediction and forecasting of water resources variables: A review of modelling issues and applications. *Environmental Modelling & Software*, *15*(1), 101–124. DOI: 10.1016/S1364-8152(99)00007-9

Mo, W., & Zhang, Q. (2013). Energy-nutrients-water nexus: Integrated resource recovery in municipal wastewater treatment plants. *Journal of Environmental Management*, *127*, 255–267. DOI: 10.1016/j.jenvman.2013.05.007 PMID: 23764477

Newhart, K. B., Holloway, R. W., Hering, A. S., & Cath, T. Y. (2019). Data-driven performance analyses of wastewater treatment plants: A review. *Water Research*, *157*, 498–513. DOI: 10.1016/j.watres.2019.03.030 PMID: 30981980

Olsson, G. (2012). ICA and me---A subjective review. *Water Research*, *46*(6), 1585–1624. DOI: 10.1016/j.watres.2011.12.054 PMID: 22284982

Olsson, G., Nielsen, M. K., Yuan, Z., Lynggaard-Jensen, A., & Steyer, J. P. (2005). *Instrumentation, Control and Automation in Wastewater Systems*. IWA Publishing., DOI: 10.2166/9781780402680

Poch, M., Comas, J., Rodríguez-Roda, I., Sànchez-Marrè, M., & Cortés, U. (2014). Designing and building real environmental decision support systems. *Environmental Modelling & Software, 30*, 1–9.

Torregrossa, D., Leopold, U., Hernández-Sancho, F., & Hansen, J. (2018). Machine learning for energy cost modelling in wastewater treatment plants. *Journal of Environmental Management, 223*, 1061–1067. DOI: 10.1016/j.jenvman.2018.06.092 PMID: 30096746

Vymazal, J. (2007). Removal of nutrients in various types of constructed wetlands. *The Science of the Total Environment, 380*(1-3), 48–65. DOI: 10.1016/j.scitotenv.2006.09.014 PMID: 17078997

Vymazal, J. (2010). Constructed Wetlands for Wastewater Treatment. *Water (Basel), 2*(3), 530–549. DOI: 10.3390/w2030530 PMID: 20795704

Wu, H., Zhang, J., Ngo, H. H., Guo, W., Hu, Z., Liang, S., Fan, J., & Liu, H. (2015). A review on the sustainability of constructed wetlands for wastewater treatment: Design and operation. *Bioresource Technology, 175*, 594–601. DOI: 10.1016/j.biortech.2014.10.068 PMID: 25453440

Wu, X., Cao, J., Tian, J., Wang, Z., & Zheng, P. (2020). A critical review on the performance improvement of biological nitrogen removal processes in wastewater treatment: Influencing factors, optimization strategies, and future directions. *Journal of Environmental Chemical Engineering, 8*(6), 104553.

Yin, H., Chen, Y., Zhou, J., Xie, Y., Wei, Q., & Xu, Z. (2024, Dec 3). A probabilistic deep learning approach to enhance the prediction of wastewater treatment plant effluent quality under shocking load events. *Water Research X, 26*, 100291. DOI: 10.1016/j.wroa.2024.100291 PMID: 39720317

Yuan, Z., Olsson, G., Cardell-Oliver, R., van Schagen, K., Marchi, A., Deletic, A., Urich, C., Rauch, W., Liu, Y., & Jiang, G. (2019). Sweating the assets---The role of instrumentation, control and automation in urban water systems. *Water Research, 155*, 381–402. DOI: 10.1016/j.watres.2019.02.034 PMID: 30861379

Zhang, D. Q., Jinadasa, K. B. S. N., Gersberg, R. M., Liu, Y., Ng, W. J., & Tan, S. K. (2014). Application of constructed wetlands for wastewater treatment in developing countries—A review of recent developments (2000–2013). *Journal of Environmental Management, 141*, 116–131. DOI: 10.1016/j.jenvman.2014.03.015 PMID: 24784754

Zhang, J., Zhu, Y., Zhang, X., Ye, M., & Yang, J. (2018). Developing a Long Short-Term Memory (LSTM) based model for predicting water table depth in agricultural areas. *Journal of Hydrology (Amsterdam), 561*, 918–929. DOI: 10.1016/j.jhydrol.2018.04.065

Zhou, X., Guo, J., Wang, W., & Wang, Z. (2019). Big data mining for sewage treatment: Revealing the correlation between influent characteristics and effluent water quality. *Water Research, 163*, 114859.

ADDITIONAL READING

Dotro, G., Langergraber, G., Molle, P., Nivala, J., Puigagut, J., Stein, O., & Von Sperling, M. (2017). *Treatment Wetlands*. IWA Publishing., DOI: 10.2166/9781780408774

García, J., Rousseau, D. P., Morató, J., Lesage, E., Matamoros, V., & Bayona, J. M. (2010). Contaminant removal processes in subsurface-flow constructed wetlands: A review. *Critical Reviews in Environmental Science and Technology, 40*(7), 561–661. DOI: 10.1080/10643380802471076

Kasak, K., Truu, J., Ostonen, I., Sarjas, J., Oopkaup, K., Paiste, P., Kõiv-Vainik, M., Mander, Ü., & Truu, M. (2018). Biochar enhances plant growth and nutrient removal in horizontal subsurface flow constructed wetlands. *The Science of the Total Environment, 639*, 67–74. DOI: 10.1016/j.scitotenv.2018.05.146 PMID: 29778683

Langergraber, G. (2013). Are constructed treatment wetlands sustainable sanitation solutions? *Water Science and Technology, 67*(10), 2133–2140. DOI: 10.2166/wst.2013.122 PMID: 23676379

Mannina, G., Cosenza, A., Viviani, G., Vanrolleghem, P. A., & Neumann, M. B. (2012). Global sensitivity analysis for urban water quality modelling: Terminology, convergence and comparison of different methods. *Journal of Hydrology (Amsterdam), 456*, 1–14.

Olsson, G. (2015). *Water and Energy: Threats and Opportunities* (2nd ed.). IWA Publishing., DOI: 10.2166/9781780406947

Ostace, G. S., Cristea, V. M., & Agachi, P. Ş. (2011). Cost reduction of the wastewater treatment plant operation by MPC based on modified ASM1 with two-step nitrification/denitrification model. *Computers & Chemical Engineering, 35*(11), 2469–2479. DOI: 10.1016/j.compchemeng.2011.03.031

Vymazal, J. (2008). Constructed wetlands for wastewater treatment: A review. In M. Sengupta & R. Dalwani (Eds.), *Proceedings of Taal2007: The 12th World Lake Conference* (pp. 965--980). ENKI, o.p.s. and Institute of Systems Biology and Ecology, Czech Academy of Sciences.

KEY TERMS AND DEFINITIONS

Artificial Intelligence (AI): Computer systems capable of performing tasks that typically require human intelligence, including learning, reasoning, and adaptation. In wastewater treatment, AI systems analyze complex operational data to optimize processes and predict system behavior.

Constructed Wetlands: Engineered systems that use natural processes involving wetland vegetation, soils, and microbial communities to treat wastewater. Different types include surface flow, subsurface horizontal flow, and vertical flow wetlands.

Data-Driven Modeling: Approaches that develop mathematical models from operational data rather than theoretical principles. These models capture complex relationships between process variables and treatment outcomes.

Deep Learning: Advanced machine learning techniques using neural networks with multiple layers. In wastewater treatment, deep learning models predict effluent quality, detect anomalies, and optimize complex biological processes.

Digital Twin: Virtual replica of a physical wastewater treatment system that integrates real-time data with simulation models, enabling scenario testing and optimization without risking actual process disruptions.

Energy Optimization: Strategies to reduce energy consumption in wastewater treatment while maintaining treatment performance. AI techniques identify energy-saving opportunities in aeration, pumping, and process configuration.

Instrumentation, Control and Automation (ICA): The application of sensors, control systems, and automated equipment to monitor and manage wastewater treatment processes. ICA systems provide the foundation for implementing AI and data-driven approaches.

Machine Learning: A subset of AI that enables systems to learn and improve from experience without explicit programming. In wastewater treatment, machine learning algorithms analyze operational data to optimize processes and predict system performance.

Nature-Based Solutions (NBS): Environmental management approaches that utilize natural processes and ecosystems to address societal challenges. In wastewater treatment, these include constructed wetlands, floating treatment wetlands, and biofiltration systems that mimic natural purification processes.

Process Control: Strategies and systems for maintaining wastewater treatment processes at desired operating conditions. Advanced control approaches include multivariable control, model predictive control, and AI-driven adaptive control.

Resource Recovery: The extraction and reuse of valuable materials from wastewater, including nutrients, energy, and water. AI systems optimize recovery processes to enhance sustainability and economic viability.

Soft Sensors: Virtual or inferential sensors that use computational models to estimate difficult-to-measure parameters based on more accessible measurements. These complement physical instrumentation in wastewater treatment monitoring.

Chapter 9
Artificial Intelligence and Machine Learning in Additive Manufacturing

Koray Özsoy
https://orcid.org/0000-0001-8663-4466
Isparta University of Applied Sciences, Turkey

Bekir Aksoy
https://orcid.org/0000-0001-8052-9411
Isparta University of Applied Sciences, Turkey

ABSTRACT

This study explores the integration of Additive Manufacturing (AM) with Artificial Intelligence (AI) and Machine Learning (ML). While AM enables complex geometries, material efficiency, and design flexibility, it faces challenges such as poor surface quality, printing defects, limited material capabilities, and slow production speeds. Addressing these issues is crucial for AM to reach its full potential. AI and ML offer promising solutions through process optimization, monitoring, defect detection, material property prediction, and generative design. Advanced sensor systems combined with ML algorithms enable early detection of defects, allowing timely interventions to reduce waste and improve quality. This study provides a comprehensive of AI and ML applications in AM, highlights key advancements, and outlines future research. The findings reveal that AI-driven technologies significantly enhance efficiency and quality control in AM. Interdisciplinary collaboration, data infrastructure, and improved AI models are essential for maximizing the impact of AI-integrated AM in Industry 4.0.

DOI: 10.4018/979-8-3373-0746-6.ch009

INTRODUCTION

Additive manufacturing (AM) has become a widely preferred method in production processes with developing technological methods and the industry 4.0 industry (Karagöz et al., 2021). Additive manufacturing (AM) is an innovative production technology that enables the production of parts by laying materials on a table in layers. AM is a technique that enables the production of parts by adding materials on top of each other in layers, instead of shaping them by subtracting or cutting them as in traditional production methods (Doğan and Kamer, 2023). Unlike traditional production methods, this technique offers significant advantages by allowing the production of more complex geometries as well as lightweight and durable parts (Bozkurt et al., 2021). This method enables the creation of parts by melting and combining layer by layer various materials such as metal, polymer and plastic (Sürmen, 2019). Offering the opportunity to produce with different materials, AM provides great convenience in production processes (Yavuz et al., 2021). In addition, it is widely used, especially in prototyping processes, in integration with advanced computer-aided design (CAD) and computer-aided manufacturing (CAM) technologies. AM provides design flexibility by enabling the production of complex geometric structures designed in a CAD environment.

The "Industrial Revolution (1.0)" is the invention of the steam engine, the "Industrial Revolution (2.0)" is the mass production because of the use of electrically operated machines, and the "Industrial Revolution (3.0)" is the widespread automation in the industry with the development of electronics and information technologies. When the historical process of industrial revolutions is examined, it is seen that there is a development that provides convenience in the production system. Therefore, there must be a new manufacturing method or production system for the Industrial Revolution (4.0). Therefore, the widespread use of 3D printer technology, which is included in the additive manufacturing method, forms the basis of the Industrial Revolution (4.0). The main purpose of the Industrial Revolution (4.0) is product development in all software and hardware, communication of production and service processes, real-time information exchange of machines and products, artificial intelligence, autonomous control and optimization, modular smart factories, cyber-physical systems for physical processes, and communication of objects with each other and with people. With the Industry (4.0) revolution, additive manufacturing systems have become rapidly widespread and have emerged as a strong alternative to traditional production methods (Erener & Boz, 2021).

AM technology is an innovative production method that enables the design and manufacturing of models with complex geometries. One of the most significant advantages of AM technology is the high mechanical strength of the produced parts. AM allows for widespread use in various engineering disciplines such as materials science, mechanics, optics, and electronics. However, the production process can vary depending on factors such as material quality, mechanical strength of the parts, printing parameters, and processing conditions. Additionally, detecting and resolving errors that may occur during the printing process can be time-consuming (Qin et al., 2022). To enhance the efficiency of production processes and optimize quality, AI and ML algorithms are increasingly being utilized. AI is defined as a technology that mimics the human brain's processes of learning and applying knowledge, while machine learning, a subfield of AI, involves training algorithms on datasets and enabling them to analyze new data (Gams et al., 2019; Arslan, 2020). AI processes input data from its environment to produce results (Bozkurt, 2023), whereas ML algorithms analyze large and complex datasets to perform classification or regression tasks. Algorithms developed in ML can classify new incoming data without requiring retraining, using pre-trained datasets (Aylak & Oral, 2021). In AM technology, AI and ML algorithms play a crucial role in error detection during the production process and post-production material quality assessment. These technologies enable early detection of potential errors, optimization of printing parameters, and acceleration of production processes. As a result, AI and ML-supported systems make it possible to use AM technology more effectively and efficiently in the industry, thereby improving the mechanical and structural integrity of the produced parts (Qin et al., 2022; Kumar et al., 2023; Jiang et al., 2022).

The concept of AI, which emerged from the idea of enabling machines to perform human-like behaviors, is now rapidly advancing and appearing in many fields alongside developing technologies. Machines trained with ML algorithms are creating systems that can learn and make decisions on their own. With the help of ML algorithms, AI technologies are integrated into various disciplines such as automotive, mechanical, electrical-electronics, computer, and industrial engineering enabling the creation of smart systems. Automotive engineering, utilizing real-time communication, has developed automatic driving systems and intervention systems suitable for the driver experience in smart vehicles (Adar & Altan, 2024). With the advancement of the Internet of Things (IoT), smart robotic systems and robots are performing tasks faster and more accurately than humans. In industrial engineering applications, robotic systems and smart conveyors on production lines contribute to increased efficiency, while drone technologies offer new possibilities for smart cargo systems and data collection processes (Genç & Tunalı, 2022). In electrical and computer engineering, AI-supported recommendation systems, smart voice response systems, and digital content platforms are used to personalize user experiences and optimize system performance (Çelik, 2021). In mechanical engineering, AI-based Predictive Maintenance (PdM) systems are being developed to predict potential errors and malfunctions in machines and robots during production in factories (Carvalho, 2019). In mechatronics engineering, AI-based optimization techniques, particularly Particle Swarm Optimization (PSO), are used to enable robotic arms to perform tasks more safely, precisely, and efficiently (Ekrem & Aksoy, 2023). In AM technologies, the use of artificial intelligence and machine learning algorithms provides significant advantages in various areas such as detecting printing errors during and after the production process, material selection, setting printing parameters, and analyzing the mechanical strength of produced parts (Ciccone et al., 2023; Büber & Taşpınar, 2023; Liu et al., 2023).

In conclusion, the integration of AI technologies into engineering disciplines offers innovative solutions across a wide range of areas, from production processes to the development of smart systems, and accelerates technological transformation.

FUNDAMENTALS OF ADDITIVE MANUFACTURING (AM)

The fundamentals of additive manufacturing were established with the stereolithography (SLA) technique, which was patented by Charles Hull in 1983. This technique operates on the principle of curing material layers using UV light (Çalışkan & Arpacıoğlu, 2020). The foundations of rapid prototyping were laid in 1951 when Munz filed a patent for a system like today's stereolithography methods. In 1968, Swainson proposed a polymerization method occurring at the intersection of two laser beams. In 1971, Ciraud described a production method based on the principle of solidifying powdered material using an energy source. In 1979, Housholder developed a technology like laser sintering systems. By the mid-1980s, 3D printing and powder-based production technologies began to be commercialized. One of the major developments in this field occurred in 1984, when Chuck Hull founded 3D Systems, launching the commercial 3D printing process. In 1987, Michael Feygin patented a system based on gluing and cutting paper layers and commercialized this technology through the Helisys company. In 1988, S. Scott Crump developed FDM (Fused Deposition Modeling) technology. In 1989, Carl Deckard received a patent for Selective Laser Sintering (SLS) technology. In 1990, the Fraunhofer Institute in Germany developed and commercialized the Selective Laser Melting (SLM) method. In 1993, Ralf Larson initiated the first studies on melting powders using an electron beam. This technology continued through a collaboration with Chalmers University in 1995 and was commercialized by ARCAM AB in 1997. The development of LENS technology also began in 1997, and its license was granted to Optomec Inc. In 2000, Concept Laser GmbH was founded by Frank Herzog, leading to the development of LaserCUSING technology. That same year, the first 3D printing systems based on PolyJet technology were introduced by Objet; in 2006, the company announced the Connex500™, the first rapid prototyping system capable of producing 3D parts using multiple materials simultaneously (Duman and Özsoy, 2018).

Materials Used in Additive Manufacturing

AM consists of various methods. These methods can be divided into three main categories based on the material to be used: solid, liquid, and powder (Dedeakayoğulları & Kaçal, 2020). The classification of these technologies according to the types of raw materials used and manufacturing methods is shown in Table 1. The technologies are denoted by their globally recognized English abbreviations in compliance with ASTM and ISO standards (Singh, Ramakrishna, & Singh, 2017).

Table 1. Types of raw materials used in AM

Additive Manufacturing Processes							
Process	Laser Based AM Processes			Electron Beam	Material Jetting	Material Adhesion	Extrusion Thermal
	Laser Melting		Laser Polymerization				
Process Figures			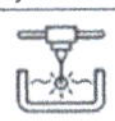		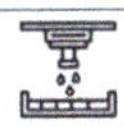	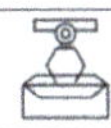	
Name / Materials	SLS	DMD	SLA	EBM	3DP	LOM	FDM
	SLM	LENS	SGC		IJP	SFP	Robocasting
	DMLS	SLC	LTP		MJM		
		LPD	BIS		BPM		
			HIS		Thermojet		
Material Type	Powder	Liquid	Solid				

(Singh, Ramakrishna, & Singh, 2017)

The process of melting material in the form of thermoplastic filament is called Fused Deposition Modeling (FDM). It is widely used today. Among the most used materials in FDM processes are acrylonitrile butadiene styrene (ABS), polylactic acid (PLA), and polyethylene terephthalate glycol (PETG) filaments. Additionally, filaments such as polyamide (PA), polycarbonate (PC), polyethylene (PE), polyvinyl alcohol (PVA), thermoplastic polyurethane (TPU), acrylonitrile styrene acrylate (ASA), and polyether ether ketone (PEEK) are also used in this method (Acierno & Patti, 2023).

Laminated Object Manufacturing (LOM) is a rapid prototyping system that was initially developed by Helisys Inc. and later continued by Cubic Technologies, the successor of this technology. LOM technology is a production method that combines materials such as metal foil, plastic foil or paper in layers using adhesives. This method, which is among the additive manufacturing technologies, is not widely used today, but can be used in areas such as prototyping, mold production and decorative designs in the furniture industry and the production of specially shaped wooden or composite parts (Aydın & Küçük, 2017).

Photopolymerization was first discovered by Charles W. Hull in 1984. Photopolymerization, which includes Stereolithography (SLA), Digital Light Processing

(DLP), and Masked Stereolithography (MSLA), is a group of 3D printing technologies based on the principle of curing liquid resin using light. The process of curing liquid material in a vat with a laser beam is referred to as Stereolithography (SLA), while curing with a lamp-style light source is called Digital Light Processing (DLP). Processes where liquid material is rapidly sprayed through multiple nozzles and then cured with ultraviolet light are named MultiJet Modeling (PolyJet/MultiJet). The working principle of the materials in these processes relies on a photoinitiator added to the resin formulation, which is designed to absorb light at specific wavelengths. When interacting with photons, it forms a reactive structure (Berberoğlu, 2019).

Through photopolymerization technology, some resins exhibiting photopolymer properties with the characteristics and strength of ABS plastic are produced. Additionally, materials used in this technology include polypropylene, epoxy, and acrylic resin-based materials. Today, acrylic polymer resins are frequently used in prosthetics and coatings. The most used type is polymethylmethacrylate (Kiraz et al., 2018; Çelik et al., 2013).

Modern manufacturing technologies, such as laser-based melting processes and electron beam manufacturing methods, are actively used in various industrial fields today (Güler, Serindağ, & Gürel, 2022). These processes can be divided into two categories: powder bed fusion (SLS, DMLS, SLM, EBM, MJF) and powder-fed fusion (NPJ). In both powder bed fusion and powder-fed fusion, the raw material is in powder form (Yalçın & Ergene, 2017). Commonly used metal powders include Ti6Al4V, AlSi10Mg, Maraging steel, 316L metal powders, and Inconel 625 and Inconel 718 alloys (Gökhan, 2020). Additionally, this method, which enables the layer-by-layer sintering of polymer powders using a high-power laser to create solid structures, stands out for its high resolution, efficient production capacity, and wide material diversity. Plastic and ceramic powders such as PA2200, PA1101, PA3200, ALUMIDE, PRIME CAST 101, PA 12, Polypropylene (PP), and TPU 90A are frequently used in the 3D printing process, where material selection determines the durability and application area of the printed part (Kozior, 2022).

Today, with the rapid advancement of additive manufacturing technologies, it is predicted that the use of ceramic materials in this field will significantly increase and become more widespread in the future (Smartech Analysis, n.d.). Figure 1 illustrates the projected economic growth of the industry between 2017 and 2028.

Figure 1. Projected market share of ceramic additive manufacturing materials between 2017 and 2028

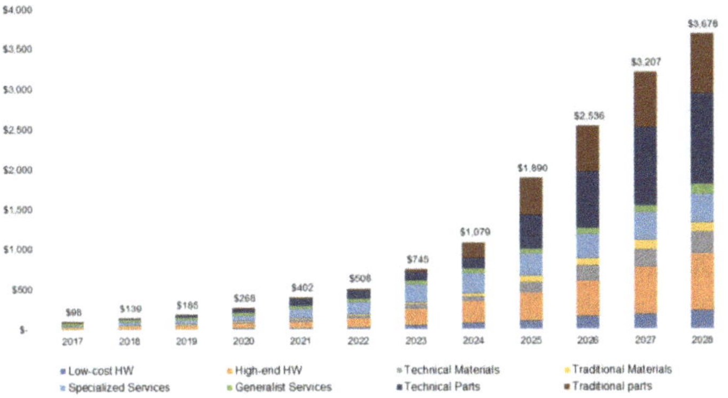

(Smartech Analysis, n.d.)

Binder jet 3D printing is also known as "powder bed and ink jet" and "drop on powder". Developed by Z Corporation in the 1990s, the binder spray technology was initially applied using starch-based materials and a water-based binder similar to household glue. In the Binder Jetting method, the production process is based on two different materials. The first is the powdered material, which serves as the main component and may consist of ceramic, polymer, composite, or metal powders. In particular, the use of metal powders for part production has become increasingly widespread in recent years. The second material is a liquid binder typically a resin that bonds the powdered material together. Today, this method, commercially offered by 3D Systems, uses gypsum-based calcium sulfate hemihydrate powder and the binder is still water-based. (Gibson et., 2021).

Challenges in Additive Manufacturing

Additive manufacturing equipment and materials require a significant investment, but in some cases, production costs can be competitive with traditional methods (Thomas & Gilbert, 2014). Although additive manufacturing offers significant advantages in product-based production, particularly in rapid prototyping, it has considerable disadvantages compared to traditional manufacturing processes when it comes to large-scale and high-volume production (Tofail et al., 2018; Ford & Despeisse, 2016). These disadvantages can primarily be attributed to the limited build volume of the processing equipment and slower processing speeds (Özsoy et al., 2019). Despite these limitations, ongoing advancements in multi-material printing and high-speed additive manufacturing technologies continue to enhance the feasibility of large-scale production.

Additive manufacturing processes offer significant advantages, such as material savings and the ability to produce structures with complex geometries. However, due to current limitations in material properties, various challenges are encountered in production for specific application areas (Przekop et al., 2025). Additionally, factors such as the thermal expansion coefficient and thermal conductivity of metal materials lead to different stress conditions during use. These challenges indicate that high-performance materials are not yet fully optimized for additive manufacturing (Jin et al., 2025). Advanced material development, including the design of novel alloys and composite materials tailored for additive processes, is essential to overcoming these limitations. Additionally, improvements in process parameter optimization and real-time monitoring technologies can enhance the consistency and reliability of printed parts, making them more suitable for demanding industrial applications.

Additive manufacturing may offer lower surface quality and precision compared to traditional manufacturing methods in some applications. This necessitates the application of post-processing techniques (e.g., heat treatment, polishing, and surface coating) to ensure that the produced parts meet functional and aesthetic requirements (Nichita, 2007). Advanced finishing techniques, such as chemical vapor smoothing and laser polishing, have been developed to enhance surface properties without compromising mechanical integrity. Furthermore, integrating automated post-processing systems into production workflows can improve efficiency and consistency in achieving high-quality surface finishes.

The layer deposition direction and orientation angle directly affect the mechanical strength of the produced part. This poses a significant challenge in design and manufacturing for applications requiring uniform strength in different directions. Therefore, determining optimal printing parameters is critical for achieving homogeneous mechanical performance (Yeşildağ & Günay, 2019). Computational modeling and experimental validation play a crucial role in identifying the ideal layer orientation to enhance mechanical reliability. Furthermore, hybrid manufacturing approaches, combining additive and subtractive techniques, can help mitigate these anisotropic effects and improve overall part performance. The challenges encountered in additive manufacturing processes and their solution methods are presented in Table 2.

Table 2. Challenges and solution approaches in additive manufacturing

Challenges of Additive Manufacturing	Solutions to the Challenges of Additive Manufacturing	References
High investment cost: Additive manufacturing equipment and materials are expensive.	**Cost Optimization:** Costs can be reduced through economies of scale and shared production models.	(Thomas & Gilbert, 2014).
Disadvantage in Large-Scale and High-Volume Production: Additive manufacturing is slower and has a limited build volume compared to traditional methods.	**Advanced Equipment:** Development of printers with larger build volumes and faster printing speeds.	(Tofail et al., 2018; Ford & Despeisse, 2016; Özsoy et al., 2019)
Limitations in Material Properties: High-performance materials are not yet fully optimized for additive manufacturing.	**Material Research:** Development and optimization of new materials specifically suited for additive manufacturing.	(Przekop et al., 2025)
Thermal and Mechanical Stress Issues: Cracks may occur due to the material's thermal expansion coefficient and thermal conductivity.	**Thermal Process Optimization:** Application of thermal treatments during and after the printing process.	(Jin et al., 2025)
Low Surface Quality and Precision: Increased need for post-processing of parts.	**Post-Processing Technologies:** Integration of methods such as automated polishing, surface coating, and CNC machining.	(Nichita, 2007)
Layer Deposition Direction and Orientation Issues: Achieving uniform mechanical strength may be challenging.	**Parameter Optimization:** Optimization of parameters such as print orientation, layer thickness, and raster angle.	(Yeşildağ & Günay, 2019)
Slow Production Speed: Production of large parts is time-consuming.	**Parallel Production:** Simultaneous operation of multiple printers and accelerated printing techniques.	(Özsoy et al., 2019)
Waste Material Management: Support structures and failed prints can generate waste.	**Waste Reduction Strategies:** Optimization of support structures and the use of recyclable materials.	(Nichita, 2007)

OVERVIEW OF ARTIFICIAL INTELLIGENCE AND MACHINE LEARNING

Key Concepts in AI and ML

Artificial Intelligence (AI) is defined as the ability of computer systems to mimic human intelligence by utilizing pre-learned information to produce outcomes. This definition encompasses AI's capability to perform cognitive functions such as learning, problem-solving, reasoning, object recognition, and language processing (Okada et al., 2023; Sarma et al., 2025).

AI systems typically build their models using statistical graphs, optimization algorithms and probabilistic approaches. Thanks to the methods they use, they can quickly realize decision-making processes without human intervention with the models they develop. Intelligent systems produced with AI reduce the need for manpower. AI is applied in a wide range of fields, including expert systems, natural language processing, visual recognition, and the development of autonomous systems. Especially in industries such as healthcare, finance, manufacturing and automotive, AI-based solutions play critical roles in areas such as process optimization, error detection and personalized service delivery. ML is a sub-discipline of AI, works through algorithms designed to automatically extract patterns and relationships from data. ML algorithms are trained using large data sets and can be integrated into decision support systems. In this way, it is frequently used in different fields such as predictive, risk management, decision-making and recommendation systems. These algorithms analyze large datasets using methods such as labeled data (supervised learning), unlabeled data (unsupervised learning), or reward-punishment mechanisms (reinforcement learning) to perform tasks like classification, regression, and clustering (Chen et al., 2023; Sarma et al., 2025). While supervised learning is used in classification and regression applications by training algorithms using historical data sets, unsupervised learning is a method based on grouping data using unknown data structures. Additionally, during the training of machine learning models, mathematical equations based on data are used, enabling the model to make predictions on new data and integrate into decision support systems.

Inspired by the layered neurons in the human brain, Artificial Neural Networks (ANN) are used to model data and solve complex problems. ANNs process data through input, hidden, and output layers; their multi-layered structures enable high accuracy rates in applications such as data analysis and pattern recognition (Abdolrasol et al., 2023; Sarma et al., 2025). Deep Learning (DL), a branch of Machine Learning (ML), is based on multi-layered artificial neural networks and can extract features from data to create more complex structural models. In DL models, transformations performed at each layer enhance the representational power of the data, improving learning accuracy and overall performance (Dargan et al., 2020; Mohammed & Kora, 2023; Sarma, 2025). Compared to traditional machine learning algorithms, the deep learning method can achieve more accurate and effective results by using larger and more complex data sets. With the transfer learning method, it performs a fast and easy training process on small data sets. As shown in Figure 2, ML algorithms are categorized based on their application methods, including Support Vector Machines (SVM), Gaussian Processes (GP), Decision Trees (DT), Random Forests (RF), Naive Bayes (NB), AdaBoost (AB), and Gradient Boosting

(GB). These algorithms use different approaches to analyze and predict data for solving problems such as classification and regression. In addition, these algorithms often use statistical and probabilistic methods to determine the relationships between data and work with different optimization techniques depending on the data set. One of these algorithms, SVM, creates a hyperplane to separate two or more classes. In addition, these algorithms often use statistical and probabilistic methods to determine the relationships between data and work with different optimization techniques depending on the data set. It searches for the optimal margin line within this hyperplane to distinguish between classes.

Figure 2. Machine learning algorithms

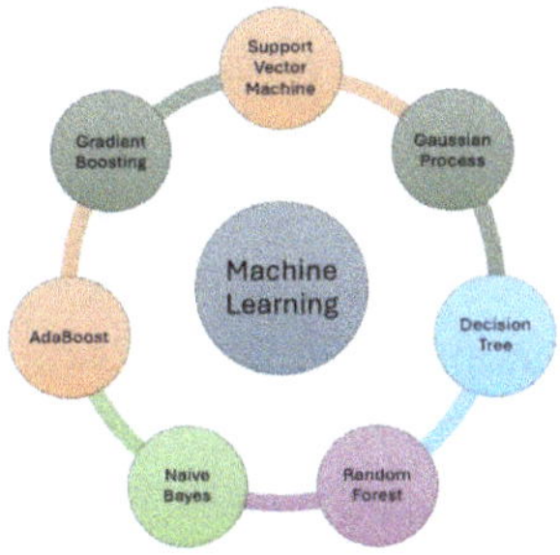

SVM is a machine learning algorithm used in classification and regression applications due to its ability to achieve high accuracy on labeled data (Roy & Chakraborty, 2023). GP offers flexible solutions for regression and classification problems in multivariate and nonlinear datasets. The GP method performs predictions using mean and covariance functions, thereby modeling uncertainties in the data as well. When modeling uncertain data structures in engineering, the GP method is often preferred because it offers a non-parametric approach to the distributions and relationships in the data set. (Deringer et al., 2021). DT divides the dataset into sub-samples consisting of a tree, branch, and root structure based on the rules it creates according to the features of the labeled dataset. Since the results of DT algorithms are easy to interpret, they are frequently used in decision support systems. During the training process, data is sent to the next sub-branch according to the rules determined at these node points. Due to its high accuracy, it is widely used in regression and classification applications. However, DT in its pure form is prone to overfitting. To prevent this, pruning techniques or ensemble learning methods are combined with DT to increase its generalization capability. (Özkan et al., 2023). RF is a powerful machine learning algorithm obtained by combining multiple decision trees through an ensemble learning approach. It provides high accuracy in both classification and regression problems while being resistant to overfitting. The success of RF is due to the fact that different decision trees are trained on random subsets, which increases the diversity, performance and generalization capability. By using bagging and classification and regression trees (CART) techniques, it enhances prediction accuracy. Thanks to random sampling and ensemble strategies, it has a high generalization capability (Özkan et al., 2023). NB performs classification based on Bayes' theorem, determining the class with the highest probability for the data under the assumption of feature independence. This method offers fast and effective results, especially in high-dimensional datasets. The NB method, which uses conditional probability calculation, is frequently used in powerful natural language processing algorithms such as text mining, spam filtering, etc. (Chen et al., 2021). AB is an ensemble learning method that aims to build a strong model by sequentially training and combining weak learners. In the learning method, model errors are minimized by increasing the weight of the learner that failed in the previous stage. It improves the overall performance of the model by assigning weights to misclassified examples at each step. AB achieves high accuracies even with different types of datasets because it is adaptive in correcting the errors of different types of weak learners (Güler, 2024). GB is a method that sequentially minimizes errors and achieves successful results in regression problems with complex datasets. At each stage, it corrects the errors of previous models, enhancing prediction accuracy by combining multiple weak learners. GB applies a progressive learning method, gradually increasing model success by focusing on the model's failures in the previous stage (Bentejan et al., 2021).

Supervised, Unsupervised, and Reinforcement Learning

Today, AI and ML are among the fundamental technologies, especially in applications such as image processing, object detection, and image segmentation. AI and ML technologies are widely used in medicine, healthcare, automotive, security and industry. It also has a major role in processing and interpreting data from different fields. These technologies use various learning methods to extract meaningful insights from data. ML is a method that performs classification and regression tasks by analyzing data and identifying patterns (Kızılkaya & Oğuzlar, 2018). Machine learning, supported by statistical methods and optimization techniques, extracts patterns from large data sets and develops predictive models. Nowadays, with the increasing computational power and accessibility of large datasets, ML methods are enabling the solution of large-scale and complex problems. The methods used within the scope of ML include supervised learning, unsupervised learning, and reinforcement learning. In supervised learning, each example in the dataset is assigned predefined labels during the training process. In this process, the model learns the labels that correspond to the data set and gains the ability to make predictions about new incoming data. First, the data must be labeled and preprocessed; unnecessary data is cleaned and divided into training and test sets. In the pre-processing stage, the accuracy of the model is improved by removing missing data, data scaling and feature selection. Supervised learning performs classification (categorical outcomes) and regression (prediction of continuous values) tasks based on the labeled data. Regression models relationships between two or more variables to generate predictions, while classification aims to categorize data into predefined classes (Kızılkaya & Oğuzlar, 2018; Calayır & Kabak, 2021). Regression techniques in particular are widely used in statistical data type applications such as time series analysis, economic forecasting and engineering. Classification methods have an important place in categorical applications such as face recognition, text mining and biometric security systems. ML models developed today are supported by deep learning techniques to make the classification and regression processes faster and more efficient.

Unsupervised learning is preferred in situations where creating labeled data is difficult and costly. In the method, the model analyzes the raw data without human intervention, revealing patterns and textures and the relationships between patterns and textures. Unsupervised learning directly trains on raw data without intervention in the dataset. This method uses techniques such as clustering to uncover hidden structures and relationships within the data It implements an effective method for subdividing data, especially in large-scale and semantically complex data structures (Kızılkaya & Oğuzlar, 2018; Calayır & Kabak, 2021). Reinforcement learning aims for an agent to develop an optimal strategy by interacting with its environment through reward and punishment mechanisms. This method allows the agent to develop an optimal policy to achieve a given goal and is widely used in game theory, robotics, autonomous systems and dynamic decision-making processes. The agent operates through a trial-and-error process in different situations and optimizes its decision-making ability based on feedback from past experiences (Sarıcan & Engin, 2024). Among the other prominent methods in the field of ML in recent years are techniques such as transfer learning and semi-supervised learning. Transfer learning minimizes training time and data requirements by reusing previously learned models for similar tasks (Erol & İnkaya, 2024), while semi-supervised learning involves the combined use of both labeled and unlabeled data (Turan & Polat, 2024). Transfer learning is usually performed by modifying the final layers of a pre-trained model to adapt it to a different problem and is highly effective in areas such as image processing and natural language processing (NLP). Semi-supervised learning improves the learning performance of the model by combining small amounts of labeled data with large amounts of unlabeled data when full labeling is costly. It is widely used in areas such as biomedical image analysis and text classification. (Erol & İnkaya, 2024; Turan & Polat, 2024).

The methods used in machine learning enable the analysis of complex datasets to derive meaningful results in applications such as image classification, object detection, and image segmentation. Moreover, in many areas such as big data analytics, financial forecasting, anomaly detection and personalized recommendation systems, ML techniques empower data-driven decision-making processes. Thus, machine learning methods have become pioneers of technological advancements by offering innovative solutions in both academic and industrial fields. Moreover, in many areas such as big data analytics, financial forecasting, anomaly detection and personalized recommendation systems, ML techniques empower data-driven decision-making processes. Figure 3 presents the titles of applications of machine learning algorithms in additive manufacturing technologies.

Figure 3. Machine learning applications in additive manufacturing methods

AI AND ML APPLICATIONS IN ADDITIVE MANUFACTURING

Process Optimization in Additive Manufacturing

The optimization of additive manufacturing processes aims to make production processes more efficient, precise, and cost-effective. By optimizing various parameters, the mechanical strength, surface roughness, and production speed of manufactured parts can be improved. Effective optimization strategies not only enhance the mechanical properties of printed components but also contribute to energy efficiency in production. Additionally, process optimization can minimize waste generation. Process optimization can be examined under three different categories:

design optimization, material optimization, and machine parameter optimization (Canlıdinç, 2023).

Design optimization is a process optimization method that aims to produce lighter and more durable products by optimizing topology and lattice structures. Topology optimization seeks to determine the optimal material distribution or structural layout within a defined design area by optimizing a target parameter under specific constraints (Canlıdinç, 2023). This method is widely used in aerospace, automotive, and biomedical applications to achieve high-performance structures with minimal material usage. Moreover, advancements in computational algorithms and simulation tools have significantly improved the efficiency and accuracy of topology optimization processes. Lattice structures are three-dimensional structures composed of one or more repeating unit cells. Each cell consists of interconnected struts joined at nodal points. By optimizing unit lattice structures, the goal is to enhance their mechanical, thermal, physical, and functional properties (Gülcan, 2021). In particular, the fine-tuning of lattice parameters can lead to improvements in impact resistance, energy absorption, and weight reduction, making them highly beneficial for structural applications.

Additive manufacturing technologies can be applied to metal, ceramic, and polymer materials. Artificial intelligence (AI) technologies have recently gained significant importance in the design and optimization of composite materials. AI-based modeling and optimization methods accelerate material selection processes in industries such as automotive, aerospace, and defense, enabling the development of more durable and sustainable materials. Additionally, AI-assisted material design holds great potential for both improving efficiency and enhancing material properties in additive manufacturing processes (Özsolak, 2019; Bozkurt et al., 2021). Studies have demonstrated that Adaptive Neuro-Fuzzy Inference System (ANFIS) models and advanced non-dominated sorting genetic algorithm (NSGA-III) optimization techniques provide more precise production capabilities for γ-TiAl raw materials used in Electron Beam Melting (EBM) processes, significantly improving the material's mechanical strength properties (Nasr & Anwar, 2025). Similarly, in studies using Invar 36 raw material in Wire Arc Additive Manufacturing (WAAM), artificial neural networks (ANN) have successfully reduced specific energy consumption and enhanced mechanical strength properties (Mattera et al., 2025). These studies highlight the critical role of AI-based modeling and optimization techniques in improving material performance and production efficiency in additive manufacturing processes.

In machine parameter optimization processes, optimizing parameters such as layer thickness, infill density, and print speed for the Fused Deposition Modeling (FDM) method (Doshi et al., 2022); scan speed, scan direction, build orientation, and laser diameter for Selective Laser Sintering (SLS) and Selective Laser Melting (SLM) methods (Hesselvig et al., n.d.); and layer thickness, infill density, lift speed, and exposure time for Stereolithography (SLA) (Turan et al., 2022) can enhance the mechanical strength, surface roughness, and production speed of manufactured parts. Moreover, the optimization of these parameters can significantly reduce residual stresses and internal defects, leading to improved part reliability and longevity. Additionally, integrating real-time monitoring and adaptive control systems into the manufacturing process can further enhance precision and repeatability. Recent advancements in artificial intelligence and machine learning have also enabled predictive modeling for optimal parameter selection, reducing trial-and-error approaches in additive manufacturing.

AI-Driven Design and Generative Design

Additive Manufacturing (AM) technology offers significant advantages in terms of time and cost for part production. In this process, design optimization and manufacturability analysis play an important role, and mistakes made in the design phase can directly affect production costs and time management. However, traditional design processes can be time-consuming before transitioning to the production phase. In this context, integrating AI-Driven Design and AI Generative Design methods into the design process enables the creation of more innovative and widely appealing products.AI-Driven Design involves analyzing datasets, finding solutions tailored to user profiles, and incorporating optimization methods. This approach accelerates the designer's decision-making process and analyzes errors during the design phase. By evaluating user data, it provides personalized design recommendations for individuals or specific profiles. And by integrating big data analytics and machine learning algorithms, the design process is more predictable and more alert to errors. (Loeffler et al., 2024). AI Generative Design, on the other hand, generates optimal design proposals based on predefined design parameters. This process generates many alternative solutions in accordance with the specified design criteria and optimizes them by selecting the most appropriate one among these solutions. It conducts design tests by determining the type of material to be used, material quantity, cost, and production processes. In particular, generative design algorithms are being integrated with sustainable engineering approaches to increase material efficiency and minimize energy consumption. These two methods contribute to the optimization of design and production stages in additive manufacturing, enabling more efficient and effective production processes (Saadi & Yang, 2023). Together, these AI-driven approaches streamline the transition from design to production, fostering innovation and efficiency in additive manufacturing. As a result, the integration of AI-supported design approaches with AM technologies paves the way for the development of smarter, innovative and environmentally friendly solutions in manufacturing processes.

Real-Time Process Monitoring and Quality Control

Real-time process monitoring and quality control methods are critical approaches that involve real-time data collection, analysis, monitoring, and improvement during the production process. Traditional process monitoring methods rely on detecting defects in parts after production, which can lead to time loss and material waste. Systems developed with AI and ML algorithms monitor the production process in real time, enabling early detection of errors and instant interventions. High-performance sensor technologies and IoT methods, in particular, provide significant advantages in error detection and quality control processes by enabling real-time data flow during production. In AM technology, errors such as layer shifting, filament jamming, and warping are commonly encountered. Such defects are particularly common in the production of parts with complex geometries and can be difficult to detect by conventional inspection methods. The use of real-time process monitoring and quality control methods allows for early detection of these errors, optimizing the production process and improving the quality of the final product. The real-time process monitoring and quality control application begins with data collection from sensors, cameras, and IoT devices during the production process.

The data collected often includes critical parameters such as temperature, pressure, material flow and layer alignment, and continuous monitoring of these parameters is essential to maintain production quality. Furthermore, advanced computer vision techniques can be used to analyze the surface defects of manufactured parts in real time. For example, thanks to high-resolution cameras and deep learning-based image processing algorithms, even micron-level distortions during production can be detected instantaneously. The collected data is then analyzed by ML algorithms trained on historical datasets to predict potential defects throughout the process. Based on the identified predictions, real-time feedback is provided, and interventions are made to address errors, enabling optimization of the process. In this way, unexpected failures in the production process are prevented and production continuity is ensured. (McCann et al., 2021; Meng et al., 2020; Qin et al., 2022). As a result, AI and ML-based real-time process monitoring and quality control systems make significant contributions to industry 4.0 transformation by enabling higher efficiency, lower defect rates and cost optimization in manufacturing processes. In the future, integrating these systems with more advanced predictive maintenance and autonomous decision-making mechanisms will make it possible for production lines to become fully autonomous.

Defect Detection and Prediction

In the AM method, the material is melted at high temperature and then flowed through a nozzle to produce the part layer by layer. The AM manufacturing process is a great advantage in manufacturing technologies for the production of parts with complex geometries. The production of parts in different geometries by melting plastic materials such as PLA, ABS and PETG at high temperatures offers a more efficient and flexible alternative to traditional production methods in terms of time and cost. However, problems caused by various factors during the production process can negatively affect the production process. These problems are caused by various factors such as layer slippage, layer lift, lack of support, low nozzle temperature. The main causes of defects are interlayer bonding weaknesses, non-uniform spreading of the material, thermal irregularities such as excessive or insufficient heating, and mechanical problems such as notch formation, especially in the production of parts with complex and large geometries. Since parts are produced layer by layer in AM, the parts produced can be inaccurate, defective or uneven. Especially in the production process of parts with complex and large geometries, defects due to weak connections between layers and notches often occur. In manufactured parts, surface quality, mechanical strength and overall performance are critical issues in the evaluation of product quality. AI and ML algorithms make predictions by performing data-based analysis. By training the defects that occur during or after the production process with ML algorithms, it provides a great advantage for the flawless production of newly produced parts. In particular, deep learning-based image processing methods are integrated with advanced sensor and camera technologies, enabling instant defect detection in the production process. In traditional defect detection methods, inspection with measuring devices, visual analysis and non-destructive testing methods are used. Compared to traditional quality control methods, AI-powered systems provide faster and highly accurate forecasts, increasing production efficiency and minimizing defective production. Deep learning-based image processing methods classify defects by analyzing images from sensors and high-resolution cameras. ML algorithms trained with data generated during the production process are trained using supervised learning and deep learning methods and perform instant or post-production defect detection. It also implements a real-time process monitoring strategy that enables intervention by anticipating defects that may occur throughout the process. In particular, deep learning architectures, such as convolutional neural networks (CNN) and evolved neural networks, enable the early elimination of production defects by detecting defects with high precision in the production process. Real-time process monitoring systems analyze data from high-sensitivity sensors, IoT devices and cameras instantaneously and control the situation with systems trained with AI and ML. With systems controlled by algorithms, instant errors that may occur in the production process are intervened in a timely manner. This prevents material wastage in the AM process, reduces costs and ensures the production of high quality parts. Furthermore, defect detection and prediction not only predicts existing defects but also makes recommendations by predicting defects that may occur. AI-supported quality control systems analyze the current state of the quality of the materials produced, while evaluating past errors and providing warnings and suggestions against possible errors. (Meng et al, 2020; Qin et al, 2022; Li et al, 2019).

AI for Material Property Prediction

Machine learning algorithms trained using data obtained within the scope of process optimization can enable faster and more accurate prediction of material properties. This way, the mechanical, thermal, and chemical properties of different material combinations can be determined in advance, allowing for the selection of the most suitable material. Additionally, the use of artificial intelligence-based models reduces dependency on experimental studies, offering significant savings in both time and cost. This approach holds great potential, particularly in the development of new products and the improvement of the performance of existing products. (Ruiz et al., 2020; Zhao & Yang, 2025)

FUTURE TRENDS AND RESEARCH DIRECTIONS

AI-Driven Autonomous Manufacturing Systems

Artificial intelligence algorithms can make industrial production processes smarter, autonomous, and more efficient. These technologies can enhance production efficiency by enabling real-time data analysis, fault detection, process optimization, and predictive modeling on production lines. Particularly, autonomous systems allow for continuous monitoring and improvement of processes without the need for human intervention. Additionally, these approaches optimize resource utilization and reduce costs by minimizing waste in production. Furthermore, AI-driven decision-making processes facilitate adaptive manufacturing, where production parameters are dynamically adjusted based on real-time data. The integration of digital twins with AI further enhances system performance by simulating and predicting potential failures before they occur. AI-powered autonomous production systems offer significant transformative potential, especially in industries requiring high precision and complexity, such as aerospace, automotive, and pharmaceuticals. (Aksoy, 2021; Iqbal et al., 2019)

AI-Integrated Smart Factories

Artificial intelligence is used in smart factories across various areas such as predictive maintenance, quality control, production planning, and supply chain management. Machine learning algorithms can analyze sensor data to predict equipment failures in advance, identify inefficiencies in production lines, and continuously improve production processes. Deep learning and image processing techniques automate defect detection in quality control processes, minimizing production errors (Şekkeli & Bakan, 2018; Iqbal et al., 2019). Furthermore, AI-driven production planning systems optimize scheduling and resource allocation by analyzing historical data and demand patterns. The integration of AI with Internet of Things (IoT) technologies enhances real-time monitoring, enabling more precise and adaptive manufacturing processes. Additionally, AI-powered supply chain management systems improve inventory forecasting and logistics efficiency, reducing operational costs and enhancing overall productivity.

CONCLUSION

The integration of Artificial Intelligence (AI) and Machine Learning (ML) into Additive Manufacturing (AM) represents a paradigm shift in modern production systems. AM technologies, known for their capabilities in producing complex geometries, material efficiency, and design flexibility, face inherent challenges such as surface defects, material constraints, and slower production rates compared to traditional manufacturing. However, the adoption of AI and ML has significantly mitigated these limitations, enabling smarter, faster, and more reliable production. Throughout this study, it has been demonstrated that AI and ML methods contribute to nearly every stage of the AM workflow. From design optimization and parameter tuning to real-time process monitoring, defect detection, and material property prediction, these technologies enhance the precision, reduce waste, and improve the mechanical properties of printed parts. For instance, the use of deep learning-based image analysis techniques in defect detection allows for early error identification during the printing process, drastically reducing the need for post-processing and minimizing production downtime. Likewise, AI-driven generative design accelerates product development while simultaneously enabling personalized and cost-effective solutions.

Another important contribution of this study is the discussion on future directions of AI-powered AM. The emergence of autonomous manufacturing systems and AI-integrated smart factories indicates that the future lies in fully adaptive, data-driven, and self-regulating production ecosystems. These advancements are particularly impactful in high-precision industries such as aerospace, automotive, and biomedical sectors, where performance, reliability, and customization are critical. In summary, the convergence of AM with AI and ML not only enhances current manufacturing capabilities but also opens the door to unprecedented levels of innovation and efficiency. For these benefits to be fully realized, interdisciplinary collaboration, investment in data infrastructure, and further research on AI model robustness and explainability will be essential. As Industry 4.0 continues to evolve, AI-enhanced additive manufacturing will play a central role in shaping the next generation of intelligent production systems.

REFERENCES

Abdolrasol, M. G., Hussain, S. S., Ustun, T. S., Sarker, M. R., Hannan, M. A., Mohamed, R., Ali, J. A., Mekhilef, S., & Milad, A. (2021). Artificial neural networks based optimization techniques: A review. *Electronics (Basel), 10*(21), 2689. DOI: 10.3390/electronics10212689

Abdolrasol, M. G., Hussain, S. S., Ustun, T. S., Sarker, M. R., Hannan, M. A., Mohamed, R., Ali, J. A., Mekhilef, S., & Milad, A. (2021). Artificial neural networks based optimization techniques: A review. *Electronics (Basel), 10*(21), 2689. DOI: 10.3390/electronics10212689

Acierno, D., & Patti, A. (2023). Fused Deposition Modelling (FDM) of Thermoplastic-Based Filaments: Process and Rheological Properties—An Overview. *Materials (Basel), 16*(24), 7664. https://www.mdpi.com/1996-1944/16/24/7664. DOI: 10.3390/ma16247664 PMID: 38138805

Adar, U. G., & Altan, Z. (2024). AUTOMIND: Otomobil Ne Kadar "Oto"? İnsan-Otonom Araç Etkileşimi Üzerine Bir İnceleme. *Uludağ Üniversitesi Mühendislik Fakültesi Dergisi, 29*(2), 639–654.

Aksoy, B. (2021). Estimation of energy produced in hydroelectric power plant industrial automation using deep learning and hybrid machine learning techniques. *Electric Power Components and Systems, 49*(3), 213–232. DOI: 10.1080/15325008.2021.1937401

Arslan, K. (2020). Eğitimde yapay zekâ ve uygulamaları. *Batı Anadolu Eğitim Bilimleri Dergisi, 11*(1), 71–88.

Aydın, L., & Küçük, S. (2017). Üç boyutlu yazıcı ve tarayıcı ile hastaya özel medikal ortez tasarımı ve geliştirilmesi. *Politeknik Dergisi, 20*(1), 1–8.

Aylak, B. L., & Oral, O. (2021). Yapay zeka ve makine öğrenmesi tekniklerinin lojistik sektöründe kullanımı. *El-Cezeri, 8*(1), 74–93.

Bentéjac, C., Csörgő, A., & Martínez-Muñoz, G. (2021). A comparative analysis of gradient boosting algorithms. *Artificial Intelligence Review, 54*(3), 1937–1967. DOI: 10.1007/s10462-020-09896-5

Berberoğlu, B. (2019). Ultraviyole ışıkla kürlenebilen organik-inorganik hibrit malzemelerin sentezi ve karakterizasyonu [Yüksek lisans tezi, Marmara Üniversitesi].

Bozkurt, A. (2023). ChatGPT, üretken yapay zeka ve algoritmik paradigma değişikliği. *Alanyazın, 4*(1), 63–72. DOI: 10.59320/alanyazin.1283282

Bozkurt, Y., Gülsoy, H., & Karayel, E. (2021). Eklemeli imalat teknolojilerinin tıbbi ekipmanların üretiminde kullanımı. *El-Cezeri*, *8*(2), 962–980.

Büber, E. D. M. Y. M., & Taşpınar, Y. S. (2023). 3 boyutlu yazıcılardaki baskı hatalarının hibrit görüntü işleme yöntemleri ile tespiti.

Calayır, G. N., & Kabak, M. (2021). Bakım için makine öğrenme tekniklerinin analizi ve bir uygulama. *Journal of Turkish Operations Management*, *5*(1), 662–675.

Canlıdinç, M. (2023). Topoloji Optimizasyonu ile Metal Eklemeli İmalat Yönteminin Endüstriyel Uygulamaları. *Kirklareli University Journal of Engineering and Science*, *9*(2), 552–565.

Carvalho, T. P., Soares, F. A., Vita, R., Francisco, R. D. P., Basto, J. P., & Alcalá, S. G. (2019). A systematic literature review of machine learning methods applied to predictive maintenance. *Computers & Industrial Engineering*, *137*, 106024. DOI: 10.1016/j.cie.2019.106024

Çelik, İ., Karakoç, F., Çakır, M. C., & Duysak, A. (2013). HIZLI PROTOTİPLEME TEKNOLOJİLERİ VE UYGULAMA ALANLARI. *Dumlupınar Üniversitesi Fen Bilimleri Enstitüsü Dergisi*, (031), 53–70.

Çelik, K. (2021). Bulut bilişim teknolojileri. *Bartın Üniversitesi İktisadi ve İdari Bilimler Fakültesi Dergisi*, *12*(24), 436–450. DOI: 10.47129/bartiniibf.1019898

Chen, H., Hu, S., Hua, R., & Zhao, X. (2021). Improved naive Bayes classification algorithm for traffic risk management. *EURASIP Journal on Advances in Signal Processing*, *2021*(1), 30. DOI: 10.1186/s13634-021-00742-6

Chen, W., Milosevic, Z., Rabhi, F. A., & Berry, A. (2023). Real-time analytics: Concepts, architectures, and ML/AI considerations. *IEEE Access : Practical Innovations, Open Solutions*, *11*, 71634–71657. DOI: 10.1109/ACCESS.2023.3295694

Ciccone, F., Bacciaglia, A., & Ceruti, A. (2023). Optimization with artificial intelligence in additive manufacturing: A systematic review. *Journal of the Brazilian Society of Mechanical Sciences and Engineering*, *45*(6), 303. DOI: 10.1007/s40430-023-04200-2

Dargan, S., Kumar, M., Ayyagari, M. R., & Kumar, G. (2020). A survey of deep learning and its applications: A new paradigm to machine learning. *Archives of Computational Methods in Engineering*, *27*(4), 1071–1092. DOI: 10.1007/s11831-019-09344-w

Dedeakayoğulları, H., & Kaçal, A. (2020). Eklemeli imalat teknolojileri ve kullanılan talaşlı imalat yöntemleri üzerine yapılan çalışmaların değerlendirilmesi. *MATECA*, *1*(1), 1–12.

Deringer, V. L., Bartók, A. P., Bernstein, N., Wilkins, D. M., Ceriotti, M., & Csányi, G. (2021). Gaussian process regression for materials and molecules. *Chemical Reviews*, *121*(16), 10073–10141. DOI: 10.1021/acs.chemrev.1c00022 PMID: 34398616

Doğan, O., & Kamer, M. S. (2023). Farklı imalat parametreleri kullanılarak eriyik yığma modelleme ile imal edilen test numunelerinin sürünme davranışlarının deneysel olarak incelenmesi. *Journal of the Faculty of Engineering & Architecture of Gazi University*, *38*(3).

Doshi, M., Mahale, A., Singh, S. K., & Deshmukh, S. (2022). Printing parameters and materials affecting mechanical properties of FDM-3D printed Parts: Perspective and prospects. *Materials Today: Proceedings*, *50*, 2269–2275. DOI: 10.1016/j.matpr.2021.10.003

Duman, B., & Özsoy, K. (2018). *Üç Boyutlu Baskı Teknolojileri*. Gece Yayınevi.

Ekrem, Ö., & Aksoy, B. (2023). Trajectory planning for a 6-axis robotic arm with particle swarm optimization algorithm. *Engineering Applications of Artificial Intelligence*, *122*, 106099. DOI: 10.1016/j.engappai.2023.106099

Erener, Ş., & Boz, S. (2021). Modern Üretim Tekniklerinde Eklemeli Imalat Sistemlerinin Yeri Ve Kullanim Alanlari. *Turkish Journal of Fashion Design and Management*, *3*(1), 47–56.

Erol, B., & İnkaya, T. (2024). Satış tahmini için uzun kısa-süreli bellek ağı tabanlı derin transfer öğrenme yaklaşımı. *Gazi Üniversitesi Mühendislik Mimarlık Fakültesi Dergisi*, *39*(1), 191–202. DOI: 10.17341/gazimmfd.1089173

Ford, S., & Despeisse, M. (2016). Additive manufacturing and sustainability: An exploratory study of the advantages and challenges. *Journal of Cleaner Production*, *137*, 1573–1587. DOI: 10.1016/j.jclepro.2016.04.150

Gams, M., Gu, I. Y. H., Härmä, A., Muñoz, A., & Tam, V. (2019). Artificial intelligence and ambient intelligence. *Journal of Ambient Intelligence and Smart Environments*, *11*(1), 71–86. DOI: 10.3233/AIS-180508

Genç, E., & Tunalı, İ. (2022). ENDÜSTRİ 4.0 VE LOJİSTİK 4.0 KAPSAMINDA AKILLI DEPO SİSTEMLERİ. *Avrasya Sosyal ve Ekonomi Araştırmaları Dergisi*, *9*(4), 194–215.

Gibson, I., Rosen, D., Stucker, B., Khorasani, M., Gibson, I., Rosen, D., ... & Khorasani, M. (2021). Binder jetting. *Additive manufacturing technologies*, 237-252.

Gökhan, Ö. (2020). Eklemeli üretim teknolojileri üzerine bir derleme. *Niğde Ömer Halisdemir Üniversitesi Mühendislik Bilimleri Dergisi*, *9*(1), 606–621.

Gülcan, O. (2021). Eklemeli imalatla üretilen kafes yapıların mekanik özellikleri. *Makina Tasarım ve İmalat Dergisi*, *19*(1), 17–34.

Güler, O. (2024). Turbofan motorlarının kestirimci bakımında makine öğrenimi algoritmaları performanslarının karşılaştırılması. *Niğde Ömer Halisdemir Üniversitesi Mühendislik Bilimleri Dergisi*, *13*(1), 99–106.

Güler, S., Serindağ, H. T., & Gürel, Ç. A. M. (2022). Tel ark eklemeli imalat: Son gelişmeler ve değerlendirmeler. *Mühendis ve Makina*, *63*(706), 82–116. DOI: 10.46399/muhendismakina.995979

Hesselvig, J., Nygaard, R. T., Budzik, M. K., & Sandberg, M. Low-fidelity modelling and analysis of polymer SLS: Insights into particle sintering and densification in straight and corner scanning passes. SSRN. Retrieved March 10, 2023 DOI: 10.2139/ssrn.5170448

Iqbal, R., Maniak, T., Doctor, F., & Karyotis, C. (2019). Fault detection and isolation in industrial processes using deep learning approaches. *IEEE Transactions on Industrial Informatics*, *15*(5), 3077–3084. DOI: 10.1109/TII.2019.2902274

Jiang, J., Xiong, Y., Zhang, Z., & Rosen, D. W. (2022). Machine learning integrated design for additive manufacturing. *Journal of Intelligent Manufacturing*, *33*(4), 1073–1086. DOI: 10.1007/s10845-020-01715-6

Jin, H., Wei, J., Ji, R., Wang, L., Liu, Y., Li, C., Yin, Y., & Liu, P. (2025). Efficient preparation of mechanical seal face reinforcement layer by composite anode jet electrodeposition. *Journal of Materials Research and Technology*, *35*, 5908–5922. DOI: 10.1016/j.jmrt.2025.02.224

Karagöz, İ., Bekdemir, A. D., & Tuna, Ö. (2021). 3B yazıcı teknolojilerinde kullanılan yöntemler ve gelişmeler üzerine bir derleme. *Düzce Üniversitesi Bilim ve Teknoloji Dergisi*, *9*(4), 1186–1213. DOI: 10.29130/dubited.877423

Kiraz, C., Sezer, H. K., & Şahin, İ. (2018). Kuyumculuk sektöründe 3b baski tasarim özgürlüğünden faydalanmaya ilişkin bir perspektif. *International Journal of 3D Printing Technologies and Digital Industry*, *2*(2), 46-58.

Kızılkaya, Y. M., & Oğuzlar, A. (2018). BAZI DENETİMLİ ÖĞRENME ALGORİTMALARININ R PROGRAMLAMA DİLİ İLE KIYASLANMASI. *Karadeniz Uluslararası Bilimsel Dergi, 37*(37), 90–98. DOI: 10.17498/kdeniz.405746

Kozior, T. (2022). Polymer powders. In *Polymers for 3D Printing* (pp. 271–306). William Andrew Publishing. DOI: 10.1016/B978-0-12-818311-3.00020-3

Kumar, S., Gopi, T., Harikeerthana, N., Gupta, M. K., Gaur, V., Krolczyk, G. M., & Wu, C. (2023). Machine learning techniques in additive manufacturing: A state of the art review on design, processes and production control. *Journal of Intelligent Manufacturing, 34*(1), 21–55. DOI: 10.1007/s10845-022-02029-5

Li, Z., Zhang, Z., Shi, J., & Wu, D. (2019). Prediction of surface roughness in extrusion-based additive manufacturing with machine learning. *Robotics and Computer-integrated Manufacturing, 57*, 488–495. DOI: 10.1016/j.rcim.2019.01.004

Liu, C., Wang, R. R., Ho, I., Kong, Z. J., Williams, C., Babu, S., & Joslin, C. (2023). Toward online layer-wise surface morphology measurement in additive manufacturing using a deep learning-based approach. *Journal of Intelligent Manufacturing, 34*(6), 2673–2689. DOI: 10.1007/s10845-022-01933-0

Loeffler, H. H., He, J., Tibo, A., Janet, J. P., Voronov, A., Mervin, L. H., & Engkvist, O. (2024). Reinvent 4: Modern AI–driven generative molecule design. *Journal of Cheminformatics, 16*(1), 20. DOI: 10.1186/s13321-024-00812-5 PMID: 38383444

Mattera, G., Caggiano, A., & Nele, L. (2025). Energy Efficiency Optimisation in Wire arc Additive Manufacturing of Invar 36 Alloy via Intelligent Data-Driven Techniques. International Journal of Precision Engineering and Manufacturing-Green Technology, 1-13.

McCann, R., Obeidi, M. A., Hughes, C., McCarthy, É., Egan, D. S., Vijayaraghavan, R. K., Joshi, A. M., Acinas Garzon, V., Dowling, D. P., McNally, P. J., & Brabazon, D. (2021). In-situ sensing, process monitoring and machine control in Laser Powder Bed Fusion: A review. *Additive Manufacturing, 45*, 102058. DOI: 10.1016/j.addma.2021.102058

Meng, L., McWilliams, B., Jarosinski, W., Park, H. Y., Jung, Y. G., Lee, J., & Zhang, J. (2020). Machine learning in additive manufacturing: A review. *Journal of the Minerals Metals & Materials Society, 72*(6), 2363–2377. DOI: 10.1007/s11837-020-04155-y

Mohammadi, M. G., Mahmoud, D., & Elbestawi, M. (2021). On the application of machine learning for defect detection in L-PBF additive manufacturing. *Optics & Laser Technology, 143*, 107338. DOI: 10.1016/j.optlastec.2021.107338

Mohammed, A., & Kora, R. (2023). A comprehensive review on ensemble deep learning: Opportunities and challenges. *Journal of King Saud University. Computer and Information Sciences*, *35*(2), 757–774. DOI: 10.1016/j.jksuci.2023.01.014

Nasr, M. M., & Anwar, S. (2025). Developing an intelligent approach based on ANFIS and advanced NSGA-III for improving the turning performance of additively manufactured γ-TiAl alloy. *International Journal of Advanced Manufacturing Technology*, *137*(7-8), 1–26. DOI: 10.1007/s00170-025-15366-5

NICHITA, G. G. (2007). An Review About Rapid Manufacturing. communications, 6(7), 9.

Okada, Y., Mertens, M., Liu, N., Lam, S. S. W., & Ong, M. E. H. (2023). AI and machine learning in resuscitation: Ongoing research, new concepts, and key challenges. *Resuscitation Plus*, *15*, 100435. DOI: 10.1016/j.resplu.2023.100435 PMID: 37547540

Özkan, B., Parim, C., & Çene, E. (2023). Ülkelerin gelişmişlik düzeylerinin karar ağacı ve rastgele orman yöntemleriyle tahmin edilmesi. *EKOIST Journal of Econometrics and Statistics*, *0*(38), 87–104. DOI: 10.26650/ekoist.2023.38.1172190

Özsolak, O. (2019). Eklemeli imalat yöntemleri ve kullanılan malzemeler. *International Journal of Innovative Engineering Applications*, *3*(1), 9–14.

Özsoy, K., Duman, B., & Gültekin, D. İ. (2019). Havacılık, uzay ve savunma sanayi için eklemeli imalat ile metal parça imalatı. 4th International Congress on 3D Printing (Additive Manufacturing) Technologies and Digital Industry (ss. 462-471).

Przekop, R. E., Konieczna, R., Głowacka, J., Sztorch, B., Głowacki, M., & Kotecka, B. (2025). 3D printing in art: Use of thermoplastics and ceramics—the current state and limitations resulting from the possibilities of technology. *Progress in Additive Manufacturing*, *10*(9), 1–31. DOI: 10.1007/s40964-025-01015-6

Qin, J., Hu, F., Liu, Y., Witherell, P., Wang, C. C., Rosen, D. W., Simpson, T. W., Lu, Y., & Tang, Q. (2022). Research and application of machine learning for additive manufacturing. *Additive Manufacturing*, *52*, 102691. DOI: 10.1016/j.addma.2022.102691

Roy, A., & Chakraborty, S. (2023). Support vector machine in structural reliability analysis: A review. *Reliability Engineering & System Safety*, *233*, 109126. DOI: 10.1016/j.ress.2023.109126

Ruiz, E., Ferreño, D., Cuartas, M., López, A., Arroyo, V., & Gutiérrez-Solana, F. (2020). Machine learning algorithms for the prediction of the strength of steel rods: An example of data-driven manufacturing in steelmaking. *International Journal of Computer Integrated Manufacturing, 33*(9), 880–894. DOI: 10.1080/0951192X.2020.1803505

Saadi, J. I., & Yang, M. C. (2023). Generative design: Reframing the role of the designer in early-stage design process. *Journal of Mechanical Design, 145*(4), 041411. DOI: 10.1115/1.4056799

Sarıcan, B., & Engin, O. (2024). Makine Çizelgeleme Problemlerinin Çözümünde Pekiştirmeli Öğrenme Etkisinin Analizi. *ALKÜ Fen Bilimleri Dergisi, 6*(2), 116–140. DOI: 10.46740/alku.1390397

Sarma, D., Rali, A. S., & Jentzer, J. C. (2025). Key Concepts in Machine Learning and Clinical Applications in the Cardiac Intensive Care Unit. *Current Cardiology Reports, 27*(1), 30. DOI: 10.1007/s11886-024-02149-9 PMID: 39831916

Şekkeli, Z. H., & Bakan, İ. (2018). AKILLI FABRİKALAR. *Journal of Life Economics, 5*(4), 203–220. DOI: 10.15637/jlecon.270

Singh, S., Ramakrishna, S., & Singh, R. (2017). Material issues in additive manufacturing: A review. *Journal of Manufacturing Processes, 25*, 185–200. DOI: 10.1016/j.jmapro.2016.11.006

Smartech Analysis. (n.d.). Ceramics additive manufacturing markets: 2017-2028. Retrieved March 10, 2025, from http://www.smartechanalysis.com/reports/ceramics-additive-manufacturing-markets-2017-2028/

Sürmen, H. K. (2019). EKLEMELİ İMALAT (3B BASKI): TEKNOLOJİLER VE UYGULAMALAR. Uludağ Üniversitesi mühendislik fakültesi dergisi, 24(2), 373-392. DOI: 10.17482/uumfd.519147

Thomas, D. S., & Gilbert, S. W. (2014). Costs and cost effectiveness of additive manufacturing (NIST Special Publication No. 1176, p. 12). National Institute of Standards and Technology.

Tofail, S. A., Koumoulos, E. P., Bandyopadhyay, A., Bose, S., O'Donoghue, L., & Charitidis, C. (2018). Additive manufacturing: Scientific and technological challenges, market uptake and opportunities. *Materials Today, 21*(1), 22–37. DOI: 10.1016/j.mattod.2017.07.001

Turan, A. K., & Polat, H. (2024). Yarı denetimli makine öğrenmesi yöntemini kullanarak müzik türlerinin tespiti. Gazi University Journal of Science Part C: Design and Technology, 1-1.

Turan, S. R., Ülkir, O., Kuncan, M., & Buldu, A. (2022). STEREOLİTHOGRAFİ EKLEMELİ İMALAT YÖNTEMİYLE FARKLI DOLULUK ORANLARINDA ÜRETİLEN NUMUNELERİN MEKANİK ÖZELLİKLERİNİN İNCELENMESİ. International Journal of 3D Printing Technologies and Digital Industry, 6(3), 399-407.

Yalçın, B., & Ergene, B. (2017). Endüstride yeni eğilim olan 3-d eklemeli imalat yöntemi ve metalurjisi. *Uluslararası Teknolojik Bilimler Dergisi*, *9*(3), 65–88.

Yavuz, İ., Erçek, E., & Yuran, A. F. (2021). Ulaştırma sektöründe imdat çekici tasarımı ve 3B yazıcı ile üretimi. *International Journal of Multidisciplinary Studies and Innovative Technologies*, *5*(1), 46–49.

Yeşildağ, İ., & Günay, M. (2019). Eklemeli imalatin metal malzemelere uygulanması üzerine bir değerlendirme. 4th International Congress on 3D Printing (Additive Manufacturing) Technologies and Digital Industry (ss. 881-892)

Zhao, X., & Yang, W. (2025). Prediction of mechanical property and costs of high-strength steel using machine learning. *Canadian Metallurgical Quarterly*, 1–13. DOI: 10.1080/00084433.2025.2471616

Chapter 10
A Digital Twin and AI-Driven Framework for Optimizing Firefighter Escape Routes in Wildfire Scenarios: FAST

Walisa Romsaiyud
https://orcid.org/0000-0002-4529-8715
Sukhothai Thammathirat Open University, Thailand

Pirom Konglerd
Sukhothai Thammathirat Open University, Thailand

Sabsathit Takaew
https://orcid.org/0009-0000-6493-5356
Sukhothai Thammathirat Open University, Thailand

ABSTRACT

Wildfires pose critical threats to firefighter safety, requiring rapid decisions under hazardous, dynamic conditions. Existing AI-based evacuation systems often lack real-time prediction and integration with physical environments. This chapter introduces FAST (Firefighter AI Safety Digital Twin), an AI-driven system bridging that gap by combining deep learning and digital twin technology for time-sensitive wildfire response. FAST integrates four components: (1) a simulation environment modeling evolving wildfire scenarios; (2) an AI engine using multivariate LSTM and DQN to predict fire spread and optimize escape routes; (3) a bidirectional sync

DOI: 10.4018/979-8-3373-0746-6.ch010

layer for real-time data exchange between virtual and physical systems; and (4) a performance assessment module. Experimental results show a prediction loss of 0.014, an 84% escape success rate, and a 260-second average evacuation time, confirming FAST's effectiveness in enhancing situational awareness, reducing response time, and improving firefighter safety.

INTRODUCTION

Wildland firefighters are responsible for controlling and extinguishing wildfires in forests, grasslands, and mountainous regions. They also engage in prevention and rehabilitation efforts to protect lives, property, and the environment (Penney, Habibi, & Cattani, 2019). Their work can span hours, days, or weeks in hazardous, physically demanding conditions, with constant risk of injury or fatality. To address these risks, firefighters undergo extensive training in risk management and safety, including burnout operations and mopping up hotspots. Certification programs ensure proficiency in tools and technologies (National Wildfire Coordinating Group, 2014). Despite such preparation, fatalities during wildfire operations remain a concern. Common causes include burn overs, heart attacks, strokes, and suffocation from smoke. Wu et al. (2019) found higher urinary mutagenicity and oxidative changes in Midwestern U.S. firefighters due to prolonged smoke exposure. Caban-Martínez, Hughes, and Bator (2021) reported increased skin cancer risks due to long-term fire response exposure. Esteves et al. (2024) noted the toll of extended hours, harsh conditions, and health hazards like burns, fractures, and inhalation, leading to acute and chronic health impacts, including cancer. Between 2013 and 2022, there were 96 wildland firefighter fatalities in the U.S., rising from 2% to 10% of total firefighter deaths (Glover et al., 2024; USFA, 2022). In 2024, California reported 8,024 wildfires burning 1,050,012 acres. Australia had 65 bushfire-related deaths from 2010 to 2020. Hu and Shao (2020) analyzed Chinese firefighter deaths, highlighting risks related to duty, age, and emergency response. In Thailand, over 23,000 wildfires occurred between 2018 and 2023, causing damage and pollution (Phantanaikasem et al., 2024). Differences in technology access and funding create disparities in wildfire management, leaving many nations struggling with firefighter injuries and deaths.

Numerous studies have explored AI technologies to support wildland firefighter rescue operations. These approaches integrate machine learning with IoT sensors, satellite imagery, and drones for accurate support in wildfire scenarios. Kulhandjian et al. (2023) used IR cameras, micro-Doppler sensors, and DCNNs to detect humans in smoke-obscured environments with 98% accuracy. Viseras et al. (2016) developed a real-time monitoring application using deep Q-learning (DQN) to track fire fronts. Garrity and Yusuf (2021) introduced a wearable device with ANN to predict and warn of dangerous temperature conditions. Marić et al. (2023) used the YOLO algorithm retrained on 30,000 images and deployed on Edge AI for real-time fire detection and automated alerts. James et al. (2023) combined CNNs for image analysis and RNNs for time-series analysis of satellite data to enable real-time fire detection and classification. Although AI models offer high predictive accuracy, many lack dynamic simulation and real-time visualization. This limits their ability to support decisions in fast-changing fire environments, highlighting the need to integrate AI with digital twins (DTs), which enable continuous awareness and adaptive decision-making.

A digital twin is not just a virtual model but a dynamic system synchronized with its real-world counterpart. Osama (2024) defines it as a dynamic digital asset reflecting the real environment. Van Schalkwyk (2022) describes it as a virtual representation continuously updated using real-time and historical data. Digital twins (DT) simulate past and present states, forecast future outcomes, and support decision-making. Their core capabilities include real-time synchronization, AI integration, and predictive simulation—key for enhancing decisions in complex situations. From an engineering perspective, implementing digital twins involves combining AI, control theory, real-time systems, and data fusion to ensure reliability, safety, and scalability in dynamic environments like wildfires. Huang et al. (2022) proposed a DT framework for building fire safety, integrating simulations and real-time monitoring with SuRF models to predict fire spread and adjust safety design. Zhang et al. (2022) developed the AID-Fire framework in a 20 m^2 fire test room, using IoT sensors, cloud computing, and Conv-LSTM models for accurate prediction. Their system achieved over 85% accuracy, 0.1-second computation, and 1-second delay in real-time simulations. In wildfire management, DTs allow simulation of virtual environments and AI modeling of fire behavior. Giannakidou et al. (2024) emphasized their role in enabling better understanding and informed decisions. Yizhou et al. (2024) reviewed DTs for fire management, noting their usefulness in remote monitoring, fire spread forecasting, and firefighter health tracking. Huang, Li, and Zheng (2024) proposed the Wildfire Digital Twin (WFDT) model, which integrates real-time data with simulations to predict and replicate wildfire dynamics.

The main contributions of this chapter are as follows:

1. FAST is introduced as an AI-powered digital twin framework that integrates a 10×10 m^2 simulation environment with adaptive models to predict optimal firefighter escape routes across all wildfire phases in real time.
2. A virtual wildfire simulation replicates realistic fire behavior using diverse data sources. The AI model applies reinforcement learning and deep learning techniques—including YOLO, CNNs, LSTMs, and algorithms such as DQN,

PPO, and A2C—to dynamically identify safe escape routes as conditions change. Both the simulation and AI continuously update to enhance wildfire response.
3. Experimental results from fast-moving wildfire scenarios confirm high prediction accuracy, reduced response time, and improved firefighter safety, demonstrating FAST's practical value in wildfire management.

The remainder of this paper is organized as follows: Section 2 reviews state-of-the-art methods, presents related work, and outlines the problem statement. Section 3 provides a detailed description of the methodology and the FAST framework used in this study. Section 4 presents the experimental results and discussion. Finally, Section 5 concludes the paper.

RELATED WORKS

This section presents a comprehensive review of foundational works and current trends relevant to the study of Digital Twin technologies and the application of Deep Learning and Reinforcement Learning algorithms. The following subsections include discussions on Digital Twin systems and various deep learning and reinforcement learning algorithms, such as YOLOv5, Convolutional Neural Networks (CNN), multivariate Long Short-Term Memory (LSTM) networks, Deep Q-Networks (DQN), Proximal Policy Optimization (PPO), and Advantage Actor-Critic (A2C). Each of these approaches plays a pivotal role in enhancing the capabilities of intelligent systems, particularly in dynamic and complex environments.

Digital Twin

The concept of digital twins has been evolving over the past two decades. In 2002, Professor Michael Grieves at the University of Michigan, USA, introduced the conceptual model of Product Lifecycle Management (PLM), which incorporated the ideas of virtual and real spaces into the mirror space and information mirror models. In 2009, the U.S. Air Force Laboratory coined the term "aircraft digital twin," and by 2010, NASA integrated the digital twin concept into its technology roadmap (Xu et al., 2025). Since 2014, the term digital twin has been widely applied across industries, such as aerospace, intelligent manufacturing, smart transportation, smart cities, precision medicine, and smart agriculture. These applications focus on leveraging the digital twin's capabilities for monitoring, detecting, predicting, controlling, automating, optimizing, decision-making, and lifecycle management (Xu et al., 2024).

The digital twin concept has been defined by researchers and organizations. Madni et al. (2019) describe it as a set of virtual information structures that comprehensively represent potential or actual physical manufacturing products, ranging from microscopic atomic levels to macroscopic geometric levels. It serves as a virtual counterpart to a physical system, continuously updating performance metrics, maintenance records, and status information throughout the lifecycle of the tangible entity. Van der Valk et al. (2022) describe the digital twin as a virtual construct that mirrors its physical counterpart, aggregating diverse data streams to facilitate data processing and manipulation. This construct establishes a bidirectional data bridge between the virtual and physical domains, fostering a symbiotic relationship of information exchange. Wu et al. (2021) highlight that the digital twin integrates information from both the physical and virtual realms, connecting the material world with the virtual world. Through multi-dimensional, multi-scale, multi-disciplinary, and multi-physical parameters, it establishes dynamic virtual models that simulate and describe the attributes, behaviors, and rules of physical entities in real-world environments.

In conclusion, the digital twin can be described as a dynamic and evolving model that bridges the gap between the physical and virtual worlds, enhancing decision-making, optimizing performance, and enabling predictive analysis through the integration of AI technologies.

The Core Components of the Digital Twin

According to Kreuzer et al. (2024), the digital twin consists of three main components: the simulation environment, AI models, and bidirectional communication.

Simulation Environment. The simulation environment acts as a virtual representation of the physical system, providing a platform to model and simulate its behavior, interactions, and conditions. It typically includes models that simulate physical, chemical, mechanical, and environmental processes affecting the system. This environment supports real-time monitoring and testing under various conditions, enabling accurate performance prediction and analysis. It is essential for visualizing and understanding complex systems, such as machinery or infrastructure, allowing scenario testing without impacting the actual system.

AI Models. AI models process data to make predictions, decisions, and optimizations based on insights from both the simulation environment and real-world data. These models use machine learning, deep learning, or reinforcement learning techniques to analyze trends and patterns, facilitating autonomous decision-making. They enhance the digital twin's capabilities by enabling predictive maintenance, anomaly detection, and real-time optimization. Continuously learning from incoming data, AI models allow the digital twin to adapt dynamically to changing conditions, improving system efficiency and performance over time.

Bidirectional Communication. Bidirectional communication integrates the physical and virtual systems, ensuring continuous data exchange between them. It collects real-time data from sensors, actuators, and IoT devices embedded in the physical system to update the digital twin's virtual model. It also enables the digital twin to send commands or adjustments back to the physical system based on AI-driven analyses and predictions. This two-way communication keeps the digital twin synchronized with the real system, allowing ongoing monitoring, adaptive learning, and timely decision-making throughout the system's lifecycle.

Deep Learning and Reinforcement Learning Algorithms

This study applies five algorithms, encompassing deep learning and reinforcement learning techniques, to identify optimal escape routes in wildfire scenarios, with the goal of reducing evacuation time and enhancing firefighter safety.

You Only Look Once (YOLO)

Huyen (2024) describes YOLO as a central object detection model in real-time environments, known for its impressive accuracy and speed in the field of computer vision. The YOLO model's coordinates consist of five components: *w* and *h*, representing the width and height of the input image; *x* and *y*, representing the center coordinates of the bounding box; and the confidence score; *c*, which corresponds to the intersection over union (IoU). The equation for YOLO is shown in Equation (1):

$$L =_{coord} \sum_{i=0}^{S^2} \sum_{j=0}^{B} 1_{ij}^{obj} \left[(x_i + \hat{x}_i)^2 + (y_i + \hat{y}_i)^2 \right] + \ldots \tag{1}$$

Where:

S: grid size,
B: number of bounding boxes per grid cell,
1_{ij}^{obj}: indicator function for the presence of an object,
(x_i, y_i): ground truth center coordinates of the object (actual label),
$(\hat{x}_i, \hat{y}_i)$: predicted center coordinates of the bounding box.

As demonstrated in Equation 1, YOLO divides an image into a grid and predicts bounding boxes and class probabilities directly from the entire image in a single evaluation. The core components of YOLO include three key elements: (1) the backbone, which is responsible for extracting essential features from images or videos; (2) the neck, which performs refinement and aggregation of features from the backbone; and (3) the head, which is the final object detection component that predicts the final output based on the processed features, as described by Vijayakumar et al. (2024), and shown in Figure 1.

Figure 1. The core components of YOLO

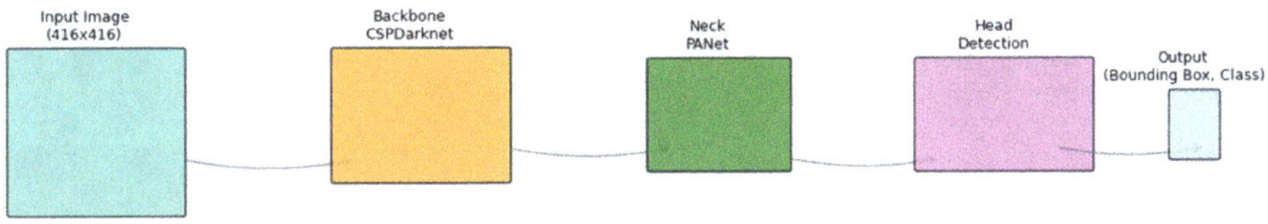

In Figure 1, the input image is of size 416x416 pixels. The backbone extracts feature from the image, which are then forwarded to the neck component for refinement and computation. These processed features are subsequently passed to the head component for setting up the bounding box and predicting the class results.

The example of the YOLO model for object detection in wildfire images and class prediction with precision values is shown in Figure 2.

Figure 2. Example of the YOLO model for object detection in wildfire images and class prediction

In Figure 2, the YOLOv5 model is used for object detection in wildfire images and class prediction. It detects and draws bounding boxes around the full body of a person and other objects such as a backpack, shoes, and a water tank and pump.

Convolutional Neural Network (CNN)

Huyen (2024) and Indolia et al. (2028) describe the convolutional neural network (CNN) as a feature extraction method from images, video streams, and spatial data, which can be used for tasks such as image classification, segmentation, and object detection. Examples of CNN applications include image classification, face detection, speech recognition, and vehicle recognition. The 2D convolution equation used in CNNs is presented in Equation (2):

$$y_{i,j} = \sum_{m=0}^{M-1} \sum_{n=0}^{N-1} x_{i+m,\, j+n} * w_{m,\, n} \qquad (2)$$

Where:

x: input image,
w: kernel or filter,
M x N: size of the kernel,
$w_{m,n}$: weight of the kernel at position *(m, n)*,
$y_{i,j}$: output feature map at position *(i, j)*.

Kunathur et al. (2018) describe the CNN architecture as consisting of a series of convolutional layers, activation layers (such as ReLU or Sigmoid), and subsampling layers, optionally followed by fully connected layers. The input to a convolutional layer is an $w \times h \times r$ image, where w represents the width, h represents the height, and r denotes the number of channels (for an RGB image, $r = 3$). The convolutional layer contains k filters (or kernels) of size $n \times n \times q$, where n is smaller than the dimensions of the image, and q can either be the same as the number of channels (r) or smaller, varying for each kernel. After the convolution operation, the output passes through an activation function, commonly ReLU (Rectified Linear Unit), which introduces non-linearity to the network, allowing it to learn more complex patterns. Each feature map is then subsampled, typically using mean or max pooling over $p \times p$ contiguous regions, where p typically ranges from 3 for smaller images to no more than 5 for larger images. After several convolutional and pooling layers, the output is flattened and passed through one or more fully connected layers. These layers are used to combine the features extracted by the convolutional layers and perform the final classification task. The output of the fully connected layers is passed through a softmax function (in the case of classification) to predict the class labels, as shown in Figure 3.

Figure 3. The core components of CNN

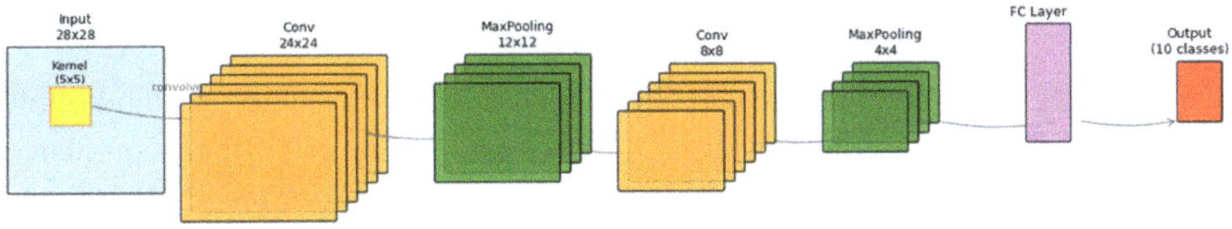

In Figure 3, the input image of size 28×28 pixels is first processed by a 5×5 convolutional kernel, reducing the spatial dimensions to 24×24. A subsequent 2×2 max pooling operation further reduces the size to 12×12. Another 5×5 convolutional layer is then applied, reducing the size to 8×8. The output is flattened into a one-dimensional vector and passed through fully connected layers. Finally, a softmax function is applied to predict the class label for the classification task.

Multivariate Long-Short Term Memory (LSTM)

Diniz et al. (2024) describe Long Short-Term Memory (LSTM) networks as being well-suited for problems involving large data sequences. LSTMs are capable of capturing both short-term and long-term dependencies within temporal sequences. An LSTM consists of a set of recurrently connected subnets, known as memory blocks, which include three multiplicative gates: the forget gate, input gate, and output gate. The mathematical formulation of LSTM is presented in Equation (3):

$$f_t = \sigma(W_f * [h_{t-1,} x_t]+b_f) \tag{3}$$

$$i_t = \sigma(W_i * [h_{t-1,} x_t]+b_i)$$

$$\widetilde{C}_t - \tanh(W_C * [h_{t-1,} x_t]+b_C)$$

$$C_t = f_i * C_{t-1} + i_t * \widetilde{C}_t$$

$$o_t = \sigma(W_0 * [h_{t-1,} x_t]+b_0)$$

$$h_t = o_i * \tanh(C_t)$$

Where:

x_t: input at time t,
h_t: hidden state,

C_t: cell state,
σ: sigmoid activation.

According to the architecture and operational pipeline explained in Goodfellow et al. (2016), the long-term memory, also referred to as the cell state - is regulated by the forget gate f_t, which determines which parts of the previous cell state should be discarded. The input gate i_t controls which information should be added to the cell state, while the output gate o_t determines which parts of the memory are used to generate the current output. The core structure of an LSTM cell is illustrated in Figure 4.

Figure 4. The core components of LSTM

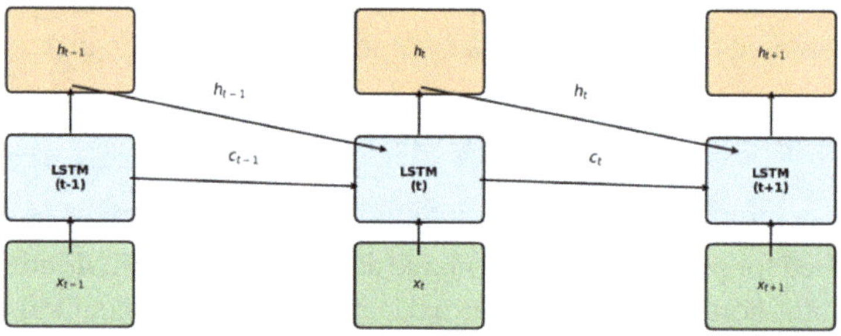

In Figure 4, x_t represents the input signal, C_t and C_{t-1} denote the current and previous cell states, and h_t and h_{t-1} represent the current and previous hidden states, respectively. In this study, a multivariate LSTM model is employed to handle multiple features at each time step, such as temperature, wind speed, flame direction, and object coordinates.

Deep Q-Networks (DQN)

Zhang et al. (2020) describe Deep Q-Networks as an approach based on Q-learning algorithms that approximate the state-action value function (Q-function) Q(s, a; θ) using a neural network parameterized by θ. The optimal Q function is learned by minimizing the mean squared error between the predicted and target Q-values, where the target values are computed based on the observed rewards and the maximum predicted Q-value of the next state, as shown in Equation (4):

$$Q(S_t, a_t) = r_t + \gamma max_{\acute{a}} Q(S_{t+1}, \acute{a}) \quad (4)$$

Where:

γ: discount factor,
S_t: state in time t,
a_t: action in time t,
r_t: reward in time t.

Q-learning aims to learn a value function $Q(S_t,a_t)$, which represents the expected return of taking action a_t in state S_t, and subsequently following the optimal policy. In the context of the discussed model, the action aa refers to movement in one of four directions—left, right, up, or down—or remaining in place (De la Fuente et al., 2024).

Proximal Policy Optimization (PPO)

Jiang (2024) describes Proximal Policy Optimization (PPO) as a sophisticated actor-critic deep reinforcement learning algorithm that has gained prominence for addressing challenges in training RL agents in complex environments. In PPO, the actor learns the policy, which determines the agent's actions in specified states, while the critic evaluates the effectiveness of selected actions in particular states. PPO improves traditional policy gradient methods by introducing a clipped surrogate objective, which limits the policy update to a predefined threshold. This prevents drastic policy changes that could lead to instability during training, as detailed in Equation (5):

$$L^{CLIP}(\theta) = E[\min(r_t(\theta)\widehat{A}_t, clip(r_t(\theta), 1-\varepsilon, 1-\varepsilon,)\widehat{A}_t)] \quad (5)$$

Where:

$r_t(\theta) = \frac{\pi_{(\theta)}(a_t|s_t)}{\pi\theta_{old}(a_t|s_t)}$: probability ration,
E_t: expected value at time t,
$\widehat{A}_t$: advantage estimate,
ε: clipping parameter.

Equation (5) defines the clipped surrogate objective function used in PPO. It calculates the expected value of the minimum between the probability ratio $r_t()$ and the clipped probability ratio, multiplied by the advantage estimate . The probability ratio $r_t()$ is the ratio of the probabilities of taking action a_t under the current policy $\pi_{()}$ to the old policy $\pi_{()}| \pi_{\theta_{old}}$. The clipping parameter (6)

Where:

$\nabla_\theta J(\theta)$: gradient of objective function,
θ: parameter of policy,
$\nabla_\theta \log \pi_\theta(a_t | s_t)$: gradient of log-probability of action a_t in state s_t.

Equation (6) represents the gradient of the objective function $J()$in the A2C algorithm. The equation calculates the gradient of the log-probability of the action a_t taken at time t under the current policy, weighted by the advantage estimate $\widehat{A}_t$. This gradient is used to update the policy parameters θ in the direction that maximizes the expected return. The expectation E_t indicates that the gradient is averaged over multiple time steps, ensuring the policy is optimized based on observed actions and their effectiveness. This approach allows A2C to efficiently update the policy through both actor and critic networks.

From equations (4), (5), and (6), it is found that DQN improves the policy by learning from the immediate outcomes of its actions, which are straightforward and tightly coupled with the actions taken. In contrast, the policy methods of PPO and A2C directly estimate the policy. The policy $\pi(a|s)$ represents the probability of taking action a_t in state s_t, and it is adjusted in the direction suggested by the policy gradient.

FAST FRAMEWORK

The Firefighter AI Safety Digital Twin (FAST) framework was developed to optimize firefighter escape routes within a digital-twin-based wildfire simulation environment. The framework was evaluated across three distinct wildfire scenarios, demonstrating capabilities for rapid response and adaptive decision-making. FAST comprises four key components: (1) a simulation environment representing dynamic wildfire conditions, (2) an AI-driven model for predicting fire spread and escape paths, (3) a bidirectional data communication layer enabling real-time synchronization between physical and virtual systems, and (4) an assessment module for evaluating firefighter escape routes under evolving environmental hazards. The components and their interactions are illustrated in Figure 5.

Figure 5. Overview of FAST framework

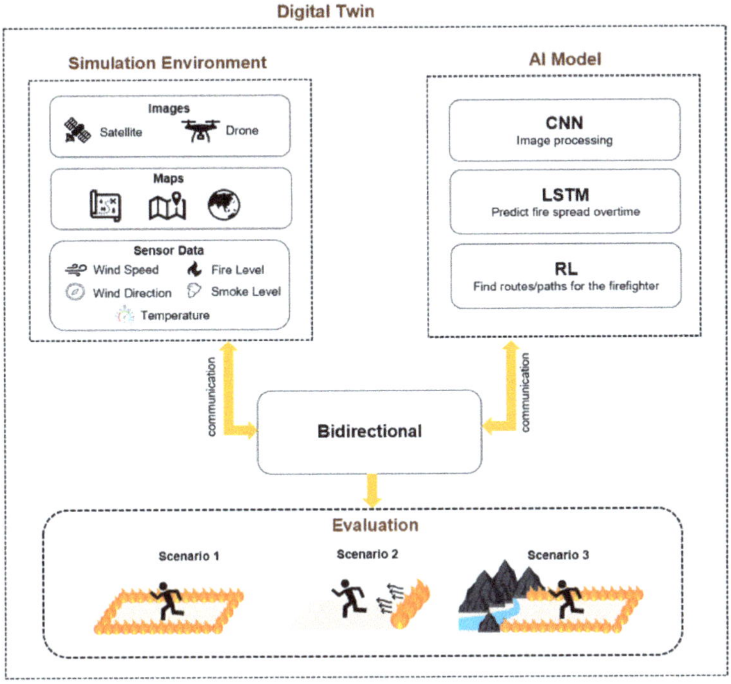

THE SIMULATION ENVIRONMENT

In the first component of the FAST framework, a digital twin environment was developed to simulate the physical landscape of Op Luang National Park, located in Chiang Mai, northern Thailand. This simulation analyzed recurring wildfire incidents and firefighter fatalities, which typically occur annually from February through the end of May. Each official firefighter and volunteer wildland firefighter was assigned responsibility for a 10 × 10-meter area, with the objective of preventing and extinguishing wildfires before moving to the next designated zone. The 10 × 10-meter grid was selected for its operational practicality, representing the typical area a single firefighter can effectively monitor and control under challenging wildfire conditions. However, the terrain of Op Luang National Park is highly diverse, comprising steep mountains, slopes, rivers, deep ravines, and flat terrain. The accumulation of dry leaves further increases the risk of ignition and rapid fire spread. Moreover, firefighters are required to carry limited water supplies along with wildfire protection jackets and essential firefighting equipment, which poses significant risks to their safety. To replicate realistic field conditions, three high-risk wildfire scenarios were simulated, each characterized by limited or nonexistent escape routes for firefighters (US Forest Service, 1999; O'Connor et al., 2017):

1. Fire surrounding the firefighter: In the first scenario, the fire encircles the firefighter on all sides, making escape difficult and requiring immediate decision-making.
2. Fire parallel to the firefighter along 10 meters with backdraft wind: In the second scenario, the fire ran parallel to the firefighter's position (for 10 meters), and strong backdraft winds posed a significant threat to survival, necessitating strategic escape planning.
3. Fire surrounding three sides, with a river or steep mountain on the fourth side: In the third scenario, the fire surrounds the firefighter on three sides, leaving only one potential escape route—either toward a river or up a steep mountain. This scenario requires the firefighter to reach higher ground or a safer route as quickly as possible, as illustrated in Figure 6.

Figure 6. 2D simulation of a 10 × 10-meter area (in square meters, m²) for the three wildfire scenarios

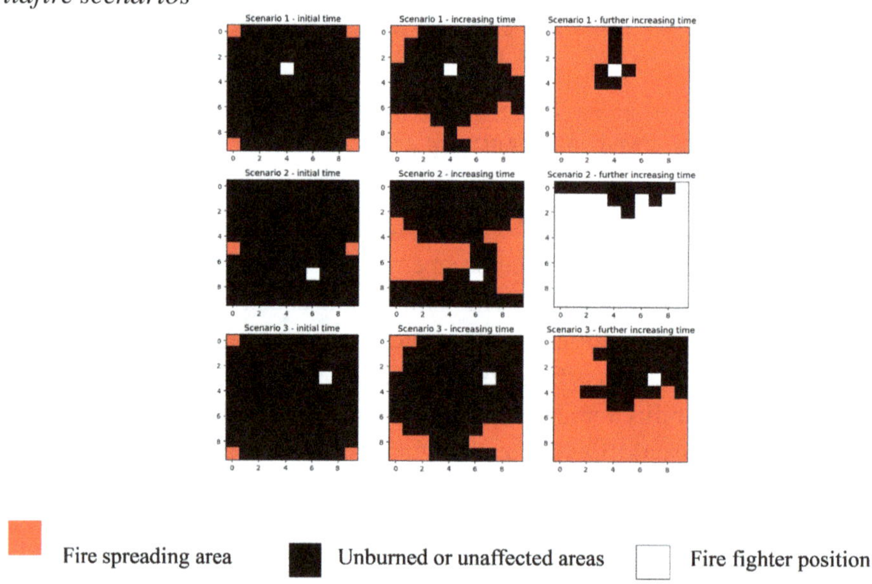

In Figure 6, a simulation of a 10×10-meter area using two-dimensional visualization to depict the progression of wildfire in three distinct scenarios over time.

The simulation is divided into three temporal phases: (1) the initial time, denoted as t_0; (2) an intermediate phase, where time increases progressively ($t_0 < t_1 < t_2 < \ldots < t_n$); and (3) an advanced phase of further time progression ($t_n < t_{n+1} < t_{n+2} < \ldots$). In Scenario 1, at t_0, the firefighter is initially surrounded by fire on all four sides. As time progresses, the fire spreads rapidly and aggressively, ultimately enclosing the firefighter completely. This scenario represents an extreme threat to life and necessitates immediate evacuation. In Scenario 2, the fire spreads in the direction of the wind, running parallel to the firefighter over a 10-meter stretch. As a result of the backdraft effect, the firefighter becomes engulfed in smoke, posing a high risk of suffocation. In such a life-threatening situation, evacuation must be carried out in the direction opposite to the wind flow to avoid smoke inhalation. In Scenario 3, the firefighter is surrounded by fire on three sides at t_0. As the fire continues to spread over time ($t_0 < t_1 < t_2 < \ldots < t_n$), the available safe zones—indicated by black regions—gradually diminish. White areas may still indicate viable escape routes during this phase. However, in the final stage ($t_n < t_{n+1} < t_{n+2} < \ldots$), the fire has spread across almost the entire area, leaving only a minimal portion as a possible exit. The black and white regions are interpreted as potential escape routes, possibly leading to higher ground or a nearby river. The firefighter must make a swift and informed decision before all exits are fully obstructed by the fire.

The data for the simulation environment was collected from multiple sources, including satellites, videos, images, maps, and sensors, during the wildfire at Op Luang National Park in Chiang Mai, northern Thailand, from February 20 to 28, 2023, covering over 200 rai (approximately 80 acres),

1. The dynamic imagery was obtained from NASA FIRMS, using Landsat (30 m resolution) and VIIRS (S-NPP, NOAA-20 & NOAA-21 with 375 m resolution) satellites, amounting to approximately 20,000 images. Additionally, public drone footage from firefighter volunteers of the Mirror Foundation was utilized, extracting 25 frames per second over a 30-minute duration, resulting in approximately 45,000 frames.
2. Maps were sourced from GISTDA (Geo-Informatics and Space Technology Development Agency) and Google Maps to provide terrain views of the landscape.
3. Sensor data included six key features: timestamp, wind speed, fire level, wind direction, smoke level, and temperature. This data was collected from land-based sensors and the MODIS dataset (Moderate Resolution Imaging Spectroradiometer) as part of Thailand's fire monitoring system, totaling 10,500 records (generated at a rate of 100 instances per 20 minutes, operating 7 hours per day over 5 days) as show in Figure 7.

Figure 7. Examples of data sources used for the simulation environment

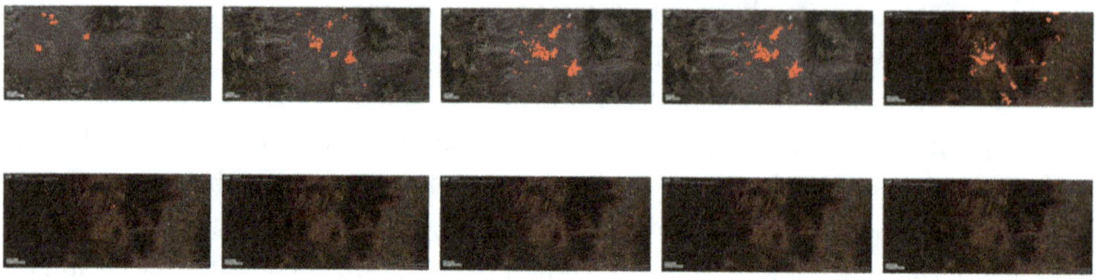

(a) Satellite images of wildfire and non-wildfire areas.

(b) Terrain view of the landscape

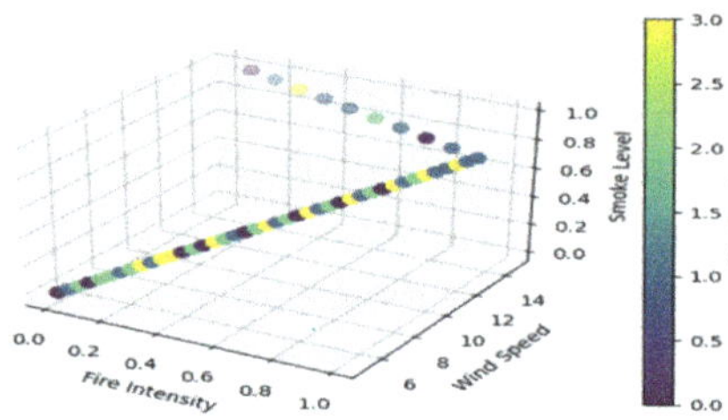

(c) 3D visualizations in the wildfire simulation

In Figure 7, the data sources used to construct the simulation environment include: (a) satellite imagery from NASA FIRMS, depicting wildfire and non-wildfire regions to provide a bird's-eye view of fire spread across the landscape; (b) terrain maps from Google Maps, illustrating key topographical features; and (c) 3D visualizations representing fire intensity, wind speed, and smoke levels within the simulated environment over a twenty-minute period.

THE AI-DRIVEN MODEL

In the second component of the FAST framework, data from diverse sources is integrated to sequentially develop three AI models: YOLOv5 and CNNs for object detection and feature extraction, a Multivariate Long Short-Term Memory (LSTM) model for fire spread prediction, and reinforcement learning algorithms for determining the optimal escape route based on the predicted fire dynamics. This modeling pipeline facilitates effective real-time decision-making within a constrained timeframe (i.e., less than the available safe time), thereby significantly enhancing firefighter survivability across a range of emergency scenarios.

Object Detection and Feature Extraction Using YOLOv5 and CNN

For the initial model development, YOLOv5 was utilized to identify key objects in images, including firefighters, backpacks, fire, and smoke. The model was also pre-trained on specific classes, such as steep slopes and rivers, to detect these features across the entire image and return bounding boxes for each detected object. Each region of interest (ROI) was subsequently extracted for further feature analysis using a Convolutional Neural Network (CNN). The CNN was responsible for feature extraction and environmental analysis by applying convolutional layers to capture spatial features, max pooling to reduce the dimensionality of the feature maps, and fully connected layers to aggregate features for class prediction. The outputs from the CNN for each ROI were concatenated into a single vector, which served as input to the subsequent model—a Multivariate Long Short-Term Memory (LSTM) network—as illustrated in Algorithm 1.

Algorithm 1. Object Detection and Feature Extraction Using YOLOv5 and CNN

1: **Input:** Load images and class labels (l, λ) = load_image_labels (I, L)
2: Predict the bounding box using YOLOv5:

$$\widehat{B} = (x, y, w, h, c_1, c_2, \dots c_n)$$

and update in an Equation (1).

3: Compute the regions of interest (ROIs):

$$ROI_{fire}, ROI_{firefighter}, ROI_{smoke}, ROI_{terrain}$$

4: Compute the feature extraction using CNN:

$$X_{conv} = Conv2D(I_{input}, W, b)$$

and update in an Equation (2).

5: Compute the convolution operation:

$$X_{conv}(\mathrm{i}, \mathrm{j}) = \sum_{m,n} I_{input}(i + m, j + n) * W(m, n) + b$$

6: **Output:** $X_{combined}$ #combined CNN

As described in Algorithm 1, the images *I* and their corresponding labels *L* are loaded, and object detection is performed using YOLOv5. The ability of YOLOv5 can detect fire, smoke, and other relevant in real-time with multi-class detection. The bounding box $\widehat{B}$ is predicted, which includes the coordinates *(x, y)*, as well as the width *w* and height *h* of the bounding box, based on class probabilities c_1, c_2,..., c_n. The regions of interest (ROIs) are then identified for further processing by the CNN. These ROIs are passed through the CNN for feature extraction, where each image undergoes a series of convolutional operations. The convolution operation is performed at each position *(i, j)*, followed by max pooling to reduce spatial dimensions. The resulting feature map is then flattened through fully connected layers. Finally, the CNN outputs corresponding to fire, firefighters, smoke, and terrain are concatenated into a single vector, which is then used as input to the multivariate LSTM model for temporal analysis.

Fire Spread Prediction Using Multivariate LSTM

For the second model, a Multivariate Long Short-Term Memory (LSTM) network was employed to predict fire spread, with its parameters fine-tuned to optimize performance based on the provided dataset. This model enhancement supports real-time decision-making in dynamic wildfire scenarios, where environmental conditions evolve rapidly. A core challenge in this task involved the integration of multiple spatial and temporal features into the multivariate LSTM architecture. These features—such as fire intensity, wind speed, wind direction, smoke level, temperature, and terrain indicators—were extracted from satellite or drone imagery using a Convolutional Neural Network (CNN) and encapsulated in a combined feature vector.

The Multivariate LSTM is particularly suited for this task, as it is capable of processing sequences of time-series data in which each time step includes multiple feature variables. Algorithm 2 details data preparation, LSTM-based prediction, and model optimization for estimating fire spread intensity.

Algorithm 2. Fire Spread Prediction Using Multivariate LSTM

1: **Input:** $X_{combined}$

2: Preprocess the input: #Reshape $X_{combined}$ into a sequence format suitable for a multivariate LSTM.

X_{LSTM} = reshape ($X_{combined}$, (T, F))

3: Compute hidden states h_t: and cell states c_t at each time step t:

$$h_t, c_t = \text{LSTM}\,(X_{LSTM}, hidden_{size}, num_{layers}, dropout)$$

4: Predict fire spread at final time step:

$$\hat{Y} = (W_y h_T + b_y)$$

5: Compute the loss function:

$$L = \frac{1}{N}\sum_{i=1}^{N} (Y_i - \hat{Y}_i)^2$$

6: Apply the gradient descent optimization:

$$\theta \leftarrow \theta - \eta \nabla_\theta L$$

7: **Output:** fire_spread_predictions

As described in Algorithm 2, after extracting spatial features using CNN, the combined feature vector $X_{combined}$ is reshaped into a multivariate sequence format for LSTM processing, represented as *(T, F)*, where *T* denotes the time steps and *F* represents the feature dimension at each time step. The multivariate LSTM processes this input sequentially to capture temporal dependencies and model the progression of fire spread over time. The hidden state h_t retains information from previous time steps and serves as the output, while the cell state c_t stores long-term dependencies within the sequence. To enhance prediction accuracy, optimization techniques are applied, including a loss function to minimize the error in fire spread prediction and gradient descent to refine the model parameters. During the multivariate LSTM forward pass, the predicted fire spread intensity $\hat{Y}$ at future time steps is computed, where W_y represents the weight matrix and b_y is the bias term in the final output layer. At the final time step *T*, the hidden state h_T is used in the output layer for the prediction. The loss function minimizes the difference between the predicted fire spread $\hat{Y}$ and the actual fire spread *Y*. Additionally, model parameters, including weights, biases (θ), the learning rate (η), and the gradient of the loss function ($_\theta L$) are updated using the gradient descent method. This iterative process reduces the loss function, optimizing the model and improving prediction accuracy. This ensures a reliable approach to help firefighters escape hazardous fire zones in the shortest possible time.

Reinforcement Learning with DQN, PPO and A2C

In the final stage of model development, reinforcement learning (RL) algorithms were implemented to find optimal escape routes quickly, aiming to enhance firefighter safety. Three RL algorithms—Deep Q-Network (DQN), Proximal Policy Optimization (PPO), and Advantage Actor-Critic (A2C)—were evaluated for their suitability. These algorithms were tested across three wildfire scenarios to identify the most effective method for generating safe and efficient escape paths. The implementation steps are summarized in Algorithm 3.

Algorithm 3. Reinforcement Learning with DQN, PPO and A2C

1: **Input:** fire_spread_predictions
2: Compute the RL model (model_type, state, action_space): *#model_type('DQN','PPO','A2C')*
3: Compute the Q-function for DQN using Equation (4):
4: Compute the policy gradient for PPO and A2C using Equation (5) to (6):
5: Return action, reward, next_state, done
6: Compute reward r_{total} based on the current state:

r_{total}= + R_s if s = safe #reward for reaching safety

-R_d if d = 0 #panelty for standing in fire

-1-1 * d if 0 < d <= 1

-1-2 * d if 1 < d <= 2

-2- d if d > 2

0 if t <= T_{safe}

-R_t if t > T_{safe}

7: Repeat:
8: Update the policy based on the reward signal
9: Compute the next escape step:
10: S_{safe} = move_firefighter (action)
11: Until convergence or termination criteria met
12: **Output:** Escape route

As described in Algorithm 3, the input vector generated by the multivariate LSTM model, which captures the temporal predictions of fire spread, is used as the current state C_t for the reinforcement learning (RL) model. Based on this state, the RL agent selects the next action N_t, such as a movement direction, using its current policy $\pi (N_t| C_t)$, and transitions to the subsequent state C_{t+1}.

The reward function R_t is dynamically computed based on both the firefighter's proximity to fire and elapsed time. This function encourages the agent to move toward safe zones while penalizing proximity to hazards and delays. Specifically:

- If the firefighter reaches a designated safe zone (S_{safe}), a positive reward $+R_s$ is granted.
- If the firefighter is located at the same position as the fire (i.e., $d=0$), a significant penalty - R_d is applied.
- Additional penalties are applied based on tunable parameters α_1, α_2, β_1, β_2 that account for the distance d between the firefighter and the fire.
- If the maximum escape time (T_{max}) does not exceed the safety threshold (T_{safe}), no penalty or reward is assigned.
- If $t>T_{safe}$, a time-based penalty $-R_t$ is applied to discourage delays.

The model iteratively updates its policy based on the reward signal until it converges to an optimal policy maximizing firefighter safety—reflected by achieving a net positive reward. The optimal policy is reached when the model consistently selects actions leading to safe escape routes with maximum survival reward.

BIDIRECTIONAL COMMUNICATION IN AI AND DIGITAL TWIN SYSTEMS

Bidirectional communication is a critical component of digital twin systems, enabling continuous data exchange and real-time interaction with AI models. This mechanism supports the dynamic reception of new data, integration of updated knowledge, and ongoing refinement of predictions and decision-making processes. Within the FAST framework, bidirectional communication is established once the digital twin and AI model are fully deployed, facilitating continuous learning, validation, and adaptation under real-world wildfire conditions. Given the highly dynamic nature of wildfires—driven by shifting wind directions, evolving fire spread patterns, firefighter mobility, and real-time sensor updates—bidirectional integration is essential for maintaining synchronization between the physical environment and its virtual representation. This process significantly enhances the system's reliability, responsiveness, and predictive accuracy over time.

For example, consider a scenario where a firefighter is surrounded by fire on three sides, with only three possible escape routes:

- A narrow route with thick smoke (uncertain safety)
- A slope leading to higher ground (safer but slow)
- A partially burning area that may soon become impassable (high risk)

The AI must determine the optimal escape route in real time. However, several factors—such as fire spread, wind direction, fire intensity, and temperature—change unpredictably over time. Therefore, the AI must continuously adapt in real-time to maximize the firefighter's chances of survival by dynamically responding to environmental changes. The details of the bidirectional process are outlined in Algorithm 4.

Algorithm 4. Bidirectional in AI and Digital Twin System

Initialize Q (s, a) *#Initialize Q-values*
Initialize Policy (π(s)) *# Initialize policy for selecting actions from state*
Initialize Environment (simulation) *# Initialize digital twin environment*
Initialize YOLO, CNN and LSTM
Input: Real-time environment and digital twin data
Compute current environment state:
$S_t = \{P_t, F_t, W_t, T_t\}$ #S_t = current state
Update AI modules:
YOLO (I_t) = detect (I_t) #Update YOLO for object detection from imagery
CNN (I_t) = feature_map (I_t) #Update CNN for spatial feature extraction
h_t = LSTM (x_t, h_{t-1}) #Update LSTM for temporal fire spread modeling
Compute the simulation environment interaction with RL:
Simulate interaction with RL agent:
$a_t = \pi (s_t)$ #Select action based on the policy
Compute the transition from state to next state:
S_{t+1}, r_t = Environment.step (a_t) #Transition to the next state
Compute the reward:
r_t = R (S_t, S_{t+1}, a_t) #Calculate reward based on state transition
Update the Q-value using Equation (4), and update the Policy using Equation (5) and (6):
Continue the loop: *#Repeat steps until termination*
Output: Firefighter escapes successfully

As described in Algorithm 4, the initialization process involves setting up the key components required for the digital twin and AI system. These include the Q-values, policy, the digital twin environment, and the YOLO, CNN, and LSTM models. These models are critical for detecting and processing real-time data. Following the initialization, real-time data is collected from various sources such as satellites, drones, maps, sensors, and weather stations. The state S_r is computed using the current environmental data, including the firefighter's position P_t, fire intensity F_t, wind speed and direction W_t, temperature T_t, and smoke levels. This integrated data is used to update both the simulation and the AI model in the digital twin system, providing new states and rewards that drive the learning process. The bidirectional communication within the reinforcement learning (RL) framework simulates the interaction between the policy function and the environment. The state transition from S_t to S_{t+1} is determined by taking action a_t, and the corresponding reward r_t is computed. Based on this reward and transition, both the Q-values and the policy are updated according to their respective update rules (Equations (4), (5), and (6)). The iterative loop continues until the simulation is complete. The bidirectional communication mechanism enables the AI model to continuously adapt and improve decision-making over time, thereby enhancing the firefighter's chances of survival.

EVALUATION

The proposed digital twin system is evaluated under three wildfire scenarios: (1) fire surrounding the firefighter completely, (2) fire running parallel with intense backdraft winds along 10 meters, and (3) fire surrounding on three sides with only one narrow escape route, a scenario linked to firefighter fatalities. The FAST system was deployed on an HPE ProLiant DL380 Gen11 server equipped with an Intel Xeon-Gold 6426Y 16-core CPU, 64GB DDR5 RAM, and 3x 960GB SATA SSDs for fast data access. The goal in each scenario is to escape within the safe time (T_safe) and identify the optimal escape route. The digital twin continuously processes real-time data, including fire spread, wind vectors, terrain, temperature, and firefighter location, to recommend adaptive actions that optimize survival. It integrates reinforcement learning agents with multivariate LSTM-based fire prediction and visual recognition modules (YOLO, CNN), ensuring responsiveness to rapidly changing conditions.

RESULT AND DISCUSSION

The FAST framework was applied to the digital twin system through the collection of multi-source data. Satellite imagery was obtained from NASA FIRMS, incorporating data from Landsat (30 m resolution) and VIIRS (S-NPP, NOAA-20, and NOAA-21 with 375 m resolution), yielding approximately 20,000 images. Additionally, public drone footage from firefighter volunteers of the Mirror Foundation contributed around 45,000 frames. Geospatial maps from GISTDA and Google Maps added approximately 12,000 images. Sensor data, comprising six key environmental features—timestamp, wind speed, fire level, wind direction, smoke level, and temperature—were collected from land-based sensors and MODIS, resulting in 10,500 records. The dataset was divided into training and testing sets using an 80:20 ratio. For instance, the sensor data captured 100 instances over a 20-minute period. Details of the FAST framework are illustrated in Figure 8.

Figure 8. Example data from sensors in 3D, displaying features such as fire intensity, wind speed, and smoke levels

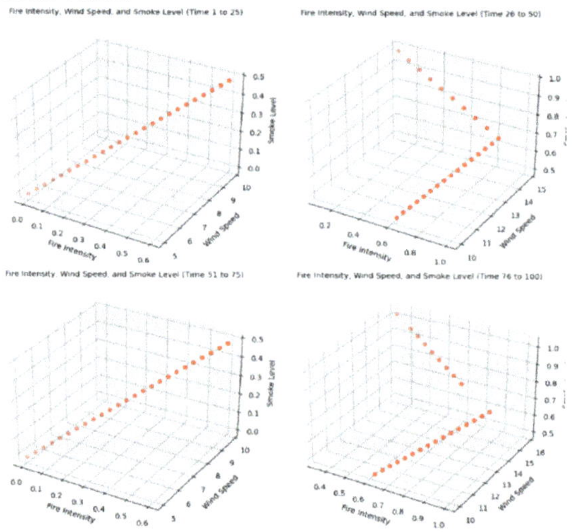

In Figure 8, presents a series of four 3D plots illustrating the progression of fire intensity, wind speed, and smoke level across four-time intervals: timestamps 1–25, 26–50, 51–75, and 76–100. These plots provide a comprehensive visualization of wildfire dynamics and environmental changes during a simulated fire event. The first plot (Time 1–25), the fire intensity remains consistently at 0.0, indicating the absence of active fire during this initial period. Wind begins to blow from the west (W) at a speed of 5.0 meters per second and subsequently intensifies, which corresponds with the early signs of fire emergence. The smoke level also starts to increase progressively, suggesting deteriorating air quality as fire conditions begin to develop. In the second time window (Time 26–50), a notable rise in fire intensity is observed in conjunction with accelerating wind speed. This phase represents the ignition and early spread of the wildfire, where both fire and smoke begin to escalate in severity. During Time 51–75, all three variables—fire intensity, wind speed, and smoke level—demonstrate a steady upward trend. This indicates a critical stage of wildfire development, where the environment becomes increasingly hazardous and difficult to manage. In the final phase (Time 76–100), fire intensity shows signs of decline, possibly due to partial containment or fuel exhaustion. However, smoke levels remain high, highlighting the lingering effects of the fire and the continued risk to visibility and respiratory health.

SIMULATION ENVIRONMENT ACROSS THREE WILDFIRE SCENARIOS

Wildfire environments were simulated using data derived from a digital twin, and a comparative visualization of three distinct scenarios was presented within a 10×10-meter grid. Each scenario illustrates the dynamic spread of fire over three-time steps (t = 0, 10, 19), with a blue dot representing the initial position of a firefighter. Fire intensity is indicated by a color scale ranging from low (black) to high (bright red). These visualizations are designed to model complex and high-risk wildfire conditions, providing test environments for AI-driven decision-making systems and digital twin frameworks aimed at enhancing firefighter safety, as shown in Figure 9.

Figure 9. Simulation of environment in three scenarios

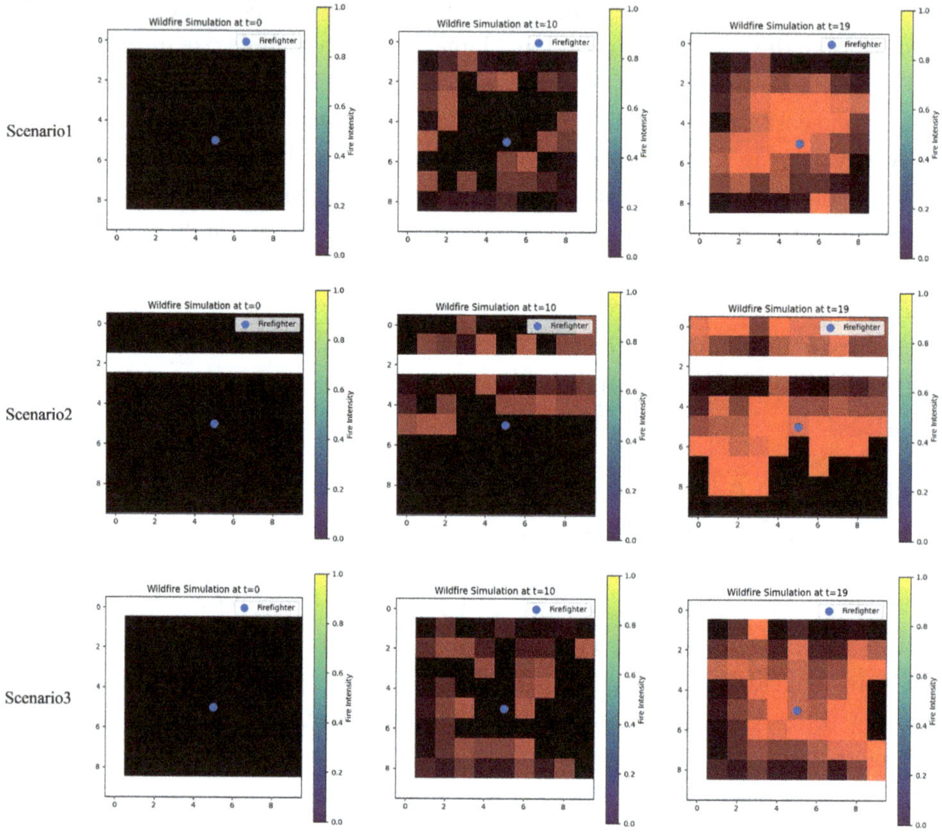

Figure 10. Comparison of fire spread

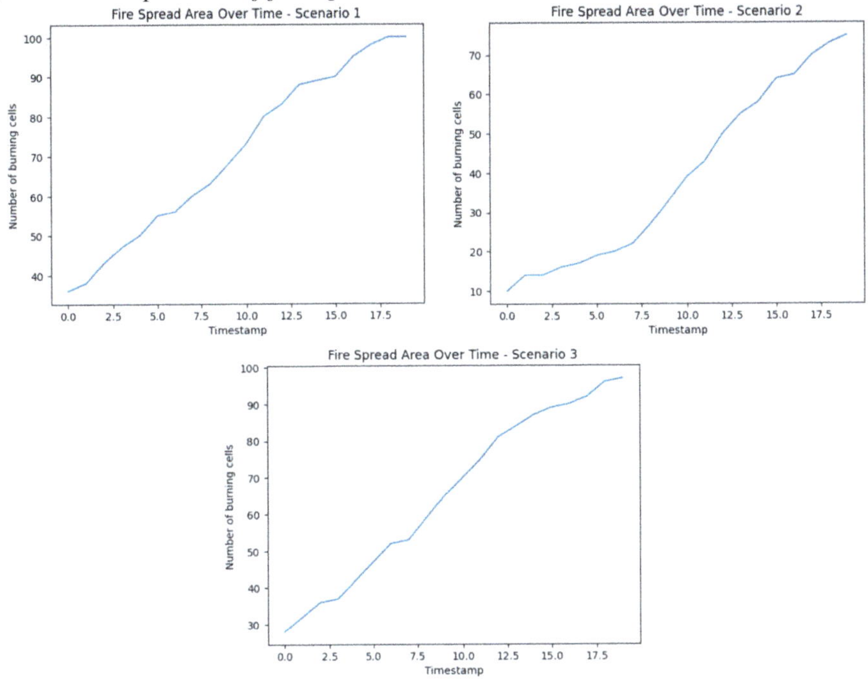

Figure 9(a) illustrates three distinct fire spread scenarios with a firefighter in a grid environment. In Scenario 1, the firefighter is centered and gradually surrounded by fire. At $t = 0$, no fire is present; by $t = 10$, flames spread inward from the edges; and by $t = 19$, the grid is nearly fully engulfed, trapping the firefighter. Scenario 2 shows a fire line running parallel to the firefighter, driven by backdraft winds. The fire begins at the upper edge at $t = 0$, spreads downward by $t = 10$, and by $t = 19$, significantly threatens the firefighter. Scenario 3 depicts a constrained escape scenario with fire on three sides and a clear edge on the right, representing a river or steep terrain. At $t = 0$, escape is only to the right; by $t = 10$, fire spreads moderately, shrinking the safe zone; and by $t = 19$, most of the grid is on fire, leaving a narrow corridor. This tests AI-based escape planning in tight spaces.

Figure 9(b) quantitatively compares fire spread by burning cells over time. Scenario 1 has the fastest spread, increasing from ~35 to 100 cells by timestamp 18, reflecting highly flammable or windy conditions. Scenario 2 shows the slowest growth, with low spread until timestamp 8, then moderate increase to ~75 cells, suggesting environmental controls or firebreaks. Scenario 3 shows steady growth reaching ~96 cells by timestamp 19, similar to Scenario 1 but with slightly slower spread, indicating less favorable fire propagation conditions.

YOLO5 AND CNN

YOLOv5 and CNN were integrated to construct the model, with hyperparameter tuning performed over 1,000 training epochs. A random search strategy was employed to identify the optimal parameters, as illustrated in Figure 10.

Figure 11. Evaluation of YOLOv5, CNN accuracy, and combined accuracy

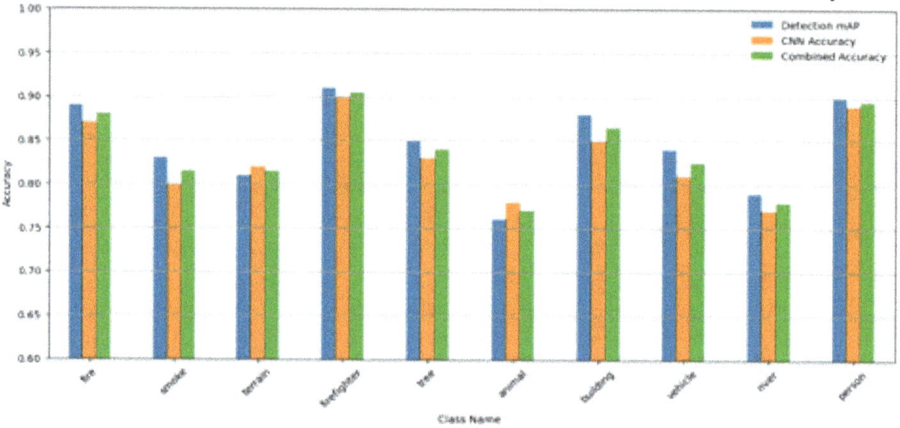

In Figure 10, the ten evaluated classes include fire, smoke, terrain, firefighter, tree, animal, building, vehicle, river, and person. The firefighter class achieved the highest performance, with a detection mAP of 0.91, a CNN accuracy of 0.90, and a combined accuracy of 0.90. The person and fire classes ranked second and third, with combined accuracies of 0.895 and 0.880, respectively. Notably, eight out of the ten classes achieved a combined accuracy exceeding 0.80, with the exceptions being the animal and river classes.

Multivariate LSTM

The Fire Spread Prediction model using multivariate LSTM was trained for 1,000 epochs, with the loss decreasing from 0.225 in the first epoch to approximately 0.025 at the 1,000th epoch. Hyperparameter tuning was conducted using a random search strategy, adjusting key parameters such as hidden layer size (from 128 to 256), learning rate (0.0005 to 0.001), and batch size (8 to 16).

Figure 12. A multivariate LSTM model loss across epochs before and after hyper-parameter tuning

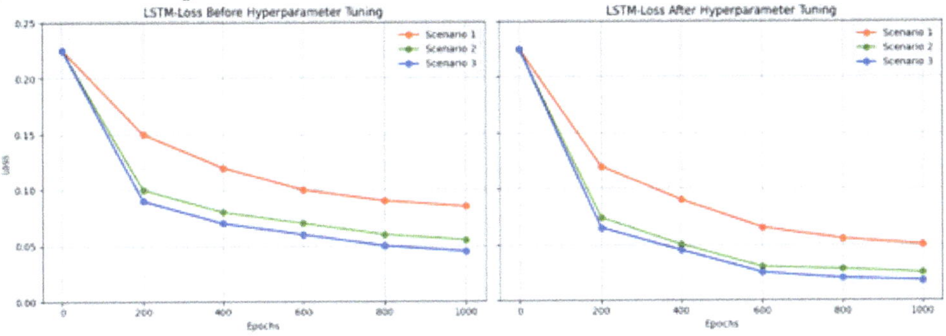

In Figure 11, the loss decreased consistently across all three scenarios. On average, the loss reduction indicates effective model learning. Scenario 3 exhibited the best performance after 1,000 epochs, achieving a final loss around 0.04 with early stopping applied. This value is close to that of Scenario 2, which ended with a loss of 0.059, while Scenario 1 showed a higher loss of 0.09, reflecting its greater complexity.

After hyperparameter tuning via random search, loss values further improved: Scenario 3 achieved a loss of 0.025, closely followed by Scenario 2 at 0.027, and Scenario 1 at 0.05. These results demonstrate that the multivariate LSTM model is effective and robust in predicting fire spread across all three scenarios, adapting well to varying levels of difficulty. Therefore, multivariate LSTM is a suitable approach for modeling fire dynamics in these diverse wildfire situations.

MULTIVARIATE LSTM AND REINFORCEMENT LEARNING ALGORITHMS

The multivariate LSTM model was evaluated in conjunction with reinforcement learning algorithms—Deep Q-Network (DQN), Proximal Policy Optimization (PPO), and Advantage Actor-Critic (A2C)—to determine the optimal escape routes under three wildfire scenarios. The evaluation metrics included Fire Spread Prediction Loss, Total Reward (average per episode), Escape Success Rate (%), Average Escape Time (seconds), and Training Time (steps).

Table 1. Performance comparison of multivariate LSTM combined with reinforcement learning algorithms (DQN, PPO, A2C) across wildfire escape scenarios

Metrics	Multivariate LSTM+DQN			Multivariate LSTM+PPO			Multivariate LSTM+A2C		
	Scenario1	Scenario2	Scenario3	Scenario1	Scenario2	Scenario3	Scenario1	Scenario2	Scenario3
Fire spread Prediction Loss	0.017	0.015	0.014	0.023	0.025	0.021	0.027	0.030	0.033
Total Reward (avg/ep)	150	160	160	120	140	155	120	110	130
Escape Success Rate (%)	80%	83%	84%	75%	78%	78%	69%	72%	73%
Average Escape Time (s)	310	280	260	360	330	320	390	380	370
Training Time (steps)	14,336	14,336	14,336	17,669	17,669	17,669	19,432	19,432	19,432

In Table 1, the multivariate LSTM + DQN model outperformed the multivariate LSTM + PPO and A2C models. Specifically, Scenario 3 yielded the lowest Fire Spread Prediction Loss of 0.012, while Scenarios 2 and 3 achieved the highest Total Reward (average per episode) of 160. Scenario 3 also demonstrated the highest Escape Success Rate at 84% and the shortest Average Escape Time of 260 seconds. Furthermore, the multivariate LSTM + DQN model required the least Training Time across all scenarios, with only 14,336 steps. These results suggest the multivariate LSTM + DQN model is both effective and practical for real-world wildfire escape route optimization.

BIDIRECTIONAL IN AI AND DIGITAL TWIN SYSTEM

The bidirectional interaction between AI and digital twin systems aims to enhance real-time decision-making and overall system accuracy. The experiments evaluate key performance metrics—including Real-time Sync Accuracy, Fire Spread Prediction Accuracy, Escape Plan Update Delay, Escape Success Rate, Average Escape Time, and System Latency per Cycle—to determine the optimal update intervals for efficient and effective system performance, as illustrated in Figure 12.

Figure 13. Comparison of bidirectional update frequencies

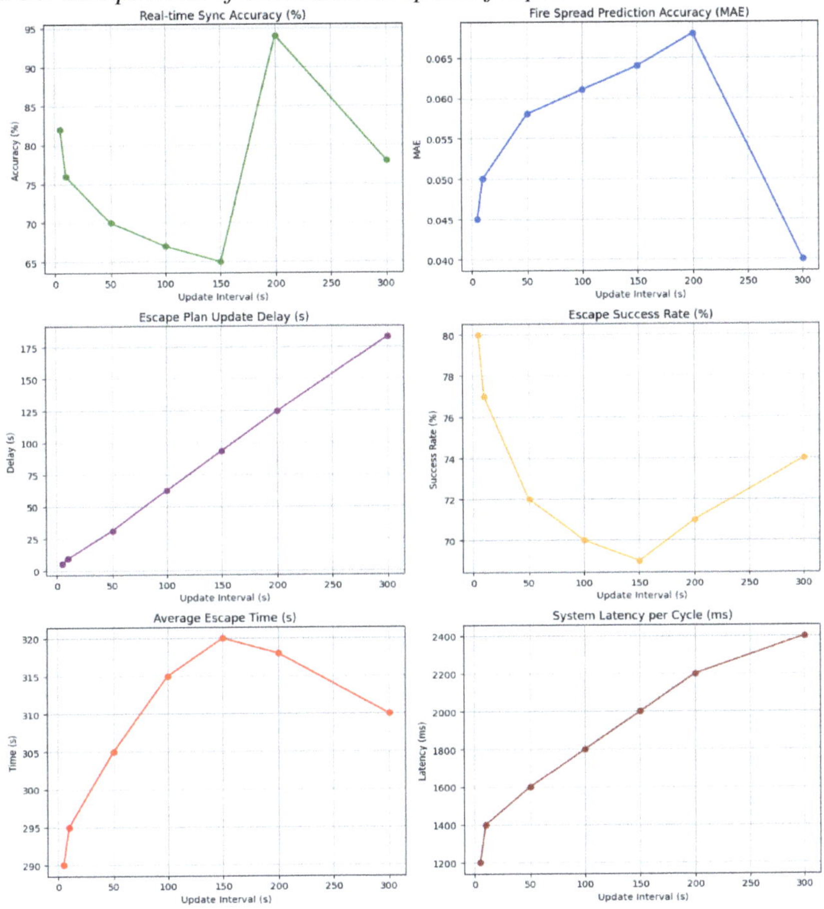

In Figure 12, Real-time Sync Accuracy and Fire Spread Prediction Accuracy reach their highest levels at a 200-second update interval. This performance likely results from improved data stabilization and more comprehensive aggregation over extended intervals, reducing noise caused by frequent updates. In contrast, Escape Plan Update Delay and System Latency per Cycle exhibit optimal values at a 300-second interval, where reduced update frequency lowers computational overhead and latency. The best outcomes for Escape Success Rate and Average Escape Time occur at a 150-second interval. Notably, shorter update intervals (0–100 seconds) yield comparatively poorer results, suggesting that the update mechanism may require further optimization to support rapid cycles and enhance evacuation outcomes.

CONCLUSION

This chapter presented the development and evaluation of the FAST (Firefighter AI Safety Digital Twin) system, an AI-driven digital twin framework designed to enhance firefighter safety and support real-time decision-making during all wildfire phases. The FAST system integrates a 10×10 m^2 virtual simulation with advanced AI models—specifically reinforcement learning algorithms and deep neural networks—to predict wildfire spread and identify optimal, safe escape routes for firefighters under rapidly changing conditions.

Among evaluated configurations, the combination of multivariate Long Short-Term Memory (LSTM) networks and Deep Q-Network (DQN) reinforcement learning showed superior performance. This model achieved the lowest fire spread prediction loss (0.014–0.017), highest escape success rate (up to 84%), and shortest average escape time (260 seconds). It also required the fewest training steps (14,336), confirming its efficiency and suitability for real-time, high-risk wildfire scenarios. These results validate the robustness and reliability of the LSTM+DQN framework for precise, time-sensitive evacuation planning in dynamic environments.

Additionally, the FAST system's bidirectional synchronization between AI agents and the digital twin environment improved operational responsiveness and accuracy. Evaluation showed update intervals of 150–200 seconds optimally balanced prediction accuracy, synchronization efficiency, and evacuation effectiveness, while intervals below 100 seconds revealed processing overhead issues, indicating a key area for future optimization.

In summary, the FAST framework provides a scalable, practical, and state-of-the-art solution for wildfire response management. By improving predictive accuracy, reducing response times, and enhancing coordination, this research establishes a strong foundation for future innovations in AI-driven digital twin systems, real-time hazard mitigation, and critical life-saving support for firefighters and emergency responders in high-risk environments.

REFERENCES

Caban-Martínez, A. J., Hughes, J., & Bator, C. (2021). A total worker health approach to skin exposure assessment: Experiences from the firefighter cancer initiative. *Annals of Work Exposures and Health*, *65*(2), 143–147. DOI: 10.1093/annweh/wxaa066 PMID: 32618337

De la Fuente, N., & Vidal Guerra, D. A. (2024). A comparative study of deep reinforcement learning models: DQN vs PPO vs A2C. *Proceedings of the ACM KDD 2024* (pp. 1–8). ACM. DOI: 10.1145/nnnnnnn.nnnnnnn

Diniz, A. P. M., Ciarelli, P. M., Salles, E. O. T., & Coco, K. F. (2024). Use of deep neural networks for clogging detection in the Submerged Entry Nozzle of the continuous casting. *Expert Systems with Applications*, *238*(Part C), 121963. DOI: 10.1016/j.eswa.2023.121963

Esteves, F., Madureira, J., Teixeira, J. P., & Costa, S. (2024). Wildland firefighters: A crucial weapon for forest fire management. Which health risks do they face? In Rodrigo-Comino, J., & Salvati, L. (Eds.), *Fire hazards: Socio-economic and regional issues* (pp. 165–189). Springer., DOI: 10.1007/978-3-031-50446-4_9

Garrity, D., & Yusuf, S. (2021). A predictive decision-aid device to warn firefighters of catastrophic temperature increases using an AI-based time-series algorithm. *Safety Science*, *138*, 105237. DOI: 10.1016/j.ssci.2021.105237

Giannakidou, S., Radoglou-Grammatikis, P., Lagkas, T., Argyriou, V., Goudos, S., Markakis, E., & Sarigiannidis, P. (2024). Leveraging the power of internet of things and artificial intelligence in forest fire prevention, detection, and restoration: A comprehensive survey. *Internet of Things : Engineering Cyber Physical Human Systems*, *26*, 101171. DOI: 10.1016/j.iot.2024.101171

Glover, K. G., Kahn, S. A., & Mittal, R. (2024). 104 Wildland firefighters suffer increasing risk of job-related death. *Journal of Burn Care & Research; Official Publication of the American Burn Association*, *45*(Suppl 1), 84. DOI: 10.1093/jbcr/irae036.103

Goodfellow, I., Bengio, Y., & Courville, A. (2016). *Deep learning*. MIT Press.

Hu, Y., & Shao, J. Z. (2020). Chinese firefighter fatalities 2007–2016. In G. Y. Wu, K. C. Tsai, & W. K. Chow (Eds.), *In 2018 11th Asia-Oceania Symposium on Fire Science and Technology (AOSFST 2018) (pp. 821–830)*. Springer. DOI: 10.1007/978-981-32-9139-3_60

Huang, X., Wu, X., & Usmani, A. (2022). Perspectives of using artificial intelligence in building fire safety. In Naser, M., & Corbett, G. (Eds.), *Handbook of cognitive and autonomous systems for fire resilient infrastructures* (pp. 139–159). Springer., DOI: 10.1007/978-3-030-98685-8_6

Huang, Y., Li, J., & Zheng, H. (2024). Modeling of wildfire digital twin: Research progress in detection, simulation, and prediction techniques. *Fire (Basel, Switzerland), 7*(1), 1–25. DOI: 10.3390/fire7110412

Huyen, C. (2024). *AI engineering: Building applications with foundation models.* O'Reilly Media.

Indolia, S., Goswami, A. K., Mishra, S. P., & Asopa, P. (2018). Conceptual understanding of convolutional neural network: A deep learning approach. *Procedia Computer Science, 132*, 679–688. DOI: 10.1016/j.procs.2018.05.069

James, G. L., Ansaf, R. B., Al Samahi, S. S., Parker, R. D., Cutler, J. M., Gachette, R. V., & Ansaf, B. I. (2023). An efficient wildfire detection system for AI-embedded applications using satellite imagery. *Fire (Basel, Switzerland), 6*(4), 169. DOI: 10.3390/fire6040169

Jiang, X. (2024). Comparison of deep reinforcement learning algorithms for trading strategy. In *Proceedings of the 2023 International Conference on Data Science, Advanced Algorithm and Intelligent Computing (DAI 2023)* (pp. 4–14). Atlantis Press. https://doi.org/DOI: 10.2991/978-94-6463-370-2_2

Kreuzer, T., Papapetrou, P., & Zdravkovic, J. (2024). Artificial intelligence in digital twins—A systematic literature review. *Data & Knowledge Engineering, 151*, 102304. DOI: 10.1016/j.datak.2024.102304

Kulhandjian, H., Davis, A., Leong, L., Bendot, M., & Kulhandjian, M. (2023). AI-based human detection and localization in heavy smoke using radar and IR camera. *In 2023 IEEE Radar Conference (RadarConf23)*, San Antonio, TX, USA, 1-6. DOI: 10.1109/RadarConf2351548.2023.10149735

Kunathur Thikshaja, U., & Paul, A. (2018). A brief review on deep learning and types of implementations for deep learning. In *Deep learning innovations and their convergence with big data* (p. 13). IGI Global., DOI: 10.4018/978-1-5225-3015-2.ch002

Lewis, R. H., Jiao, J., Seong, K., Farahi, A., Navrátil, P., Casebeer, N., & Niyogi, D. (2024). Fire and smoke digital twin – A computational framework for modeling fire incident outcomes. *Computers, Environment and Urban Systems, 110*, 102093. DOI: 10.1016/j.compenvurbsys.2024.102093

Madni, A. M., Madni, C. C., & Lucero, S. D. (2019). Leveraging digital twin technology in model-based systems engineering. *Systems*, *7*(1), 7. DOI: 10.3390/systems7010007

Marić, L., Bjelopera, M., Mandić, D., Periša, M., Mandić, N., & Knezović, M. (2023). Advancement of an integrated technological platform for wildfire management through edge computing. *In 2023 8th International Conference on Smart and Sustainable Technologies (SpliTech)* (pp. 1–6). IEEE. https://doi.org/DOI: 10.23919/SpliTech58164.2023.10193111

Moradian, A. (2024). Digital twins for sustainable semiconductor manufacturing. In M. Grieves & E. Y. Hua (Eds.), *Digital twins, simulation, and the metaverse: Simulation foundations, methods and applications* (pp. [page range]). Springer. DOI: 10.1007/978-3-031-69107-2_12

NASA. (n.d.). *NASA wildfire digital twin pioneers new AI models and streaming data techniques for forecasting fire and smoke*. NASA Science. https://science.nasa.gov/science-research/science-enabling-technology/nasa-wildfire-digital-twin-pioneers-new-ai-models-and-streaming-data-techniques-for-forecasting-fire-and-smoke/

National Wildfire Coordinating Group. (2014). *Wildland fire incident management field guide.*

O'Connor, S. M., Bell, J., & Morrison, S. (2017). Lessons Learned from the South Canyon Fire. *Fire Management Today*, *77*(2), 32–38. https://www.fs.fed.us/fire/publications/fire-management-today/2017/summer/lessons-learned-south-canyon-fire.shtml

Osama, Z. (2024). The digital twin framework: A roadmap to the development of user-centred digital twin in the built environment. *Journal of Building Engineering*, *98*, 111081. DOI: 10.1016/j.jobe.2024.111081

Penney, G., Habibi, D., & Cattani, M. (2019). Firefighter tenability and its influence on wildfire suppression. *Fire Safety Journal*, *106*, 38–51. DOI: 10.1016/j.firesaf.2019.03.012

Phantanaikasem, P., Pannawan, A., Jamroen, C., & Surinkaew, T. (2024). Potential impacts of wildfire on distribution networks in Thailand: A data-driven assessment using hotspot analysis. *In 2024 IEEE 22nd Student Conference on Research and Development (SCOReD)*, 39–43. DOI: 10.1109/SCOReD64708.2024.10872680

U.S. Fire Administration. (2022, Nov). *Fire-related firefighter injuries (2018–2020)*. https://www.usfa.fema.gov/statistics/reports/firefighters-departments/firefighter-injuries-v22i4.html

US Forest Service. (1999). *Mann Gulch Fire Disaster*. USDA Forest Service. https://www.fs.usda.gov/manngulchfire

Van der Valk, H., Haße, H., Möller, F., & Otto, B. (2022). Archetypes of digital twins. *Business & Information Systems Engineering*, 1–17. DOI: 10.1007/s12599-022-00763-2

Van Schalkwyk, P. (2022). *Digital Twin Capabilities Periodic Table: A Digital Twin Consortium User Guide. Digital Twin Consortium*. https://www.digitaltwinconsortium.org/wp-content/uploads/sites/3/2022/06/Digital-Twin-Capabilities-Periodic-Table-User-Guide.pdf

Vijayakumar, A., & Vairavasundaram, S. (2024). YOLO-based object detection models: A review and its applications. *Multimedia Tools and Applications*, *83*(17), 83535–83574. DOI: 10.1007/s11042-024-18872-y

Viseras, M., Meissner, M., & Marchal, J. (2016). Wildfire front monitoring with multiple UAVs using deep Q-learning. *IEEE Access : Practical Innovations, Open Solutions*, *1*, 1–14. DOI: 10.1109/ACCESS.2021.3055651

Wu, C., Zhou, Y., Pessôa, M. V. P., Peng, Q., & Tan, R. (2021). Conceptual digital twin modeling based on an integrated five-dimensional framework and TRIZ function model. *Journal of Manufacturing Systems*, *58*, 79–93. DOI: 10.1016/j.jmsy.2020.07.006

Wu, C. M., Adetona, A., Song, C., Naeher, L., & Adetona, O. (2019). Measuring acute pulmonary responses to occupational wildland fire smoke exposure using exhaled breath condensate. *Archives of Environmental & Occupational Health*, *75*(1), 65–69. DOI: 10.1080/19338244.2018.1562413 PMID: 30668286

Xu, L., Yu, H., Qin, H., Chai, Y., Yan, N., Li, D., & Chen, Y. (2024). Digital twin for aquaponics factory: Analysis, opportunities, and research challenges. *IEEE Transactions on Industrial Informatics*, *20*(4), 5060–5073. DOI: 10.1109/TII.2023.3331135

Xu, Q., Wang, J., Gao, W., Ren, S., & Li, Z. (2025). Digital twin: State-of-the-art and future perspectives. In *2024 5th International Conference on Computer Science and Management Technology (ICCSMT '24)* (pp. 731–740). Association for Computing Machinery. DOI: 10.1145/3708036.3708159

Yizhou, L., Zhang, T., Ding, Y., Wadhwani, R., & Huang, X. (2024). Review and perspectives of digital twin systems for wildland fire management. *Journal of Forestry Research*, *36*(1), 1–24. DOI: 10.1007/s11676-024-01810-x

Zhang, T., Wang, Z., Zeng, Y., Wu, X., Huang, X., & Xiao, F. (2022). Building Artificial-Intelligence Digital Fire (AID-Fire) system: A real-scale demonstration. *Journal of Building Engineering*, *62*, 105363. DOI: 10.1016/j.jobe.2022.105363

Zhang, Z., Zohren, S., & Roberts, S. (2020). Deep reinforcement learning for trading. *The Journal of Financial Data Science*, *2*(2), 25–40. DOI: 10.3905/jfds.2020.1.030

Compilation of References

Abdolrasol, M. G., Hussain, S. S., Ustun, T. S., Sarker, M. R., Hannan, M. A., Mohamed, R., Ali, J. A., Mekhilef, S., & Milad, A. (2021). Artificial neural networks based optimization techniques: A review. *Electronics (Basel), 10*(21), 2689. DOI: 10.3390/electronics10212689

Achouch, M., Dimitrova, M., Ziane, K., Sattarpanah Karganroudi, S., Dhouib, R., Ibrahim, H., & Adda, M. (2022). On predictive maintenance in industry 4.0: Overview, models, and challenges. *Applied Sciences (Basel, Switzerland), 12*(16), 8081. DOI: 10.3390/app12168081

Acierno, D., & Patti, A. (2023). Fused Deposition Modelling (FDM) of Thermoplastic-Based Filaments: Process and Rheological Properties—An Overview. *Materials (Basel), 16*(24), 7664. https://www.mdpi.com/1996-1944/16/24/7664. DOI: 10.3390/ma16247664 PMID: 38138805

Adar, U. G., & Altan, Z. (2024). AUTOMIND: Otomobil Ne Kadar "Oto"? İnsan-Otonom Araç Etkileşimi Üzerine Bir İnceleme. *Uludağ Üniversitesi Mühendislik Fakültesi Dergisi, 29*(2), 639–654.

Adewale, B. A., Ene, V. O., Ogunbayo, B. F., & Aigbavboa, C. O. (2024). A Systematic Review of the Applications of AI in a Sustainable Building's Lifecycle. *Buildings (Basel, Switzerland), 14*(7), 2137. DOI: 10.3390/buildings14072137

Afachao, K., & Abu-Mahfouz, A. (2022). A review of intelligent IoT devices at the edge. *Proceedings of the 2022 International Conference on Artificial Intelligence of Things*. DOI: 10.1109/ICAIoT57170.2022.10121824

Aggarwal, A., Mittal, M., & Battineni, G. (2021). Generative adversarial network: An overview of theory and applications. *International Journal of Information Management Data Insights, 1*(1), 100004.DOI: 10.1016/j.jjimei.2020.100004

Ahmad, R., & Kamaruddin, S. (2012). An overview of time-based and condition-based maintenance in industrial application. *Computers & Industrial Engineering*, *63*(1), 135–149. DOI: 10.1016/j.cie.2012.02.002

Akanbi, T., & Zhang, J. (2021). Design information extraction from construction specifications to support cost estimation. *Automation in Construction*, *131*, 103835. DOI: 10.1016/j.autcon.2021.103835

Aksoy, B. (2021). Estimation of energy produced in hydroelectric power plant industrial automation using deep learning and hybrid machine learning techniques. *Electric Power Components and Systems*, *49*(3), 213–232. DOI: 10.1080/15325008.2021.1937401

Al Sehrawy, A., Amoudi, O., Tong, M., & Callaghan, N. (2021, November). A review of the challenges to integrating BIM and building sustainability assessment. In AIP Conference Proceedings (Vol. 2428, No. 1). AIP Publishing. DOI: 10.1063/5.0071055

Alajlan, N. N., & Ibrahim, D. M. (2022). TinyML: Enabling of inference deep learning models on ultra-low-power IoT edge devices for AI applications. *Micromachines*, *13*(4), 851. DOI: 10.3390/mi13060851 PMID: 35744466

Alemán-Gómez, Y., Arribas-Gil, A., Desco, M., Elías, A., & Romo, J. (2022). Depthgram: Visualizing outliers in high-dimensional functional data with application to fMRI data exploration. *Statistics in Medicine*, *41*(11), 2005–2024. DOI: 10.1002/sim.9342 PMID: 35118686

Alfons, A., Croux, C., & Gelper, S. (2013). Sparse least trimmed squares regression for analyzing high-dimensional large data sets. *The Annals of Applied Statistics*, *7*(1), 226–248. DOI: 10.1214/12-AOAS575

Al-Kahwati, K., Birk, W., Nilsfors, E. F., & Nilsen, R. (2022). Experiences of a digital twin based predictive maintenance solution for belt conveyor systems. *PHM Soc. Eur. Conf.*, *7*(1), 1–8. DOI: 10.36001/phme.2022.v7i1.3355

Al-Sabaawi, A., Muayad Ibrahim, H., & Mohsin, Z., & sabah Alzubaidi, L. (2020). Amended Convolutional Neural Network with Global Average Pooling for Image Classification. *20th International Conference Intelligent Systems Design and Applications*.

Alzubaidi, L., Zhang, J., Humaidi, A. J., Al-Dujaili, A., Duan, Y., Al-Shamma, O., Santamaría, J., Fadhel, M. A., Al-Amidie, M., & Farhan, L. (2021). Review of deep learning: Concepts, CNN architectures, challenges, applications, future directions. *Journal of Big Data*, *8*(1), 53. DOI: 10.1186/s40537-021-00444-8 PMID: 33816053

Amrouni Hosseini, M., Ravanshadnia, M., Rahimzadegan, M., & Ramezani, S. (2024). Next-Generation Building Condition Assessment: BIM and Neural Network Integration. *Journal of Performance of Constructed Facilities, 38*(6), 04024050. DOI: 10.1061/JPCFEV.CFENG-4828

Andrade, P., Silva, I., Signoretti, G., Da Silva, M. B. D., Dias, J., Marques, L., & Costa, D. G. (2021). An unsupervised TinyML approach applied for pavement anomalies detection under the Internet of Intelligent Vehicles. *2021 IEEE International Workshop on Metrology for Industry*. DOI: 10.1109/MetroInd4.0IoT51437.2021.9488546

Andreas Kamilaris, F. X. P.-B. (2018). Deep learning in agriculture: A survey. *Computers and Electronics in Agriculture, 147*, 70–90. DOI: 10.1016/j.compag.2018.02.016

Arbogast, T., & Wheeler, M. F. (1996). A nonlinear mixed finite element method for a degenerate parabolic equation arising in flow in porous media. *SIAM Journal on Numerical Analysis, 33*(4), 1669–1687. DOI: 10.1137/S0036142994266728

Arribas-Gil, A., & Romo, J. (2014). Shape outlier detection and visualization for functional data: The outliergram. *Biostatistics (Oxford, England), 15*(4), 603–619. DOI: 10.1093/biostatistics/kxu006 PMID: 24622037

Arslan, K. (2020). Eğitimde yapay zekâ ve uygulamaları. *Batı Anadolu Eğitim Bilimleri Dergisi, 11*(1), 71–88.

Arulkumaran, K., Deisenroth, M. P., Brundage, M., & Bharath, A. A. (2017). A Brief Survey of Deep Reinforcement Learning. *IEEE Signal Processing Magazine, 34*(6), 26–38. DOI: 10.1109/MSP.2017.2743240

ArunKumar, K. E., Kalaga, D. V., Mohan Sai Kumar, C., Kawaji, M., & Brenza, T. M.ArunKumar. (2022). Comparative analysis of Gated Recurrent Units (GRU), long Short-Term memory (LSTM) cells, autoregressive Integrated moving average (ARIMA), seasonal autoregressive Integrated moving average (SARIMA) for forecasting COVID-19 trends. *Alexandria Engineering Journal, 61*(10), 7585–7603. DOI: 10.1016/j.aej.2022.01.011

Ashraf, J., Bakhshi, A. D., Moustafa, N., Khurshid, H., Javed, A., & Beheshti, A. (2021). Novel Deep Learning-Enabled LSTM Autoencoder Architecture for Discovering Anomalous Events from Intelligent Transportation Systems. *IEEE Transactions on Intelligent Transportation Systems, 22*(7), 4507–4518. DOI: 10.1109/TITS.2020.3017882

Ashworth, D. H. (1971). Belt Drives. In *Mechanical Power Transmission* (pp. 103–110). Palgrave Macmillan UK. DOI: 10.1007/978-1-349-01197-1_11

Astrid, M., Zaheer, M. Z., Lee, J.-Y., & Lee, S.-I. (2021). Learning not to reconstruct anomalies. arXiv. https://arxiv.org/abs/2110.09742 DOI: 10.5244/C.35.205

Aydın, L., & Küçük, S. (2017). Üç boyutlu yazıcı ve tarayıcı ile hastaya özel medikal ortez tasarımı ve geliştirilmesi. *Politeknik Dergisi*, *20*(1), 1–8.

Ayesha, S., Hanif, M. K., & Talib, R. (2020). Overview and comparative study of dimensionality reduction techniques for high dimensional data. *Information Fusion*, *59*, 44–58. DOI: 10.1016/j.inffus.2020.01.005

Aylak, B. L., & Oral, O. (2021). Yapay zeka ve makine öğrenmesi tekniklerinin lojistik sektöründe kullanımı. *El-Cezeri*, *8*(1), 74–93.

Banihashemi, S., Khalili, S., Sheikhkhoshkar, M., & Fazeli, A. (2022). Machine learning-integrated 5D BIM informatics: Building materials costs data classification and prototype development. *Innovative Infrastructure Solutions*, *7*(3), 215. DOI: 10.1007/s41062-022-00822-y

Bank, D., Koenigstein, N., & Giryes, R. (2020). *Autoencoders. Machine Learning for Data Science Handbook: Data Mining and Knowledge Discovery Handbook*. Springer International Publishing.

Bardi, F. C., & Kyriakides, S. (2006). Plastic buckling of circular tubes under axial compression—part I: Experiments. *International Journal of Mechanical Sciences*, *48*(8), 830–841. DOI: 10.1016/j.ijmecsci.2006.03.005

Baresi, L., Mendonça, D. F., & Garriga, M. (2017). Empowering low-latency applications through a serverless edge computing architecture.

Bas, E., Egrioglu, E., & Cansu, T. (2024). Robust training of median dendritic artificial neural networks for time series forecasting. *Expert Systems with Applications*, *238*, 122080. DOI: 10.1016/j.eswa.2023.122080

Basharat, A., Gritai, A., & Shah, M. (2008). Learning object motion patterns for anomaly detection and improved object detection. In *2008 IEEE Conference on Computer Vision and Pattern Recognition* (pp. 1–8). IEEE. DOI: 10.1109/CVPR.2008.4587510

Bassier, M., Vergauwen, M., & Van Genechten, B. (2017). Automated classification of heritage buildings for as-built BIM using machine learning techniques. *ISPRS Annals of the Photogrammetry, Remote Sensing and Spatial Information Sciences*, *4*(W2), 25–30. DOI: 10.5194/isprs-annals-IV-2-W2-25-2017

Bassir, D., Lodge, H., Chang, H., Majak, J., & Chen, G. (2023). Application of artificial intelligence and machine learning for BIM. *International Journal for Simulation and Multidisciplinary Design Optimization, 14*, 5. DOI: 10.1051/smdo/2023005

Becker, S., Styp-Rekowski, K., Stoll, O. V. L., & Kao, O. (2022). *Federated learning for autoencoder-based condition monitoring in the industrial internet of things*. arXiv. https://arxiv.org/abs/2206.08187 DOI: 10.1109/BigData55660.2022.10020836

Bengio, Y., Courville, A., & Vincent, P. (2013). Representation learning: A review and new perspectives. *IEEE Transactions on Pattern Analysis and Machine Intelligence, 35*(8), 1798–1828. DOI: 10.1109/TPAMI.2013.50 PMID: 23787338

Benjamini, Y., & Hochberg, Y. (1995). Controlling the false discovery rate: A practical and powerful approach to multiple testing. *Journal of the Royal Statistical Society. Series B, Statistical Methodology, 57*(1), 289–300. DOI: 10.1111/j.2517-6161.1995.tb02031.x

Benjamini, Y., & Yekutieli, D. (2001). The control of the false discovery rate in multiple testing under dependency. *Annals of Statistics, 29*(4), 1165–1188. DOI: 10.1214/aos/1013699998

Bentéjac, C., Csörgő, A., & Martínez-Muñoz, G. (2021). A comparative analysis of gradient boosting algorithms. *Artificial Intelligence Review, 54*(3), 1937–1967. DOI: 10.1007/s10462-020-09896-5

Berahmand, K., Daneshfar, F., Salehi, E. S., Li, Y., & Xu, Y. (2024). Autoencoders and their applications in machine learning: A survey. *Artificial Intelligence Review, 57*(2), 28. DOI: 10.1007/s10462-023-10662-6

Berberoğlu, B. (2019). Ultraviyole ışıkla kürlenebilen organik-inorganik hibrit malzemelerin sentezi ve karakterizasyonu [Yüksek lisans tezi, Marmara Üniversitesi].

Bikku, T. (2020). Multi-layered deep learning perceptron approach for health risk prediction. *Journal of Big Data, 7*(1), 50. DOI: 10.1186/s40537-020-00316-7

Bikku, T., & Satya Sree, K. P. N. V. (2020). DEEP LEARNING APPROACHES FOR CLASSIFYING DATA: A REVIEW. *Journal of Engineering Science and Technology, 15*(4), 2580–2594.

Binson, V. A., Thomas, S., Subramoniam, M., Arun, J., Naveen, S., & Madhu, S. (2024). A review of machine learning algorithms for biomedical applications. *Annals of Biomedical Engineering, 52*(5), 1159–1183. DOI: 10.1007/s10439-024-03459-3 PMID: 38383870

Bishop, R. E. D., & Sherrington, I. (2000). *Mechanical wear prediction and analysis*. Arnold Publishers.

Blazques, D., & Domenech, J. (2018). Big Data sources and methods for social and economic analyses. *Technological Forecasting and Social Change, 130*, 99–113. DOI: 10.1016/j.techfore.2017.07.027

Bloch, H. P., & Geitner, F. K. (1999). *Practical machinery management for process plants: Machinery failure analysis and troubleshooting* (Vol. 2). Gulf Professional Publishing.

Bozkurt, A. (2023). ChatGPT, üretken yapay zeka ve algoritmik paradigma değişikliği. *Alanyazın, 4*(1), 63–72. DOI: 10.59320/alanyazin.1283282

Bozkurt, Y., Gülsoy, H., & Karayel, E. (2021). Eklemeli imalat teknolojilerinin tıbbi ekipmanların üretiminde kullanımı. *El-Cezeri, 8*(2), 962–980.

Bradlow, E. T., Gangwar, M., Kopalle, P., & Voleti, S. (2017). The role of Big Data and predictive analytics in retailing. *Journal of Retailing, 93*(1), 79–95. DOI: 10.1016/j.jretai.2016.12.004

Bridge, J. & Dodds, J.C. (2019). Managerial decision making. Routledge, Abingdon-on-Thames.

Büber, E. D. M. Y. M., & Taşpınar, Y. S. (2023). 3 boyutlu yazıcılardaki baskı hatalarının hibrit görüntü işleme yöntemleri ile tespiti.

Bučková, B., Brunovský, M., Bareš, M., & Hlinka, J. (2020). Predicting sex from EEG: Validity and generalizability of deep-learning-based interpretable classifier. *Frontiers in Neuroscience, 14*, 589303. DOI: 10.3389/fnins.2020.589303 PMID: 33192274

Caban-Martínez, A. J., Hughes, J., & Bator, C. (2021). A total worker health approach to skin exposure assessment: Experiences from the firefighter cancer initiative. *Annals of Work Exposures and Health, 65*(2), 143–147. DOI: 10.1093/annweh/wxaa066 PMID: 32618337

Cai, R., Zhang, H., Liu, W., Gao, S., & Hao, Z. (2021). Appearance-motion memory consistency network for video anomaly detection. In Proceedings of the AAAI Conference on Artificial Intelligence (Vol. 35, pp. 938–946). https://ojs.aaai.org/index.php/AAAI/article/view/16163

Calayır, G. N., & Kabak, M. (2021). Bakım için makine öğrenme tekniklerinin analizi ve bir uygulama. *Journal of Turkish Operations Management, 5*(1), 662–675.

Canlıdinç, M. (2023). Topoloji Optimizasyonu ile Metal Eklemeli İmalat Yönteminin Endüstriyel Uygulamaları. *Kirklareli University Journal of Engineering and Science*, *9*(2), 552–565.

Cao, C., Lu, Y., & Zhang, Y. (2024). Context recovery and knowledge retrieval: A novel two-stream framework for video anomaly detection. *IEEE Transactions on Image Processing*, *33*, 1810–1825. DOI: 10.1109/TIP.2024.3372466 PMID: 38451764

Capogrosso, L., Cunico, F., Cheng, D. S., Fummi, F., & Cristani, M. (2023). A machine learning-oriented survey on tiny machine learning. *arXiv preprint arXiv:2301.12345*. https://arxiv.org/abs/2301.12345

Carvalho, T. P., Soares, F. A. A. M. N., Vita, R., Francisco, R. P., Basto, J. P., & Alcalá, S. G. (2019). A systematic literature review of machine learning methods applied to predictive maintenance. *Computers & Industrial Engineering*, *137*, 106024. DOI: 10.1016/j.cie.2019.106024

Çelik, İ., Karakoç, F., Çakır, M. C., & Duysak, A. (2013). HIZLI PROTOTİPLEME TEKNOLOJİLERİ VE UYGULAMA ALANLARI. *Dumlupınar Üniversitesi Fen Bilimleri Enstitüsü Dergisi*, (031), 53–70.

Çelik, K. (2021). Bulut bilişim teknolojileri. *Bartın Üniversitesi İktisadi ve İdari Bilimler Fakültesi Dergisi*, *12*(24), 436–450. DOI: 10.47129/bartiniibf.1019898

Chen, S., & Guo, W. (2023). Auto-Encoders in Deep Learning—A Review with New Perspectives. In *Mathematics* (Vol. 11, Issue 8). MDPI. DOI: 10.3390/math11081777

Cheng, K.-W., Chen, Y.-T., & Fang, W.-H. (2015). Video anomaly detection and localization using hierarchical feature representation and Gaussian process regression. In *Proceedings of the IEEE Conference on Computer Vision and Pattern Recognition* (pp. 2909–2917). IEEE. DOI: 10.1109/CVPR.2015.7298909

Chen, H., Hu, S., Hua, R., & Zhao, X. (2021). Improved naive Bayes classification algorithm for traffic risk management. *EURASIP Journal on Advances in Signal Processing*, *2021*(1), 30. DOI: 10.1186/s13634-021-00742-6

Chen, L. P. (2024). Variable selection and estimation for misclassified binary responses and multivariate error-prone predictors. *Journal of Computational and Graphical Statistics*, *33*(2), 407–420. DOI: 10.1080/10618600.2023.2218428

Chen, R. S., Zhang, H. Y., Hao, X. K., Yu, H. X., Shi, T., Zhou, H. S., Wang, P., Zhao, Z.-F., & Wang, P. (2024, Oct). Experimental study on ultimate bearing capacity of short thin-walled steel tubes reinforced with high-ductility concrete. [). Elsevier.]. *Structures (Oxford)*, *68*, 107109. DOI: 10.1016/j.istruc.2024.107109

Chen, W., Milosevic, Z., Rabhi, F. A., & Berry, A. (2023). Real-time analytics: Concepts, architectures, and ML/AI considerations. *IEEE Access : Practical Innovations, Open Solutions*, *11*, 71634–71657. DOI: 10.1109/ACCESS.2023.3295694

Chen, Z., Duan, J., Kang, L., & Qiu, G. (2022). Supervised anomaly detection via conditional generative adversarial network and ensemble active learning. *IEEE Transactions on Pattern Analysis and Machine Intelligence*, *45*(6), 7781–7798. DOI: 10.1109/TPAMI.2022.3225476 PMID: 36445991

Chetana, V. L., Kolisetty, S. S., & Amogh, K. (2020). A Short Survey of Dimensionality Reduction Techniques. In *Recent Advances in Computer Based Systems, Processes and Applications* (pp. 3–14). CRC Press. DOI: 10.1201/9781003043980-2

Childs, P. R. N. (2013). *Mechanical design* (2nd ed.). Butterworth-Heinemann.

Chollet, F. (2024). Keras, version 3.0.0. Github repository. [https://github.com/keras-team/keras]

Chollet, F. (2018). *Deep learning with Python*. Manning Publications Co.

Chowdhury, S., & Yedavalli, R. K. (2016). Dynamics of belt-pulley-shaft systems. *Mechanism and Machine Theory*, *98*, 199–215. DOI: 10.1016/j.mechmachtheory.2015.11.011

Ciccone, F., Bacciaglia, A., & Ceruti, A. (2023). Optimization with artificial intelligence in additive manufacturing: A systematic review. *Journal of the Brazilian Society of Mechanical Sciences and Engineering*, *45*(6), 303. DOI: 10.1007/s40430-023-04200-2

Citakoglu, H. (2015). Comparison of artificial intelligence techniques via empirical equations for prediction of solar radiation. *Computers and Electronics in Agriculture*, *118*, 28–37. DOI: 10.1016/j.compag.2015.08.020

Claude Sammut, G. I. W. (n.d.). *Encyclopedia of Machine Learning*. Springer-Verlag New York Inc.

Collacott, R. A. (1979). *Mechanical fault diagnosis and condition monitoring*. Springer.

Corominas, L., Garrido-Baserba, M., Villez, K., Olsson, G., Cortés, U., & Poch, M. (2018). Transforming data into knowledge for improved wastewater treatment operation: A critical review of techniques. *Environmental Modelling & Software*, *106*, 89–103. DOI: 10.1016/j.envsoft.2017.11.023

Corona, E., Lee, L. H., & Kyriakides, S. (2006). Yield anisotropy effects on buckling of circular tubes under bending. *International Journal of Solids and Structures, 43*(22-23), 7099–7118. DOI: 10.1016/j.ijsolstr.2006.03.005

Coşar, S., Donatiello, G., Bogorny, V., Garate, C., Alvares, L. O., & Brémond, F. (2017). Toward abnormal trajectory and event detection in video surveillance. *IEEE Transactions on Circuits and Systems for Video Technology*, *27*(3), 683–695. DOI: 10.1109/TCSVT.2016.2589859

Croitoru, F.-A., Ristea, N.-C., Dascalescu, D., Ionescu, R. T., Khan, F. S., & Shah, M. (2022). Lightning fast video anomaly detection via adversarial knowledge distillation. arXiv. https://arxiv.org/abs/2211.15597

Dabiri, S., Heaslip, K. P., Abbas, M. M., Gerdes, R. M., & Ramakrishnan, N. (2018). *Application of Deep Learning in Intelligent Transportation Systems.*

Dai, W., & Genton, M. G. (2018). Multivariate functional data visualization and outlier detection. *Journal of Computational and Graphical Statistics*, *27*(4), 923–934. DOI: 10.1080/10618600.2018.1473781

Dalal, N., Triggs, B., & Schmid, C. (2006). Human detection using oriented histograms of flow and appearance. In A. Leonardis, H. Bischof, & A. Pinz (Eds.), Computer Vision – ECCV 2006: 9th European Conference on Computer Vision, Graz, Austria, May 7-13, 2006, Proceedings, Part II (pp. 428–441). Springer. DOI: 10.1007/11744047_33

Dalal, N., & Triggs, B. (2005). Histograms of oriented gradients for human detection. In *2005 IEEE Computer Society Conference on Computer Vision and Pattern Recognition (CVPR'05)* (Vol. 1, pp. 886–893). IEEE. DOI: 10.1109/CVPR.2005.177

Dargan, S., Kumar, M., Ayyagari, M. R., & Kumar, G. (2020). A survey of deep learning and its applications: A new paradigm to machine learning. *Archives of Computational Methods in Engineering*, *27*(4), 1071–1092. DOI: 10.1007/s11831-019-09344-w

De la Fuente, N., & Vidal Guerra, D. A. (2024). A comparative study of deep reinforcement learning models: DQN vs PPO vs A2C. *Proceedings of the ACM KDD 2024* (pp. 1–8). ACM. DOI: 10.1145/nnnnnnn.nnnnnnn

Dedeakayoğulları, H., & Kaçal, A. (2020). Eklemeli imalat teknolojileri ve kullanılan talaşlı imalat yöntemleri üzerine yapılan çalışmaların değerlendirilmesi. *MATECA*, *1*(1), 1–12.

Deepak, K., Chandrakala, S., & Mohan, C. K. (2021). Residual spatiotemporal autoencoder for unsupervised video anomaly detection. *Signal, Image and Video Processing*, *15*(1), 215–222. DOI: 10.1007/s11760-020-01740-1

Delaigle, A., & Hall, P. (2012). Achieving near perfect classification for functional data. *Journal of the Royal Statistical Society. Series A, (Statistics in Society), 74*, 267–286.

Deng, H., Hong, H., Luo, D., Deng, Y., & Su, C. (2020). Automatic indoor construction process monitoring for tiles based on BIM and computer vision. *Journal of Construction Engineering and Management*, *146*(1), 04019095. DOI: 10.1061/(ASCE)CO.1943-7862.0001744

Deringer, V. L., Bartók, A. P., Bernstein, N., Wilkins, D. M., Ceriotti, M., & Csányi, G. (2021). Gaussian process regression for materials and molecules. *Chemical Reviews*, *121*(16), 10073–10141. DOI: 10.1021/acs.chemrev.1c00022 PMID: 34398616

Dey, P., Gopal, M., Pradhan, P., & Pal, T. (2019). On robustness of radial basis function network with input perturbation. *Neural Computing & Applications*, *31*(2), 523–537. DOI: 10.1007/s00521-017-3086-5

Dezfooli, M. S., Abolmaali, A., & Razavi, M. (2015). Coupled nonlinear finite-element analysis of soil–steel pipe structure interaction. *International Journal of Geomechanics*, *15*(1), 04014032. DOI: 10.1061/(ASCE)GM.1943-5622.0000387

Dheeru, D., & Taniskidou, E. K. (2017). *UCI Machine Learning Repository*. University of California. [http://archive.ics.uci.edu/ml]

Dilek, E., & Dener, M. (2024). Enhancement of video anomaly detection performance using transfer learning and fine-tuning. *IEEE Access : Practical Innovations, Open Solutions*, *12*, 123456–123467. DOI: 10.1109/ACCESS.2024.3404553

Dimitrov, A., & Golparvar-Fard, M. (2014). Vision-based material recognition for automated monitoring of construction progress and generating building information modeling from unordered site image collections. *Advanced Engineering Informatics*, *28*(1), 37–49. DOI: 10.1016/j.aei.2013.11.002

Diniz, A. P. M., Ciarelli, P. M., Salles, E. O. T., & Coco, K. F. (2024). Use of deep neural networks for clogging detection in the Submerged Entry Nozzle of the continuous casting. *Expert Systems with Applications*, *238*(Part C), 121963. DOI: 10.1016/j.eswa.2023.121963

Doğan, O., & Kamer, M. S. (2023). Farklı imalat parametreleri kullanılarak eriyik yığma modelleme ile imal edilen test numunelerinin sürünme davranışlarının deneysel olarak incelenmesi. *Journal of the Faculty of Engineering & Architecture of Gazi University*, *38*(3).

Dong, G., Liao, G., Liu, H., & Kuang, G. (2018). A Review of the Autoencoder and Its Variants: A Comparative Perspective from Target Recognition in Synthetic-Aperture Radar Images. *IEEE Geoscience and Remote Sensing Magazine*, *6*(3), 44–68. DOI: 10.1109/MGRS.2018.2853555

Dong, S., Wang, P., & Abbas, K. (2021a). A survey on deep learning and its applications. In *Computer Science Review* (Vol. 40). Elsevier Ireland Ltd., DOI: 10.1016/j.cosrev.2021.100379

Doshi, K., & Yilmaz, Y. (2020). Any-shot sequential anomaly detection in surveillance videos. In *Proceedings of the IEEE/CVF Conference on Computer Vision and Pattern Recognition Workshops* (pp. 934–935). IEEE. DOI: 10.1109/CVPRW50498.2020.00475

Doshi, M., Mahale, A., Singh, S. K., & Deshmukh, S. (2022). Printing parameters and materials affecting mechanical properties of FDM-3D printed Parts: Perspective and prospects. *Materials Today: Proceedings*, *50*, 2269–2275. DOI: 10.1016/j.matpr.2021.10.003

Duman, B., & Özsoy, K. (2018). *Üç Boyutlu Baskı Teknolojileri*. Gece Yayınevi.

Duman, E., & Erdem, O. A. (2019). Anomaly detection in videos using optical flow and convolutional autoencoder. *IEEE Access : Practical Innovations, Open Solutions*, *7*, 183914–183923. DOI: 10.1109/ACCESS.2019.2960654

Duong, H.-T., Le, V.-T., & Hoang, V. T. (2023). Deep learning-based anomaly detection in video surveillance: A survey. *Sensors (Basel)*, *23*(11), 5024. DOI: 10.3390/s23115024 PMID: 37299751

Ekrem, Ö., & Aksoy, B. (2023). Trajectory planning for a 6-axis robotic arm with particle swarm optimization algorithm. *Engineering Applications of Artificial Intelligence*, *122*, 106099. DOI: 10.1016/j.engappai.2023.106099

Elghaish, F., Chauhan, J. K., Matarneh, S., Rahimian, F. P., & Hosseini, M. R. (2022). Artificial intelligence-based voice assistant for BIM data management. *Automation in Construction*, *140*, 104320. DOI: 10.1016/j.autcon.2022.104320

Emunds, C., Pauen, N., Richter, V., Frisch, J., & van Treeck, C. (2022). SpaRSE-BIM: Classification of IFC-based geometry via sparse convolutional neural networks. *Advanced Engineering Informatics*, *53*, 101641. DOI: 10.1016/j.aei.2022.101641

Erener, Ş., & Boz, S. (2021). Modern Üretim Tekniklerinde Eklemeli Imalat Sistemlerinin Yeri Ve Kullanim Alanlari. *Turkish Journal of Fashion Design and Management*, *3*(1), 47–56.

Erol, B., & İnkaya, T. (2024). Satış tahmini için uzun kısa-süreli bellek ağı tabanlı derin transfer öğrenme yaklaşımı. *Gazi Üniversitesi Mühendislik Mimarlık Fakültesi Dergisi*, *39*(1), 191–202. DOI: 10.17341/gazimmfd.1089173

Esan, D. O., & Owolawi, P. A. C. T. (2023). Generative Adversarial Networks: Applications, Challenges, and Open Issues. *IntechOpen*. https://doi.org/http://dx.doi.org/10.5772/intechopen.113098

Esteves, F., Madureira, J., Teixeira, J. P., & Costa, S. (2024). Wildland firefighters: A crucial weapon for forest fire management. Which health risks do they face? In Rodrigo-Comino, J., & Salvati, L. (Eds.), *Fire hazards: Socio-economic and regional issues* (pp. 165–189). Springer., DOI: 10.1007/978-3-031-50446-4_9

Fang, Z., Zhou, J. T., Xiao, Y., Li, Y., & Yang, F. (2020). Multi-encoder towards effective anomaly detection in videos. *IEEE Transactions on Multimedia*, *23*, 4106–4116. DOI: 10.1109/TMM.2020.3037538

Fan, J., Ke, Y., & Wang, K. (2020). Factor-adjusted regularized model selection. *Journal of Econometrics*, *216*(1), 71–85. DOI: 10.1016/j.jeconom.2020.01.006 PMID: 32269406

Fan, Y., Wen, G., Li, D., Qiu, S., Levine, M. D., & Xiao, F. (2020). Video anomaly detection and localization via Gaussian mixture fully convolutional variational autoencoder. *Computer Vision and Image Understanding*, *195*, 102920. Advance online publication. DOI: 10.1016/j.cviu.2020.102920

Fazli, H., Farooq, S., Yang, C., & Wæhrens, B. V. (2023). Proactive and reactive approaches towards sustainable practices in manufacturing companies: Emerging economies perspective. *Sustainability (Basel)*, *15*(17), 12796. DOI: 10.3390/su151712796

Filzmoser, P., & Nordhausen, K. (2021). Robust linear regression for high-dimensional data: An overview. *Wiley Interdisciplinary Reviews: Computational Statistics*, *13*(4), e1524. DOI: 10.1002/wics.1524

Fitch, E. C. (2013). *Proactive maintenance for mechanical systems* (No. 5). Elsevier.

Fordellone, M. (2019). *Statistical analysis of complex data. Dimensionality reduction and classification methods*. LAP LAMBERT Academic Publishing.

Ford, S., & Despeisse, M. (2016). Additive manufacturing and sustainability: An exploratory study of the advantages and challenges. *Journal of Cleaner Production, 137*, 1573–1587. DOI: 10.1016/j.jclepro.2016.04.150

Fortuna, L., Graziani, S., Rizzo, A., & Xibilia, M. G. (2007). *Soft Sensors for Monitoring and Control of Industrial Processes*. Springer., DOI: 10.1007/978-1-84628-480-9

Frizzo-Stefenon, S., Seman, L., & Mariani, V. (2023). Machine fault detection using a hybrid CNN-LSTM attention-based model. *Sensors (Basel), 23*(9), 4512. DOI: 10.3390/s23094512 PMID: 37177716

Gams, M., Gu, I. Y. H., Härmä, A., Muñoz, A., & Tam, V. (2019). Artificial intelligence and ambient intelligence. *Journal of Ambient Intelligence and Smart Environments, 11*(1), 71–86. DOI: 10.3233/AIS-180508

Garrity, D., & Yusuf, S. (2021). A predictive decision-aid device to warn firefighters of catastrophic temperature increases using an AI-based time-series algorithm. *Safety Science, 138*, 105237. DOI: 10.1016/j.ssci.2021.105237

Gawate, E., & Rane, P. (2023). Empowering intelligent manufacturing with the potential of edge computing with NVIDIA's Jetson Nano. *2023 International Conference on Computing, Communication and Automation*. DOI: 10.1109/ICCCIS60361.2023.10425278

Genç, E., & Tunalı, İ. (2022). ENDÜSTRİ 4.0 VE LOJİSTİK 4.0 KAPSAMINDA AKILLI DEPO SİSTEMLERİ. *Avrasya Sosyal ve Ekonomi Araştırmaları Dergisi, 9*(4), 194–215.

Gerike, B., Panachev, I., & Kuzin, E. (2017). Development of the preventive maintenance system for belt conveyors reducers. In *E3S Web of Conferences* (Vol. 15, p. 03008). EDP Sciences. DOI: 10.1051/e3sconf/20171503008

Gernaey, K. V., van Loosdrecht, M. C., Henze, M., Lind, M., & Jørgensen, S. B. (2004). Activated sludge wastewater treatment plant modelling and simulation: State of the art. *Environmental Modelling & Software, 19*(9), 763–783. DOI: 10.1016/j.envsoft.2003.03.005

Gholamalinezhad, H., & Khosravi, H. (2020). Pooling Methods in Deep Neural Networks, a Review. *ArXiv Preprint ArXiv:2009.07485*.

Giannakidou, S., Radoglou-Grammatikis, P., Lagkas, T., Argyriou, V., Goudos, S., Markakis, E., & Sarigiannidis, P. (2024). Leveraging the power of internet of things and artificial intelligence in forest fire prevention, detection, and restoration: A comprehensive survey. *Internet of Things : Engineering Cyber Physical Human Systems*, *26*, 101171. DOI: 10.1016/j.iot.2024.101171

Gibson, I., Rosen, D., Stucker, B., Khorasani, M., Gibson, I., Rosen, D., ... & Khorasani, M. (2021). Binder jetting. *Additive manufacturing technologies*, 237-252.

Glover, K. G., Kahn, S. A., & Mittal, R. (2024). 104 Wildland firefighters suffer increasing risk of job-related death. *Journal of Burn Care & Research; Official Publication of the American Burn Association*, *45*(Suppl 1), 84. DOI: 10.1093/jbcr/irae036.103

Gökhan, Ö. (2020). Eklemeli üretim teknolojileri üzerine bir derleme. *Niğde Ömer Halisdemir Üniversitesi Mühendislik Bilimleri Dergisi*, *9*(1), 606–621.

Gong, D., Liu, L., Le, V., Saha, B., Mansour, M. R., Venkatesh, S., & van den Hengel, A. (2019). Memorizing normality to detect anomaly: Memory-augmented deep autoencoder for unsupervised anomaly detection. In *Proceedings of the IEEE/CVF International Conference on Computer Vision* (pp. 1705–1714). IEEE. DOI: 10.1109/ICCV.2019.00179

Gong, T., Zhu, L., Yu, F., & Tang, T. (2023). Edge intelligence in intelligent transportation systems: A survey. *IEEE Transactions on Intelligent Transportation Systems*, *24*(1). DOI: 10.1109/TITS.2023.3275741

Goodfellow, I., Bengio, Y., & Courville, A. (2016). *Deep Learning*. MIT Pres.

Goodfellow, I., Bengio, Y., & Courville, A. (2016). *Deep learning*. MIT Press.

Greene, W. H. (2017). *Econometric analysis* (8th ed.). Pearson.

Gui, J., Chen, T., Zhang, J., Cao, Q., Sun, Z., Luo, H., & Tao, D. (2023). A survey on self-supervised learning: Algorithms, applications, and future trends. *arXiv*. https://arxiv.org/abs/2302.00487

Gülcan, O. (2021). Eklemeli imalatla üretilen kafes yapıların mekanik özellikleri. *Makina Tasarım ve İmalat Dergisi*, *19*(1), 17–34.

Güler, O. (2024). Turbofan motorlarının kestirimci bakımında makine öğrenimi algoritmaları performanslarının karşılaştırılması. *Niğde Ömer Halisdemir Üniversitesi Mühendislik Bilimleri Dergisi*, *13*(1), 99–106.

Güler, S., Serindağ, H. T., & Gürel, Ç. A. M. (2022). Tel ark eklemeli imalat: Son gelişmeler ve değerlendirmeler. *Mühendis ve Makina, 63*(706), 82–116. DOI: 10.46399/muhendismakina.995979

Gupta, V., Mitra, R., Koenig, F., Kumar, M., & Tiwari, M. K. (2023). Predictive maintenance of baggage handling conveyors using IoT. *Computers & Industrial Engineering, 177*, 109033. DOI: 10.1016/j.cie.2023.109033

Gutschi, C., Furian, N., Suschnigg, J., Neubacher, D., & Voessner, S. (2019). Log-based predictive maintenance in discrete parts manufacturing. *Procedia CIRP, 79*, 528–533. DOI: 10.1016/j.procir.2019.02.098

Gwidon, W. S. (2010). *Mechanical engineering handbook*. CRC Press.

Haghighat, A. K., Ravichandra-Mouli, V., Chakraborty, P., Esfandiari, Y., Arabi, S., & Sharma, A. (2020). Applications of Deep Learning in Intelligent Transportation Systems. *Journal of Big Data Analytics in Transportation, 2*(2), 115–145. DOI: 10.1007/s42421-020-00020-1

Haimi, H., Mulas, M., Corona, F., & Vahala, R. (2013). Data-derived soft-sensors for biological wastewater treatment plants: An overview. *Environmental Modelling & Software, 47*, 88–107. DOI: 10.1016/j.envsoft.2013.05.009

Han, H., & Siebert, J. (2022). TinyML: A systematic review and synthesis of existing research. *Proceedings of the International Conference on Artificial Intelligence*. DOI: 10.1109/ICAIIC54071.2022.9722636

Hardin, J., & Rocke, D. M. (2005). The distribution of robust distances. *Journal of Computational and Graphical Statistics, 14*(4), 928–946. DOI: 10.1198/106186005X77685

Harris, T. A., & Kotzalas, M. N. (2006). *Rolling bearing analysis* (5th ed.). CRC Press. DOI: 10.1201/9781482275148

Hasan, M., Choi, J., Neumann, J., Roy-Chowdhury, A. K., & Davis, L. S. (2016). Learning temporal regularity in video sequences. In *Proceedings of the IEEE Conference on Computer Vision and Pattern Recognition* (pp. 733–742). IEEE. DOI: 10.1109/CVPR.2016.85

Hatcher, W. G., & Yu, W. (2018). A Survey of Deep Learning: Platforms, Applications and Emerging Research Trends. *IEEE Access : Practical Innovations, Open Solutions, 6*, 24411–24432. DOI: 10.1109/ACCESS.2018.2830661

Hayajneh, A., Alasali, F., Salama, A., & Holderbaum, W. (2024). Intelligent solar forecasts: Modern machine learning models and TinyML role for improved solar energy yield predictions. *IEEE Access : Practical Innovations, Open Solutions*, *12*, 10846–10864. DOI: 10.1109/ACCESS.2024.3354703

Haydari, A., & Yilmaz, Y. (2022). Deep Reinforcement Learning for Intelligent Transportation Systems: A Survey. *IEEE Transactions on Intelligent Transportation Systems*, *23*(1), 11–32. DOI: 10.1109/TITS.2020.3008612

Hazra, T., & Anjaria, K. (2022). Applications of game theory in deep learning: A survey. *Multimedia Tools and Applications*, *81*(6), 8963–8994. DOI: 10.1007/s11042-022-12153-2 PMID: 35496996

Heinze, G., Wallisch, C., & Dunkler, D. (2017). Variable selection–A review and recommendations for the practicing statistician. *Biometrical Journal. Biometrische Zeitschrift*, *60*(3), 431–449. DOI: 10.1002/bimj.201700067 PMID: 29292533

Heisler, H. (2002). *Advanced vehicle technology* (2nd ed.). Butterworth-Heinemann.

Herb, K. (2024). *Engineering mathematics and artificial intelligence: Foundations, methods, and applications*. Routledge.

Hertel, L., de Prado, M. T., & Ferrara, E. (2023). Transformer training strategies for forecasting multiple load time series. *IEEE Transactions on Smart Grid*.

Hesselvig, J., Nygaard, R. T., Budzik, M. K., & Sandberg, M. Low-fidelity modelling and analysis of polymer SLS: Insights into particle sintering and densification in straight and corner scanning passes. SSRN. Retrieved March 10, 2023 DOI: 10.2139/ssrn.5170448

Himeur, Y., Ghanem, K., Alsalemi, A., Bensaali, F., & Amira, A. (2021). Artificial intelligence based anomaly detection of energy consumption in buildings: A review, current trends and new perspectives. *Applied Energy*, *287*, 116601. DOI: 10.1016/j.apenergy.2021.116601

Hinton, G. E., & Salakhutdinov, R. R. (2006). Reducing the dimensionality of data with neural networks. *Science*, *313*(5786), 504–507. DOI: 10.1126/science.1127647 PMID: 16873662

Hochreiter, S., Bengio, Y., Frasconi, P., & Schmidhuber, J. (2001). Gradient Flow in Recurrent Nets: the Difficulty of Learning Long-Term Dependencies. *A Field Guide to Dynamical Recurrent Neural Networks. IEEE Press, 1.*

Holder, A., & Eichholz, J. (2019). *An introduction to computational science*. Springer. DOI: 10.1007/978-3-030-15679-4

Hong, S., Ahn, S., Jo, Y., & Park, S. (2024). Making anomalies more anomalous: Video anomaly detection using a novel generator and destroyer. *IEEE Access : Practical Innovations, Open Solutions, 12*, 98765–98776. DOI: 10.1109/ACCESS.2024.3374383

Hosamo, H. H., Svennevig, P. R., Svidt, K., Han, D., & Nielsen, H. K. (2022). A Digital Twin predictive maintenance framework of air handling units based on automatic fault detection and diagnostics. *Energy and Building*, *261*, 111988. DOI: 10.1016/j.enbuild.2022.111988

Hosamo, H. H., Tingstveit, M. S., Nielsen, H. K., Svennevig, P. R., & Svidt, K. (2022). Multiobjective optimization of building energy consumption and thermal comfort based on integrated BIM framework with machine learning-NSGA II. *Energy and Building*, *277*, 112479. DOI: 10.1016/j.enbuild.2022.112479

Hu, Y., & Shao, J. Z. (2020). Chinese firefighter fatalities 2007–2016. In G. Y. Wu, K. C. Tsai, & W. K. Chow (Eds.), *In 2018 11th Asia-Oceania Symposium on Fire Science and Technology (AOSFST 2018) (pp. 821–830).* Springer. DOI: 10.1007/978-981-32-9139-3_60

Huang, F., Cui, Y., Dong, R., Wei, J., & Chen, B. (2019). Evaluation on ultimate load-carrying capacity of concrete-filled steel tubular arch structure with preload. *Advances in Structural Engineering*, *22*(13), 2755–2770. DOI: 10.1177/1369433219850091

Huang, H., & Sun, Y. (2019). A decomposition of total variation depth for understanding functional outliers. *Technometrics*, *61*(4), 445–458. DOI: 10.1080/00401706.2019.1574241

Huang, X., Wu, X., & Usmani, A. (2022). Perspectives of using artificial intelligence in building fire safety. In Naser, M., & Corbett, G. (Eds.), *Handbook of cognitive and autonomous systems for fire resilient infrastructures* (pp. 139–159). Springer., DOI: 10.1007/978-3-030-98685-8_6

Huang, Y., Li, J., & Zheng, H. (2024). Modeling of wildfire digital twin: Research progress in detection, simulation, and prediction techniques. *Fire (Basel, Switzerland)*, *7*(1), 1–25. DOI: 10.3390/fire7110412

Huber, P. J. (2009). *Robust statistics* (2nd ed.). Wiley. DOI: 10.1002/9780470434697

Huettmann, F., Craig, E. H., Herrick, K. A., Baltensperger, A. P., Humphries, G. R., Lieske, D. J., & Young, B. D. (2018). Use of machine learning (ML) for predicting and analyzing ecological and 'presence only'data: an overview of applications and a good outlook. *Machine learning for ecology and sustainable natural resource management*, 27-61.

Hu, J., Yu, G., Wang, S., Zhu, E., Cai, Z., & Zhu, X. (2022). Detecting anomalous events from unlabeled videos via temporal masked auto-encoding. In *2022 IEEE International Conference on Multimedia and Expo (ICME)* (pp. 1–6). IEEE. DOI: 10.1109/ICME52920.2022.9859873

Hu, X., Lian, J., Zhang, D., Gao, X., Jiang, L., & Chen, W. (2022). Video anomaly detection based on 3D convolutional auto-encoder. *Signal, Image and Video Processing, 16*(7), 1885–1893. DOI: 10.1007/s11760-022-02148-9

Huyen, C. (2024). *AI engineering: Building applications with foundation models*. O'Reilly Media.

Hu, Z. Z., Leng, S., Lin, J. R., Li, S. W., & Xiao, Y. Q. (2022). Knowledge extraction and discovery based on BIM: A critical review and future directions. *Archives of Computational Methods in Engineering, 29*(1), 335–356. DOI: 10.1007/s11831-021-09576-9

Iman, M., Arabnia, H. R., & Rasheed, K. (2023). A Review of Deep Transfer Learning and Recent Advancements. In *Technologies* (Vol. 11, Issue 2). MDPI. DOI: 10.3390/technologies11020040

Indolia, S., Goswami, A. K., Mishra, S. P., & Asopa, P. (2018). Conceptual understanding of convolutional neural network: A deep learning approach. *Procedia Computer Science, 132*, 679–688. DOI: 10.1016/j.procs.2018.05.069

Ionescu, R. T., Khan, F. S., Georgescu, M.-I., & Shao, L. (2019). Object-centric auto-encoders and dummy anomalies for abnormal event detection in video. In *Proceedings of the IEEE/CVF Conference on Computer Vision and Pattern Recognition* (pp. 7842–7851). IEEE. DOI: 10.1109/CVPR.2019.00803

Ionescu, R. T., Smeureanu, S., Alexe, B., & Popescu, M. (2017). Unmasking the abnormal events in video. In *Proceedings of the IEEE International Conference on Computer Vision* (pp. 2895–2903). IEEE. DOI: 10.1109/ICCV.2017.312

Iqbal, R., Maniak, T., Doctor, F., & Karyotis, C. (2019). Fault detection and isolation in industrial processes using deep learning approaches. *IEEE Transactions on Industrial Informatics, 15*(5), 3077–3084. DOI: 10.1109/TII.2019.2902274

Jacobson, N. J., Bentley, K. H., Walton, A., Wang, S. B., Fortgang, R. G., Millner, A. J., Coombs, G.III, Rodman, A. M., & Coppersmith, D. D. L. (2020). Ethical dilemmas posed by mobile health and machine learning in psychiatry research. *Bulletin of the World Health Organization, 98*(4), 270–276. DOI: 10.2471/BLT.19.237107 PMID: 32284651

James, G. L., Ansaf, R. B., Al Samahi, S. S., Parker, R. D., Cutler, J. M., Gachette, R. V., & Ansaf, B. I. (2023). An efficient wildfire detection system for AI-embedded applications using satellite imagery. *Fire (Basel, Switzerland)*, *6*(4), 169. DOI: 10.3390/fire6040169

Janiesch, C., Zschech, P., & Heinrich, K. (2021). Machine learning and deep learning. *Electronic Markets*, *31*(3), 685–695. DOI: 10.1007/s12525-021-00475-2

Jiang, J., Xiong, Y., Zhang, Z., & Rosen, D. W. (2022). Machine learning integrated design for additive manufacturing. *Journal of Intelligent Manufacturing*, *33*(4), 1073–1086. DOI: 10.1007/s10845-020-01715-6

Jiang, X. (2024). Comparison of deep reinforcement learning algorithms for trading strategy. In *Proceedings of the 2023 International Conference on Data Science, Advanced Algorithm and Intelligent Computing (DAI 2023)* (pp. 4–14). Atlantis Press. https://doi.org/DOI: 10.2991/978-94-6463-370-2_2

Jiang, Z., & Luo, F. (2021). Graph neural networks for traffic forecasting: A survey. *IEEE Transactions on Intelligent Transportation Systems*, *22*(11), 7028–7046.

Jia, W., Sun, M., Lian, J., & Hou, S. (2022). Feature dimensionality reduction : A review. *Complex & Intelligent Systems*, *8*(3), 2663–2693. DOI: 10.1007/s40747-021-00637-x

Jin, H., Wei, J., Ji, R., Wang, L., Liu, Y., Li, C., Yin, Y., & Liu, P. (2025). Efficient preparation of mechanical seal face reinforcement layer by composite anode jet electrodeposition. *Journal of Materials Research and Technology*, *35*, 5908–5922. DOI: 10.1016/j.jmrt.2025.02.224

Jung, N., & Lee, G. (2019). Automated classification of building information modeling (BIM) case studies by BIM use based on natural language processing (NLP) and unsupervised learning. *Advanced Engineering Informatics*, *41*, 100917. DOI: 10.1016/j.aei.2019.04.007

Jung, S., & Yeon, M. (2021). Comparison of Deep Learning Techniques for River Streamflow Forecasting. *IEEE Access : Practical Innovations, Open Solutions*, *9*, 71805–71820. DOI: 10.1109/ACCESS.2021.3077703

Jurečková, J., Picek, J., & Schindler, M. (2019). *Robust statistical methods with R* (2nd ed.). CRC Press. DOI: 10.1201/b21993

Juvinall, R. C., & Marshek, K. M. (2011). *Fundamentals of machine component design* (5th ed.). Wiley.

Kadlec, R. H., & Wallace, S. (2008). *Treatment Wetlands* (2nd ed.). CRC Press., DOI: 10.1201/9781420012514

Kalina, J., & Duintjer Tebbens, J. (2015). Algorithms for regularized linear discriminant analysis. Proceedings of the 6th International Conference on Bioinformatics Models Methods, and Algorithms (BIOINFORMATICS '15), Scitepress, Lisbon, 128-133. DOI: 10.5220/0005234901280133

Kalina, J. (2015). Three contributions to robust regression diagnostics. Journal of Applied Mathematics. *Statistics and Informatics*, *11*(2), 69–78.

Kalina, J. (2018). A robust pre-processing of BeadChip microarray images. *Biocybernetics and Biomedical Engineering*, *38*(3), 556–563. DOI: 10.1016/j.bbe.2018.04.005

Kalina, J., & Matonoha, C. (2020). A sparse pair-preserving centroid-based supervised learning method for high-dimensional biomedical data or images. *Biocybernetics and Biomedical Engineering*, *40*(2), 774–786. DOI: 10.1016/j.bbe.2020.03.008

Kalina, J., & Rensová, D. (2015). How to reduce dimensionality of data: Robustness point of view. *Serbian Journal of Management*, *10*(1), 131–140. DOI: 10.5937/sjm10-6531

Kalina, J., Seidl, L., Zvára, K., Grünfeldová, H., Slovák, D., & Zvárová, J. (2013). System for selecting relevant information for decision support. *Studies in Health Technology and Informatics*, *186*, 83–87. DOI: 10.3233/978-1-61499-240-0-83 PMID: 23542973

Kalina, J., Vašaničová, P., & Litavcová, E. (2019). Regression quantiles under heterosceasticity and multicollinearity: Analysis of travel and tourism competitiveness. Ekonomický časopis. *Journal of Economics*, *67*, 69–85.

Kalina, J., & Vidnerová, P. (2020). On robust training of regression neural networks. In Aneiros, G., Horová, I., Hušková, M., & Vieu, P. (Eds.), *Functional and high-dimensional statistics and related fields. IWFOS 2020. Contributions to Statistics* (pp. 145–152). Springer. DOI: 10.1007/978-3-030-47756-1_20

Kalina, J., & Zvárová, J. (2013). Decision support systems in the process of improving patient safety. In Moumtzoglou, A., & Kastania, A. (Eds.), *E-health technologies and improving patient safety: Exploring organizational factors* (pp. 71–83). IGI Global. DOI: 10.4018/978-1-4666-2657-7.ch005

Kalogridis, I., Claeskens, G., & Van Aelst, S. (2023). Robust and efficient estimation of nonparametric generalized linear models. *Test*, *32*(3), 1055–1078. DOI: 10.1007/s11749-023-00866-x PMID: 37363066

Kamalov, F., Nazir, A., Safaraliev, M., Cherukuri, A. K., & Zgheib, R. (2021). Comparative analysis of activation functions in neural networks. *2021 28th IEEE International Conference on Electronics, Circuits, and Systems, ICECS 2021 - Proceedings*. DOI: 10.1109/ICECS53924.2021.9665646

Karagöz, İ., Bekdemir, A. D., & Tuna, Ö. (2021). 3B yazıcı teknolojilerinde kullanılan yöntemler ve gelişmeler üzerine bir derleme. *Düzce Üniversitesi Bilim ve Teknoloji Dergisi*, *9*(4), 1186–1213. DOI: 10.29130/dubited.877423

Kaveh, A. (2024). *Applications of artificial neural networks and machine learning in civil engineering*. Springer. DOI: 10.1007/978-3-031-66051-1

Khalifa, R. M., Yacout, S., Bassetto, S., & Shaban, Y. (2023). Experimental vibration data collected for a belt-drive system under different operating conditions. *Data in Brief*, *48*, 109175. DOI: 10.1016/j.dib.2023.109175 PMID: 37383798

Khurmi, R. S., & Gupta, J. K. (2005). *A Textbook of Machine Design. S.* Chand Publishing.

Kiangala, K. S., & Wang, Z. (2018). Initiating predictive maintenance for a conveyor motor in a bottling plant using industry 4.0 concepts. *International Journal of Advanced Manufacturing Technology*, *97*(9–12), 3251–3271. DOI: 10.1007/s00170-018-2093-8

Kim, J., & Grauman, K. (2009). Observe locally, infer globally: A space-time MRF for detecting abnormal activities with incremental updates. In *2009 IEEE Conference on Computer Vision and Pattern Recognition* (pp. 2921–2928). IEEE. DOI: 10.1109/CVPR.2009.5206569

Kim, S., Jeong, K., Hong, T., Lee, J., & Lee, J. (2023). Deep learning–based automated generation of material data with object–space relationships for scan to BIM. *Journal of Management Engineering*, *39*(3), 04023004. DOI: 10.1061/JMENEA.MEENG-5143

Kiraz, C., Sezer, H. K., & Şahin, İ. (2018). Kuyumculuk sektöründe 3b baski tasarim özgürlüğünden faydalanmaya ilişkin bir perspektif. *International Journal of 3D Printing Technologies and Digital Industry, 2*(2), 46-58.

Kızılkaya, Y. M., & Oğuzlar, A. (2018). BAZI DENETİMLİ ÖĞRENME ALGORİTMALARININ R PROGRAMLAMA DİLİ İLE KIYASLANMASI. *Karadeniz Uluslararası Bilimsel Dergi*, *37*(37), 90–98. DOI: 10.17498/kdeniz.405746

Koo, B., Jung, R., & Yu, Y. (2021). Automatic classification of wall and door BIM element subtypes using 3D geometric deep neural networks. *Advanced Engineering Informatics*, *47*, 101200. DOI: 10.1016/j.aei.2020.101200

Koresh, H. J. D. (2024). Impact of the preprocessing steps in deep learning-based image classifications. *National Academy Science Letters*, *47*(6), 1–3. DOI: 10.1007/s40009-023-01372-2

Kozhushko, G. G., Yampolsky, D. A., & Lukashuk, M. D. (2020). Analysis of forced vibrations of conveyor belts. IOP Conf. Ser. Mater. Sci. Eng. 709 (3), 033022. DOI: 10.1088/1757-899X/709/3/033022

Kozior, T. (2022). Polymer powders. In *Polymers for 3D Printing* (pp. 271–306). William Andrew Publishing. DOI: 10.1016/B978-0-12-818311-3.00020-3

Krar, S., Gill, A., & Smid, P. (2005). *Technology of machine tools* (6th ed.). McGraw-Hill.

Kratz, L., & Nishino, K. (2009). Anomaly detection in extremely crowded scenes using spatio-temporal motion pattern models. In *2009 IEEE Conference on Computer Vision and Pattern Recognition* (pp. 1446–1453). IEEE. DOI: 10.1109/CVPR.2009.5206771

Kreuzer, T., Papapetrou, P., & Zdravkovic, J. (2024). Artificial intelligence in digital twins—A systematic literature review. *Data & Knowledge Engineering*, *151*, 102304. DOI: 10.1016/j.datak.2024.102304

Krishna, G. V., Narayanamurthy, V., & Viswanath, C. (2020). Modeling the buckling characteristics of the metal-FRP hybrid cylinder. *Composite Structures*, *250*, 112505. DOI: 10.1016/j.compstruct.2020.112505

Kropp, C., Koch, C., & König, M. (2018). Interior construction state recognition with 4D BIM registered image sequences. *Automation in Construction*, *86*, 11–32. DOI: 10.1016/j.autcon.2017.10.027

Kulhandjian, H., Davis, A., Leong, L., Bendot, M., & Kulhandjian, M. (2023). AI-based human detection and localization in heavy smoke using radar and IR camera. *In 2023 IEEE Radar Conference (RadarConf23)*, San Antonio, TX, USA, 1-6. DOI: 10.1109/RadarConf2351548.2023.10149735

Kumar, A., & Khari, M. (2023). Efficient video anomaly detection using residual variational autoencoder. In *2023 International Conference on Communication System, Computing and IT Applications (CSCITA)* (pp. 50–55). IEEE. DOI: 10.1109/CSCITA55725.2023.10104785

Kumari, S. A., & Srinivasan, S. (2019). Ash fouling monitoring and soot-blow optimization for reheater in thermal power plant. *Applied Thermal Engineering*, *149*, 62–72. DOI: 10.1016/j.applthermaleng.2018.12.031

Kumar, M., Samui, P., Kumar, D. R., & Asteris, P. G. (2024). State-of-the-art XG-Boost, RF and DNN based soft-computing models for PGPN piles. *Geomechanics and Geoengineering*, *19*(6), 975–990. DOI: 10.1080/17486025.2024.2337702

Kumar, S., Gopi, T., Harikeerthana, N., Gupta, M. K., Gaur, V., Krolczyk, G. M., & Wu, C. (2023). Machine learning techniques in additive manufacturing: A state of the art review on design, processes and production control. *Journal of Intelligent Manufacturing*, *34*(1), 21–55. DOI: 10.1007/s10845-022-02029-5

Kunathur Thikshaja, U., & Paul, A. (2018). A brief review on deep learning and types of implementations for deep learning. In *Deep learning innovations and their convergence with big data* (p. 13). IGI Global., DOI: 10.4018/978-1-5225-3015-2.ch002

Lakshmanaprabu, S. K., Mohanty, S. N., Krishnamoorthy, S., Uthayakumar, J., & Shankar, K. (2019). Online clinical decision support system using optimal deep neural networks. *Applied Soft Computing*, *81*, 105487. DOI: 10.1016/j.asoc.2019.105487

Lalmuanawma, S., Hussain, J., & Chhaakchhuak, L. (2020). Applications of machine learning and artificial intelligence for Covid-19 (SARS-CoV-2) pandemic: A review. *Chaos, Solitons, and Fractals*, *139*, 110059. DOI: 10.1016/j.chaos.2020.110059 PMID: 32834612

Latif, S., Cuayáhuitl, H., Pervez, F., Shamshad, F., Ali, H. S., & Cambria, E. (2023). A survey on deep reinforcement learning for audio-based applications. *Artificial Intelligence Review*, *56*(3), 2193–2240. DOI: 10.1007/s10462-022-10224-2

Lee, J., Bagheri, B., & Kao, H. A. (2015). A cyber-physical systems architecture for industry 4.0-based manufacturing systems. *Manufacturing Letters*, *3*, 18–23. DOI: 10.1016/j.mfglet.2014.12.001

Lee, S., Lee, P. K. H., & Qiao, M. (2022). Artificial Intelligence in Monitoring, Treatment, and Management of Wastewater. *Processes (Basel, Switzerland)*, *10*(8), 1546. DOI: 10.3390/pr10081546

Lewis, R. H., Jiao, J., Seong, K., Farahi, A., Navrátil, P., Casebeer, N., & Niyogi, D. (2024). Fire and smoke digital twin – A computational framework for modeling fire incident outcomes. *Computers, Environment and Urban Systems*, *110*, 102093. DOI: 10.1016/j.compenvurbsys.2024.102093

Le, X. H., Ho, H. V., Lee, G., & Jung, S. (2019). Application of Long Short-Term Memory (LSTM) neural network for flood forecasting. *Water (Basel)*, *11*(7), 1387. DOI: 10.3390/w11071387

Li, H., Li, J., Guan, X., Liang, B., Lai, Y., & Luo, X. (2019). Research on Overfitting of Deep Learning. *Proceedings - 2019 15th International Conference on Computational Intelligence and Security, CIS 2019*, 78–81. DOI: 10.1109/CIS.2019.00025

Li, P., & Pei, Y. J. L. (2023). A comprehensive survey on design and application of autoencoder in deep learning. *Applied Soft Computing, Elsevier*, 110176.

Li, Y., Orgerie, A.C., Rodero, I., Amersho, B.L., Parashar, M. et al. (2018). End-to-end models for edge cloud-based IoT platforms: Application to data stream analysis in IoT. [https://hal.archives-ouvertes.fr/hal-01673501]

Liao, Q.. (2021). A review of graph neural networks in power systems. *IEEE Transactions on Power Systems*, *36*(6), 5024–5039.

Li, D., Nie, X., Li, X., Zhang, Y., & Yin, Y. (2022). Context-related video anomaly detection via generative adversarial network. *Pattern Recognition Letters*, *156*, 183–189. DOI: 10.1016/j.patrec.2022.03.004

Li, K., Gan, V. J., Li, M., Gao, M. Y., Tiong, R. L., & Yang, Y. (2024). Automated generative design and prefabrication of precast buildings using integrated BIM and graph convolutional neural network. *Developments in the Built Environment*, *18*, 100418. DOI: 10.1016/j.dibe.2024.100418

Li, N., & Chang, F. (2019). Video anomaly detection and localization via multivariate Gaussian fully convolution adversarial autoencoder. *Neurocomputing*, *369*, 92–105. DOI: 10.1016/j.neucom.2019.08.044

Li, N., Chang, F., & Liu, C. (2020). Spatial-temporal cascade autoencoder for video anomaly detection in crowded scenes. *IEEE Transactions on Multimedia*, *23*, 203–215. DOI: 10.1109/TMM.2020.2984093

Lipson, C., & Juran, I. (1991). *Quality by design: Taguchi methods and U.S. industry*. McGraw-Hill.

Liu, C., Tang, D., Zhu, H., & Nie, Q. (2021). A novel predictive maintenance method based on deep adversarial learning in the intelligent manufacturing system. *IEEE Access : Practical Innovations, Open Solutions*, *9*, 49557–49575. DOI: 10.1109/ACCESS.2021.3069256

Liu, C., Wang, R. R., Ho, I., Kong, Z. J., Williams, C., Babu, S., & Joslin, C. (2023). Toward online layer-wise surface morphology measurement in additive manufacturing using a deep learning-based approach. *Journal of Intelligent Manufacturing*, *34*(6), 2673–2689. DOI: 10.1007/s10845-022-01933-0

Liu, H., & Lang, B. (2019). Machine learning and deep learning methods for intrusion detection systems: A survey. *Applied Sciences (Basel, Switzerland), 9*(20), 4396. DOI: 10.3390/app9204396

Liu, J. Z., Fang, H., & Chan, T. M. (2022). Experimental investigation on material properties and residual stresses in cold-formed high strength steel irregular octagonal hollow sections. *Journal of Constructional Steel Research, 191*, 107170. DOI: 10.1016/j.jcsr.2022.107170

Liu, X., He, D., Lodewijks, G., Pang, Y., & Mei, J. (2019). Integrated decision making for predictive maintenance of belt conveyor systems. *Reliability Engineering & System Safety, 188*, 347–351. DOI: 10.1016/j.ress.2019.03.047

Liu, X., & Zhang, Z.. (2025). A survey on diffusion models for anomaly detection. *IEEE Transactions on Neural Networks and Learning Systems.*

Li, W., Mahadevan, V., & Vasconcelos, N. (2014). Anomaly detection and localization in crowded scenes. *IEEE Transactions on Pattern Analysis and Machine Intelligence, 36*(1), 18–32. DOI: 10.1109/TPAMI.2013.111 PMID: 24231863

Li, Y. H., Harfiya, L. N., Purwandari, K., & Der Lin, Y. (2020). Real-time cuffless continuous blood pressure estimation using deep learning model. *Sensors (Basel), 20*(19), 1–19. DOI: 10.3390/s20195606 PMID: 33007891

Li, Z., Zhang, Z., Shi, J., & Wu, D. (2019). Prediction of surface roughness in extrusion-based additive manufacturing with machine learning. *Robotics and Computer-integrated Manufacturing, 57*, 488–495. DOI: 10.1016/j.rcim.2019.01.004

Locatelli, M., Seghezzi, E., Pellegrini, L., Tagliabue, L. C., & Di Giuda, G. M. (2021). Exploring natural language processing in construction and integration with building information modeling: A scientometric analysis. *Buildings (Basel, Switzerland), 11*(12), 583. DOI: 10.3390/buildings11120583

Loeffler, H. H., He, J., Tibo, A., Janet, J. P., Voronov, A., Mervin, L. H., & Engkvist, O. (2024). Reinvent 4: Modern AI–driven generative molecule design. *Journal of Cheminformatics, 16*(1), 20. DOI: 10.1186/s13321-024-00812-5 PMID: 38383444

López-Robles, J. R., Rodríguez-Salvador, M., Gamboa-Rosales, N. K., Ramirez-Rosales, S., & Cobo, M. J. (2019). The last five years of Big Data Research in economics, econometrics and finance: Identification and conceptual analysis. *Procedia Computer Science, 162*, 729–736. DOI: 10.1016/j.procs.2019.12.044

Lu, C., Shi, J., & Jia, J. (2013). Abnormal event detection at 150 fps in Matlab. In *Proceedings of the IEEE International Conference on Computer Vision* (pp. 2720–2727). IEEE. DOI: 10.1109/ICCV.2013.338

Luo, W., Liu, W., & Gao, S. (2017a). A revisit of sparse coding based anomaly detection in stacked RNN framework. In *Proceedings of the IEEE International Conference on Computer Vision* (pp. 341–349). IEEE. DOI: 10.1109/ICCV.2017.45

Luo, W., Liu, W., & Gao, S. (2017b). Remembering history with convolutional LSTM for anomaly detection. In *2017 IEEE International Conference on Multimedia and Expo (ICME)* (pp. 439–444). IEEE. DOI: 10.1109/ICME.2017.8019325

Luo, W., Liu, W., Lian, D., & Gao, S. (2019). Video anomaly detection with sparse coding inspired deep neural networks. *IEEE Transactions on Pattern Analysis and Machine Intelligence*, *43*(3), 1070–1084. DOI: 10.1109/TPAMI.2019.2944377 PMID: 31567072

Luo, W., Liu, W., Lian, D., Tang, J., Duan, L., Peng, X., & Gao, S. (2021). Future frame prediction network for video anomaly detection. *IEEE Transactions on Pattern Analysis and Machine Intelligence*, *44*(11), 7505–7520. DOI: 10.1109/TPAMI.2021.3129349 PMID: 34797762

Luo, X., Li, Y., Wang, W., Ban, X., Wang, J. H., & Zhao, W. (2020). A robust multilayer extreme learning machine using kernel risk-sensitive loss criterion. *International Journal of Machine Learning and Cybernetics*, *11*(1), 197–216. DOI: 10.1007/s13042-019-00967-w

Lv, H., Chen, C., Cui, Z., Xu, C., Li, Y., & Yang, J. (2021). Learning normal dynamics in videos with meta prototype network. In *Proceedings of the IEEE/CVF Conference on Computer Vision and Pattern Recognition* (pp. 15425–15434). IEEE. DOI: 10.1109/CVPR46437.2021.01517

Madni, A. M., Madni, C. C., & Lucero, S. D. (2019). Leveraging digital twin technology in model-based systems engineering. *Systems*, *7*(1), 7. DOI: 10.3390/systems7010007

Mahadevan, V., Li, W., Bhalodia, V., & Vasconcelos, N. (2010). Anomaly detection in crowded scenes. In *2010 IEEE Computer Society Conference on Computer Vision and Pattern Recognition* (pp. 1975–1981). IEEE. DOI: 10.1109/CVPR.2010.5539872

Mahdi, M. S., & Mohammed, A. I. (2024, Aug). The effect of bending angle and radius on wall thickness variation according to a geometric ratio in tube bending. In *AIP Conference Proceedings* (Vol. 3105, No. 1). AIP Publishing. DOI: 10.1063/5.0212722

Mahmoud, M. M. E., Rodrigues, J. J. P. C., Ahmed, S. H., Shah, S. C., Al-Muhtadi, J. F., Korotaev, V. V., & De Albuquerque, V. H. C. (2018). Enabling technologies on cloud of things for smart healthcare. *IEEE Access : Practical Innovations, Open Solutions*, *6*, 31950–31967. DOI: 10.1109/ACCESS.2018.2845399

Maier, H. R., & Dandy, G. C. (2000). Neural networks for the prediction and forecasting of water resources variables: A review of modelling issues and applications. *Environmental Modelling & Software, 15*(1), 101–124. DOI: 10.1016/S1364-8152(99)00007-9

Malaka, J., & Hetmańczyk, M. (2020). Intelligent Drive in Industry 4.0–Protection of Toothed Belt Transmission on the Basis of Its Digital Twin. In *Sympozjon Modelowanie w Mechanice* (pp. 201–210). Springer International Publishing.

Mansour, R. F., Escorcia-Gutierrez, J., Gamarra, M., Villanueva, J. A., & Leal, N. (2021). Intelligent video anomaly detection and classification using faster RCNN with deep reinforcement learning model. *Image and Vision Computing, 112*, 104229. DOI: 10.1016/j.imavis.2021.104229

Marić, L., Bjelopera, M., Mandić, D., Periša, M., Mandić, N., & Knezović, M. (2023). Advancement of an integrated technological platform for wildfire management through edge computing. *In 2023 8th International Conference on Smart and Sustainable Technologies (SpliTech)* (pp. 1–6). IEEE. https://doi.org/DOI: 10.23919/SpliTech58164.2023.10193111

Marozzi, M., Mukherjee, A., & Kalina, J. (2020). Interpoint distance tests for high-dimensional comparison studies. *Journal of Applied Statistics, 47*(4), 653–665. DOI: 10.1080/02664763.2019.1649374 PMID: 35707487

Martinez, W. L., Martinez, A. R., & Solka, J. L. (2017). *Exploratory data analysis with MATLAB* (3rd ed.). Chapman & Hall/CRC.

Mathew, V., Toby, T., Singh, V., Rao, B. M., & Kumar, M. G. (2018). "Prediction of Remaining Useful Lifetime (RUL) of turbofan engine using machine learning," in IEEE international conference on circuits and systems. *ICCS, 2017*, 306–311. DOI: 10.1109/ICCS1.2017.8326010

Mattera, G., Caggiano, A., & Nele, L. (2025). Energy Efficiency Optimisation in Wire arc Additive Manufacturing of Invar 36 Alloy via Intelligent Data-Driven Techniques. International Journal of Precision Engineering and Manufacturing-Green Technology, 1-13.

Ma, X., Wang, H., Xue, B., Zhou, M., Ji, B., & Li, Y. (2014). Depth-based human fall detection via shape features and improved extreme learning machine. *IEEE Journal of Biomedical and Health Informatics, 18*(6), 1915–1922. DOI: 10.1109/JBHI.2014.2304357 PMID: 25375688

McArthur, J. J., Shahbazi, N., Fok, R., Raghubar, C., Bortoluzzi, B., & An, A. (2018). Machine learning and BIM visualization for maintenance issue classification and enhanced data collection. *Advanced Engineering Informatics*, *38*, 101–112. DOI: 10.1016/j.aei.2018.06.007

McCann, R., Obeidi, M. A., Hughes, C., McCarthy, É., Egan, D. S., Vijayaraghavan, R. K., Joshi, A. M., Acinas Garzon, V., Dowling, D. P., McNally, P. J., & Brabazon, D. (2021). In-situ sensing, process monitoring and machine control in Laser Powder Bed Fusion: A review. *Additive Manufacturing*, *45*, 102058. DOI: 10.1016/j.addma.2021.102058

Mehran, R., Oyama, A., & Shah, M. (2009). Abnormal crowd behavior detection using social force model. In *2009 IEEE Conference on Computer Vision and Pattern Recognition* (pp. 935–942). IEEE. DOI: 10.1109/CVPR.2009.5206641

Meng, L., McWilliams, B., Jarosinski, W., Park, H. Y., Jung, Y. G., Lee, J., & Zhang, J. (2020). Machine learning in additive manufacturing: A review. *Journal of the Minerals Metals & Materials Society*, *72*(6), 2363–2377. DOI: 10.1007/s11837-020-04155-y

Merz, M., & Actuarial, S. (2023). Recurrent Neural Networks. *StatisticalFoundationsofActuarialLearninganditsApplications,Springer, Actuarial*, 381–406.

Micorek, J., Possegger, H., Narnhofer, D., Bischof, H., & Kozinski, M. (2024). Mulde: Multiscale log-density estimation via denoising score matching for video anomaly detection. In Proceedings of the IEEE/CVF Conference on Computer Vision and Pattern Recognition (pp. 18868–18877). IEEE. https://openaccess.thecvf.com/content/CVPR2024/html/Micorek_Mulde_Multiscale_Log-Density_Estimation_via_Denoising_Score_Matching_for_Video_CVPR_2024_paper.html

Mikołajczyk. Agnieszka, M. G. (2018). Data augmentation for improving deep learning in image classification problem. *IEEE*, 117–122.

Minsky, M. (1961). Steps Toward Artificial Intelligence*. *PROCEEDINGS OF THE IRE*, 8–29.

Mishra, S., & Jabin, S. (2024). Anomaly detection in surveillance videos using deep autoencoder. *International Journal of Information Technology : an Official Journal of Bharati Vidyapeeth's Institute of Computer Applications and Management*, *16*(2), 1111–1122. DOI: 10.1007/s41870-023-01659-z

Mobley, R. K. (2002). *An introduction to predictive maintenance* (2nd ed.). Butterworth-Heinemann.

Mohammadi, M. G., Mahmoud, D., & Elbestawi, M. (2021). On the application of machine learning for defect detection in L-PBF additive manufacturing. *Optics & Laser Technology, 143*, 107338. DOI: 10.1016/j.optlastec.2021.107338

Mohammed, A., & Kora, R. (2023). A comprehensive review on ensemble deep learning: Opportunities and challenges. *Journal of King Saud University. Computer and Information Sciences, 35*(2), 757–774. DOI: 10.1016/j.jksuci.2023.01.014

Moradian, A. (2024). Digital twins for sustainable semiconductor manufacturing. In M. Grieves & E. Y. Hua (Eds.), *Digital twins, simulation, and the metaverse: Simulation foundations, methods and applications* (pp. [page range]). Springer. DOI: 10.1007/978-3-031-69107-2_12

Morris, B. T., & Trivedi, M. M. (2008). A survey of vision-based trajectory learning and analysis for surveillance. *IEEE Transactions on Circuits and Systems for Video Technology, 18*(8), 1114–1127. DOI: 10.1109/TCSVT.2008.927109

Mott, R. L. (2017). *Machine Elements in Mechanical Design* (6th ed.). Pearson.

Mo, W., & Zhang, Q. (2013). Energy-nutrients-water nexus: Integrated resource recovery in municipal wastewater treatment plants. *Journal of Environmental Management, 127*, 255–267. DOI: 10.1016/j.jenvman.2013.05.007 PMID: 23764477

Mrázek, V., Sarwar, S. S., Sekanina, L., Vašíček, Z., & Roy, K. (2016). Design of power-efficient approximate multipliers for approximate artificial neural networks. *2016 IEEE/ACM International Conference on Computer-Aided Design (ICCAD)*, 1-7. DOI: 10.1145/2966986.2967021

Mu, H., Sun, R., Yuan, G., & Wang, Y. (2021). Abnormal human behavior detection in videos: A review. *Information Technology and Control, 50*(3), 522–545. DOI: 10.5755/j01.itc.50.3.27864

Mulero-Palencia, S., Álvarez-Díaz, S., & Andrés-Chicote, M. (2021). Machine learning for the improvement of deep renovation building projects using as-built BIM models. *Sustainability (Basel), 13*(12), 6576. DOI: 10.3390/su13126576

Mungoli, N. (2023a). Advancements in Deep Learning: A Comprehensive Study of the Latest Trends and Techniques in Machine Learning. *International Journal of Advanced Engineering Technologies and Innovations, 04*, 43–64.

Nabavi, A., Ramaji, I., Sadeghi, N., & Anderson, A. (2023). Leveraging Natural Language Processing for Automated Information Inquiry from Building Information Models. Journal of Information Technology in Construction, 28.

Naji, Y., Setkov, A., Loesch, A., Gouiffès, M., & Audigier, R. (2022). Spatio-temporal predictive tasks for abnormal event detection in videos. In 2022 18th IEEE International Conference on Advanced Video and Signal Based Surveillance (AVSS) (pp. 1–8). IEEE. DOI: 10.1109/AVSS56176.2022.9959669

Nanga, S., Bawah, A. T., Acquaye, B. A., Billa, M., Baeta, F. D., Odai, N. A., Obeng, S. K., & Nsiah, A. D. (2021). Review of Dimension Reduction Methods. *Journal of Data Analysis and Information Processing*. *Journal of Data Analysis and Information Processing*, *9*(3), 189–231. DOI: 10.4236/jdaip.2021.93013

Narisetty, N. N., & Nair, V. N. (2016). Extremal depth for functional data and applications. *Journal of the American Statistical Association*, *111*(516), 1705–1714. DOI: 10.1080/01621459.2015.1110033

NASA. (n.d.). *NASA wildfire digital twin pioneers new AI models and streaming data techniques for forecasting fire and smoke*. NASA Science. https://science.nasa.gov/science-research/science-enabling-technology/nasa-wildfire-digital-twin-pioneers-new-ai-models-and-streaming-data-techniques-for-forecasting-fire-and-smoke/

Nasr, M. M., & Anwar, S. (2025). Developing an intelligent approach based on ANFIS and advanced NSGA-III for improving the turning performance of additively manufactured γ-TiAl alloy. *International Journal of Advanced Manufacturing Technology*, *137*(7-8), 1–26. DOI: 10.1007/s00170-025-15366-5

National Wildfire Coordinating Group. (2014). *Wildland fire incident management field guide*.

Nayak, R., Pati, U. C., & Das, S. K. (2021). A comprehensive review on deep learning-based methods for video anomaly detection. *Image and Vision Computing*, *106*, 104078. DOI: 10.1016/j.imavis.2020.104078

Neale, M. J. (1995). *The tribology handbook* (2nd ed.). Butterworth-Heinemann.

Newhart, K. B., Holloway, R. W., Hering, A. S., & Cath, T. Y. (2019). Data-driven performance analyses of wastewater treatment plants: A review. *Water Research*, *157*, 498–513. DOI: 10.1016/j.watres.2019.03.030 PMID: 30981980

Nguyen, H., Kieu, L. M., Wen, T., & Cai, C. (2018). Deep learning methods in transportation domain: A review. *IET Intelligent Transport Systems*, *12*(9), 998–1004. DOI: 10.1049/iet-its.2018.0064

NICHITA, G. G. (2007). An Review About Rapid Manufacturing. communications, 6(7), 9.

Ning, Z., Wang, Z., Liu, Y., Liu, J., & Song, L. (2024). Memory-enhanced appearance-motion consistency framework for video anomaly detection. *Computer Communications, 216*, 159–167. DOI: 10.1016/j.comcom.2024.01.004

Nisha, C. M., & Thangarasu, N. (2023a). Deep learning algorithms and their relevance: A review. *International Journal of Data Informatics and Intelligent Computing*, *2*(4), 1–10. DOI: 10.59461/ijdiic.v2i4.78

Norton, R. L. (2020). *Design of Machinery* (6th ed.). McGraw-Hill Education.

Nunez-Morales, J. D., Hsu, S. H., Ibrahim, A., & Golparvar-Fard, M. (2023, May). New metrics to benchmark and improve bim visibility within a synthetic image generation process for computer vision progress tracking. In Canadian Society of Civil Engineering Annual Conference (pp. 209-221). Cham: Springer Nature Switzerland.

O'Connor, S. M., Bell, J., & Morrison, S. (2017). Lessons Learned from the South Canyon Fire. *Fire Management Today*, *77*(2), 32–38. https://www.fs.fed.us/fire/publications/fire-management-today/2017/summer/lessons-learned-south-canyon-fire.shtml

O'Shea, K., & Nash, R. (2015). An Introduction to Convolutional Neural Networks. *ArXiv Preprint ArXiv:1511.08458.*

Obaidur Rahman, M., Sabir Hossain, M., Shafiul Alam Forhad, M., Junaid, T.-S., & Kamal Hossen, M. (2019). Predicting Prices of Stock Market using Gated Recurrent Units (GRUs) Neural Networks. *IJCSNS International Journal of Computer Science and Network Security, 19*(1).

Okada, Y., Mertens, M., Liu, N., Lam, S. S. W., & Ong, M. E. H. (2023). AI and machine learning in resuscitation: Ongoing research, new concepts, and key challenges. *Resuscitation Plus*, *15*, 100435. DOI: 10.1016/j.resplu.2023.100435 PMID: 37547540

Olson, D. L. (2017). *Descriptive data mining*. Springer. DOI: 10.1007/978-981-10-3340-7

Olsson, G. (2012). ICA and me---A subjective review. *Water Research*, *46*(6), 1585–1624. DOI: 10.1016/j.watres.2011.12.054 PMID: 22284982

Olsson, G., Nielsen, M. K., Yuan, Z., Lynggaard-Jensen, A., & Steyer, J. P. (2005). *Instrumentation, Control and Automation in Wastewater Systems*. IWA Publishing., DOI: 10.2166/9781780402680

Olugboyega, O., Oseghale, G. E., & Aigbavboa, C. (2023). Multiple holistic case study of project-level building information modelling (BIM) adoption in Nigeria. *Construction Innovation*, *23*(3), 567–586. DOI: 10.1108/CI-10-2021-0199

Osama, Z. (2024). The digital twin framework: A roadmap to the development of user-centred digital twin in the built environment. *Journal of Building Engineering*, *98*, 111081. DOI: 10.1016/j.jobe.2024.111081

Oskar, L. (2022). Measuring and extracting data from a BIM model.

Ouerghemmi, N., Salem, S. B., Salah, M., Bacha, K., & Chaari, A. (2023). V-belt fault detection using Extended Park Vector Approach (EPVA) in centrifugal fan driven by an induction motor. *Mechanical Systems and Signal Processing*, *200*, 110566. DOI: 10.1016/j.ymssp.2023.110566

Özkan, B., Parim, C., & Çene, E. (2023). Ülkelerin gelişmişlik düzeylerinin karar ağacı ve rastgele orman yöntemleriyle tahmin edilmesi. *EKOIST Journal of Econometrics and Statistics*, *0*(38), 87–104. DOI: 10.26650/ekoist.2023.38.1172190

Özsolak, O. (2019). Eklemeli imalat yöntemleri ve kullanılan malzemeler. *International Journal of Innovative Engineering Applications*, *3*(1), 9–14.

Özsoy, K., Duman, B., & Gültekin, D. İ. (2019). Havacılık, uzay ve savunma sanayi için eklemeli imalat ile metal parça imalatı. 4th International Congress on 3D Printing (Additive Manufacturing) Technologies and Digital Industry (ss. 462-471).

Park, C., Cho, M., Lee, M., & Lee, S. (2022). FastAno: Fast anomaly detection via spatio-temporal patch transformation. In *Proceedings of the IEEE/CVF Winter Conference on Applications of Computer Vision* (pp. 2249–2259). IEEE. DOI: 10.1109/WACV51458.2022.00197

Park, J., Kim, J., Lee, D., Jeong, K., Lee, J., Kim, H., & Hong, T. (2022). Deep learning–based automation of scan-to-BIM with modeling objects from occluded point clouds. *Journal of Management Engineering*, *38*(4), 04022025. DOI: 10.1061/(ASCE)ME.1943-5479.0001055

Pawar, K., & Attar, V. (2019). Deep learning approaches for video-based anomalous activity detection. *World Wide Web (Bussum)*, *22*(2), 571–601. DOI: 10.1007/s11280-018-0582-1

Pedral Sampaio, R., Aguiar Costa, A., & Flores-Colen, I. (2022). A systematic review of artificial intelligence applied to facility management in the building information modeling context and future research directions. *Buildings (Basel, Switzerland)*, *12*(11), 1939. DOI: 10.3390/buildings12111939

Penney, G., Habibi, D., & Cattani, M. (2019). Firefighter tenability and its influence on wildfire suppression. *Fire Safety Journal, 106*, 38–51. DOI: 10.1016/j.firesaf.2019.03.012

Perez-Perez, Y., Golparvar-Fard, M., & El-Rayes, K. (2021). Scan2BIM-NET: Deep learning method for segmentation of point clouds for scan-to-BIM. *Journal of Construction Engineering and Management, 147*(9), 04021107. DOI: 10.1061/(ASCE)CO.1943-7862.0002132

Peterson, M. B., & Winer, W. O. (1980). *Wear control handbook*. ASME.

Phantanaikasem, P., Pannawan, A., Jamroen, C., & Surinkaew, T. (2024). Potential impacts of wildfire on distribution networks in Thailand: A data-driven assessment using hotspot analysis. *In 2024 IEEE 22nd Student Conference on Research and Development (SCOReD)*, 39–43. DOI: 10.1109/SCOReD64708.2024.10872680

Piciarelli, C., Micheloni, C., & Foresti, G. L. (2008). Trajectory-based anomalous event detection. *IEEE Transactions on Circuits and Systems for Video Technology, 18*(11), 1544–1554. DOI: 10.1109/TCSVT.2008.2005599

Poch, M., Comas, J., Rodríguez-Roda, I., Sànchez-Marrè, M., & Cortés, U. (2014). Designing and building real environmental decision support systems. *Environmental Modelling & Software, 30*, 1–9.

Pollak, A., Temich, S., Ptasiński, W., Kucharczyk, J., & Gąsiorek, D. (2021). Prediction of belt-drive faults in case of predictive maintenance in industry 4.0 platform. *Applied Sciences (Basel, Switzerland), 11*(21), 10307. DOI: 10.3390/app112110307

Pothuganti, S. (2018). Review on over-fitting and under-fitting problems in Machine Learning and solutions. *Article in International Journal of Advanced Research in Electrical Electronics and Instrumentation Engineering, 7*(9). Advance online publication. DOI: 10.15662/IJAREEIE.2018.0709015

Pruckovskaja, V., Weißenfeld, A., & Heistracher, C.. (2023). *Federated learning for predictive maintenance and quality inspection in industrial applications. Proceedings of PHM Europe 2023.* https://arxiv.org/abs/2304.11101 DOI: 10.1109/PHM58589.2023.00064

Przekop, R. E., Konieczna, R., Głowacka, J., Sztorch, B., Głowacki, M., & Kotecka, B. (2025). 3D printing in art: Use of thermoplastics and ceramics—the current state and limitations resulting from the possibilities of technology. *Progress in Additive Manufacturing, 10*(9), 1–31. DOI: 10.1007/s40964-025-01015-6

Pulcini, V., & Modoni, G. (2024). Machine learning-based digital twin of a conveyor belt for predictive maintenance. *International Journal of Advanced Manufacturing Technology*, *133*(11), 6095–6110. DOI: 10.1007/s00170-024-14097-3

Purwono, Ma'arif, A., Rahmaniar, W., Fathurrahman, H. I. K., Frisky, A. Z. K., & Haq, Q. M. U. (2022). Understanding of Convolutional Neural Network (CNN): A Review. *International Journal of Robotics and Control Systems, 2*(4), 739–748. DOI: 10.31763/ijrcs.v2i4.888

Pu, Y., Apel, D. B., Liu, V., & Mitri, H. (2019). Machine learning methods for rockburst prediction-state-of-the-art review. *International Journal of Mining Science and Technology*, *29*(4), 565–570. DOI: 10.1016/j.ijmst.2019.06.009

Qin, J., Hu, F., Liu, Y., Witherell, P., Wang, C. C., Rosen, D. W., Simpson, T. W., Lu, Y., & Tang, Q. (2022). Research and application of machine learning for additive manufacturing. *Additive Manufacturing*, *52*, 102691. DOI: 10.1016/j.addma.2022.102691

Qiu, L., Zhang, D., Tian, Y., & Al-Nabhan, N. (2021). Deep learning-based algorithm for vehicle detection in intelligent transportation systems. *The Journal of Supercomputing*, *77*(10), 11083–11098. DOI: 10.1007/s11227-021-03712-9

Quarteroni, A. (2018). The role of statistics in the era of big data: A computational scientist' perspective. *Statistics & Probability Letters*, *136*, 63–67. DOI: 10.1016/j.spl.2018.02.047

Rahman Ahad, M. A., Tan, J. K., Kim, H., & Ishikawa, S. (2018). Activity representation by SURF-based templates. *Computer Methods in Biomechanics and Biomedical Engineering. Imaging & Visualization*, *6*(5), 573–583. DOI: 10.1080/21681163.2017.1298472

Ramachandra, B., Jones, M. J., & Vatsavai, R. (2020). Learning a distance function with a Siamese network to localize anomalies in videos. In *Proceedings of the IEEE/CVF Winter Conference on Applications of Computer Vision* (pp. 2598–2607). IEEE. DOI: 10.1109/WACV45572.2020.9093417

Rao, J. S. (2011). *Mechanical Vibrations*. New Age International Publishers.

Ribeiro, F., Marins, M., Netto, S., & da Silva, E. A. (2017). Rotating machinery fault diagnosis using similarity-based models. XXXV Simpósio Brasileiro de Telecomunicações e Processamento de Sinais-sbrt2017.

Ristea, N.-C., Croitoru, F.-A., Ionescu, R. T., Popescu, M., Khan, F. S., & Shah, M. (2024). Self-distilled masked auto-encoders are efficient video anomaly detectors. In *Proceedings of the IEEE/CVF Conference on Computer Vision and Pattern Recognition* (pp. 15984–15995). IEEE., https://openaccess.thecvf.com/content/CVPR2024/html/Ristea_Self-Distilled_Masked_Auto-Encoders_are_Efficient_Video_Anomaly_Detectors_CVPR_2024_paper.html DOI: 10.1109/CVPR52733.2024.01513

Rossini, N. S., Dassisti, M., Benyounis, K. Y., & Olabi, A. G. (2012). Methods of measuring residual stresses in components. *Materials & Design*, *35*, 572–588. DOI: 10.1016/j.matdes.2011.08.022

Rousseeuw, P. J., & Leroy, A. M. (1987). *Robust regression and outlier detection*. Wiley. DOI: 10.1002/0471725382

Rousseeuw, P. J., Raymaekers, J., & Hubert, M. (2018). A measure of directional outlyingness with applications to image data and video. *Journal of Computational and Graphical Statistics*, *27*(2), 345–359. DOI: 10.1080/10618600.2017.1366912

Rousseeuw, P. J., & Van Driessen, K. (1999). A fast algorithm for the minimum covariance determinant estimator. *Technometrics*, *41*(3), 212–223. DOI: 10.1080/00401706.1999.10485670

Rousseeuw, P. J., & Van Driessen, K. (2006). Computing LTS regression for large data sets. *Data Mining and Knowledge Discovery*, *12*(1), 29–45. DOI: 10.1007/s10618-005-0024-4

Roy, A., & Chakraborty, S. (2023). Support vector machine in structural reliability analysis: A review. *Reliability Engineering & System Safety*, *233*, 109126. DOI: 10.1016/j.ress.2023.109126

Roy, P. K., & Om, H. (2018). Suspicious and violent activity detection of humans using HOG features and SVM classifier in surveillance videos. In Satapathy, S. C., Bhateja, V., Le, P. N. H., Nguyen, N., & Rao, B. L. (Eds.), *Advances in soft computing and machine learning in image processing* (pp. 277–294). Springer., DOI: 10.1007/978-3-319-63754-9_13

Ruiz, E., Ferreño, D., Cuartas, M., López, A., Arroyo, V., & Gutiérrez-Solana, F. (2020). Machine learning algorithms for the prediction of the strength of steel rods: An example of data-driven manufacturing in steelmaking. *International Journal of Computer Integrated Manufacturing*, *33*(9), 880–894. DOI: 10.1080/0951192X.2020.1803505

Rumin, P., Kotowicz, J., Hogg, D., & Zastawna-Rumin, A. (2023). Utilization of measurements, machine learning, and analytical calculation for preventing belt flip over on conveyor belts. *Measurement, 218*, 113157. DOI: 10.1016/j.measurement.2023.113157

Rusiecki, A. (2013). Robust learning algorithm based on LTA estimator. *Neurocomputing, 120*, 624–632. DOI: 10.1016/j.neucom.2013.04.008

Saadi, J. I., & Yang, M. C. (2023). Generative design: Reframing the role of the designer in early-stage design process. *Journal of Mechanical Design, 145*(4), 041411. DOI: 10.1115/1.4056799

Sabokrou, M., Fathy, M., & Hoseini, M. (2016). Video anomaly detection and localisation based on the sparsity and reconstruction error of auto-encoder. *Electronics Letters, 52*(13), 1122–1124. DOI: 10.1049/el.2016.0440

Sabokrou, M., Fayyaz, M., Fathy, M., Moayed, Z., & Klette, R. (2018). Deep-anomaly: Fully convolutional neural network for fast anomaly detection in crowded scenes. *Computer Vision and Image Understanding, 172*, 88–97. DOI: 10.1016/j.cviu.2018.02.006

Sadiq, M. T., Yu, X., & Yuan, Z. (2021). Exploiting dimensionality reduction and neural network techniques for the development of expert brain-computer interfaces. *Expert Systems with Applications, 164*, 114031. DOI: 10.1016/j.eswa.2020.114031

Sakib, N., & Wuest, T. (2018). Challenges and opportunities of condition-based predictive maintenance: A review. *Procedia CIRP, 78*, 267–272. DOI: 10.1016/j.procir.2018.08.318

Santhosh, K. K., Dogra, D. P., & Roy, P. P. (2020). Anomaly detection in road traffic using visual surveillance: A survey. *ACM Computing Surveys, 53*(6), 119. DOI: 10.1145/3417988

Sarıcan, B., & Engin, O. (2024). Makine Çizelgeleme Problemlerinin Çözümünde Pekiştirmeli Öğrenme Etkisinin Analizi. *ALKÜ Fen Bilimleri Dergisi, 6*(2), 116–140. DOI: 10.46740/alku.1390397

Sarker, I. H. (2021). Deep Learning: A Comprehensive Overview on Techniques, Taxonomy, Applications and Research Directions. In *SN Computer Science* (Vol. 2, Issue 6). Springer. DOI: 10.1007/s42979-021-00815-1

Sarma, D., Rali, A. S., & Jentzer, J. C. (2025). Key Concepts in Machine Learning and Clinical Applications in the Cardiac Intensive Care Unit. *Current Cardiology Reports, 27*(1), 30. DOI: 10.1007/s11886-024-02149-9 PMID: 39831916

Şekkeli, Z. H., & Bakan, İ. (2018). AKILLI FABRİKALAR. *Journal of Life Economics*, *5*(4), 203–220. DOI: 10.15637/jlecon.270

Seydgar, M., Motamedi, A., & Poirier, E. (2022). *Deep neural networks to assist in bim creation using scanned data: A review*. Transforming Construction with Reality Capture Technologies.

Shamshiri, A., Ryu, K. R., & Park, J. Y. (2024). Text mining and natural language processing in construction. *Automation in Construction*, *158*, 105200. DOI: 10.1016/j.autcon.2023.105200

Sharma, S., Kang, B., Kidambi, N. V., & Mukhopadhyay, S. (2025). HamQ: Hamming weight-based energy-aware quantization for analog compute-in-memory accelerator in intelligent sensors. *IEEE Sensors Journal*, *25*(2). DOI: 10.1109/JSEN.2024.3382479

Sharma, S., Sharma, S., & Athaiya, A. (2020). Activation Functions in Neural Networks. *International Journal of Engineering Applied Sciences and Technology*, *4*(12), 310–316. DOI: 10.33564/IJEAST.2020.v04i12.054

Shehadeh, H. K. (2024). Evaluating the Impact of Transitioning Maintenance Strategy from Reactive to Proactive in Power Generation Companies: An Empirical Analysis. *WSEAS Transactions on Business And Economics*, *21*, 820–838. DOI: 10.37394/23207.2024.21.69

Shevlyakov, G. L., & Oja, H. (2016). *Robust correlation. Theory and applications*. Wiley. DOI: 10.1002/9781119264507

Shigley, J. E., Mischke, C. R., Budynas, R. G., & Nisbett, K. (2011). *Mechanical Engineering Design* (9th ed.). McGraw-Hill Education.

Shorten, C., & Khoshgoftaar, T. M. (2019). A survey on Image Data Augmentation for Deep Learning. *Journal of Big Data*, *6*(1), 60. DOI: 10.1186/s40537-019-0197-0

Singh, A., Jones, M. J., & Learned-Miller, E. G. (2023). Eval: Explainable video anomaly localization. In *Proceedings of the IEEE/CVF Conference on Computer Vision and Pattern Recognition* (pp. 18717–18726). IEEE. DOI: 10.1109/CVPR52729.2023.01793

Singh, R., Saini, K., Sethi, A., Tiwari, A., Saurav, S., & Singh, S. (2023). STEMGAN: Spatio-temporal generative adversarial network for video anomaly detection. *Applied Intelligence*, *53*(23), 28133–28152. DOI: 10.1007/s10489-023-04940-7

Singh, S., Ramakrishna, S., & Singh, R. (2017). Material issues in additive manufacturing: A review. *Journal of Manufacturing Processes*, *25*, 185–200. DOI: 10.1016/j.jmapro.2016.11.006

Smartech Analysis. (n.d.). Ceramics additive manufacturing markets: 2017-2028. Retrieved March 10, 2025, from http://www.smartechanalysis.com/reports/ceramics-additive-manufacturing-markets-2017-2028/

Staerk, C., Byrd, A., & Mayr, A. (2024). Recent methodological trends in epidemiology: No need for data-driven variable selection? *American Journal of Epidemiology*, *193*(2), 370–376. DOI: 10.1093/aje/kwad193 PMID: 37771042

Statti, F., Sued, M., & Yohai, V. J. (2018). High breakdown point robust estimators with missing data. *Communications in Statistics. Theory and Methods*, *47*(21), 5145–5162. DOI: 10.1080/03610926.2017.1388396

Steyvers, M., & Kumar, A. (2024). Three challenges for AI-assisted decision-making. *Perspectives on Psychological Science*, *19*(5), 722–734. DOI: 10.1177/17456916231181102 PMID: 37439761

Su, E. C. Y., & Wu, H. M. (2024). Dimension reduction and visualization of multiple time series data: A symbolic data analysis approach. *Computational Statistics*, *39*(4), 1937–1969. DOI: 10.1007/s00180-023-01440-7

Sundaram, K. M., Hussain, A., Sanjeevikumar, P., Holm-Nielsen, J. B., Kaliappan, V. K., & Santhoshi, B. K. (2021). Deep Learning for Fault Diagnostics in Bearings, Insulators, PV Panels, Power Lines, and Electric Vehicle Applications - The State-of-the-Art Approaches. In *IEEE Access* (Vol. 9, pp. 41246–41260). Institute of Electrical and Electronics Engineers Inc., DOI: 10.1109/ACCESS.2021.3064360

Sun, L., Sun, H., Wang, J., Wu, S., Zhao, Y., & Xu, Y. (2021). Breast mass detection in mammography based on image template matching and CNN. *Sensors (Basel)*, *2021*(8), 2855. DOI: 10.3390/s21082855 PMID: 33919623

Sun, Y., & Genton, M. G. (2011). Functional boxplots. *Journal of Computational and Graphical Statistics*, *20*(2), 316–334. DOI: 10.1198/jcgs.2011.09224

Suriya, M., Chandran, V., & Sumithra, M. G. (2022). Enhanced deep convolutional neural network for malarial parasite classification. *International Journal of Computers and Applications*, *44*(12), 1113–1122. DOI: 10.1080/1206212X.2019.1672277

Sürmen, H. K. (2019). EKLEMELİ İMALAT (3B BASKI): TEKNOLOJİLER VE UYGULAMALAR. Uludağ Üniversitesi mühendislik fakültesi dergisi, 24(2), 373-392. DOI: 10.17482/uumfd.519147

Su, T., Li, H., & An, Y. (2021). A BIM and machine learning integration framework for automated property valuation. *Journal of Building Engineering*, *44*, 102636. DOI: 10.1016/j.jobe.2021.102636

Szrek, J., Wodecki, J., Błazej, R., & Zimroz, R. (2020). An inspection robot for belt conveyor maintenance in underground mine-infrared thermography for overheated idlers detection. *Applied Sciences (Basel, Switzerland)*, *10*(14), 4984–5017. DOI: 10.3390/app10144984

Tabatabaeian, A., Ghasemi, A. R., Shokrieh, M. M., Marzbanrad, B., Baraheni, M., & Fotouhi, M. (2022). Residual stress in engineering materials: A review. *Advanced Engineering Materials*, *24*(3), 2100786. DOI: 10.1002/adem.202100786

Talaei Khoei, T., Ould Slimane, H., & Kaabouch, N. (2023). Deep learning: systematic review, models, challenges, and research directions. In *Neural Computing and Applications* (Vol. 35, Issue 31, pp. 23103–23124). Springer Science and Business Media Deutschland GmbH. DOI: 10.1007/s00521-023-08957-4

Tao, F., Cheng, Y., Qi, Q., Zhang, M., Zhang, H., & Sui, F. (2018). Digital twin-driven product design, manufacturing and service with big data. *International Journal of Advanced Manufacturing Technology*, *94*(9-12), 3563–3576. DOI: 10.1007/s00170-017-0233-1

Tarwani, M., K., & Edem, S. (. (2017). Survey on Recurrent Neural Network in Natural Language Processing. *International Journal of Engineering Trends and Technology*, *48*(6), 301–304. DOI: 10.14445/22315381/IJETT-V48P253

Taye, M. M. (2023). Understanding of machine learning with deep learning: Architectures, workflow, applications and future directions. *Computers*, *12*(5), 91. DOI: 10.3390/computers12050091

Tegmark, M. (2017). *Life 3.0: Being human in the age of artificial intelligence*. Alfred A. Knopf.

Teng, Y., Xu, J., Pan, W., & Zhang, Y. (2022). A systematic review of the integration of building information modeling into life cycle assessment. *Building and Environment*, *221*, 109260. DOI: 10.1016/j.buildenv.2022.109260

Teuwen, J., & Moriakov, N. (2019). Convolutional neural networks. In *Handbook of Medical Image Computing and Computer Assisted Intervention* (pp. 481–501). Elsevier., DOI: 10.1016/B978-0-12-816176-0.00025-9

Thomas, D. S., & Gilbert, S. W. (2014). Costs and cost effectiveness of additive manufacturing (NIST Special Publication No. 1176, p. 12). National Institute of Standards and Technology.

Thompson, R. (2000). *Power transmission: Drives and belts.* Industrial Press.

Tian, Y., Jin, X., Zhu, G., Hu, Y., Wang, Y., & Huang, Z. (2025). Data-driven dimension reduction for high-dimensional random vibration systems with cubic nonlinearity. *Nonlinear Dynamics, 113*(5), 4163–4177. DOI: 10.1007/s11071-024-10039-6

Tlusty, J. (2000). *Manufacturing processes and equipment.* Prentice Hall.

Tofail, S. A., Koumoulos, E. P., Bandyopadhyay, A., Bose, S., O'Donoghue, L., & Charitidis, C. (2018). Additive manufacturing: Scientific and technological challenges, market uptake and opportunities. *Materials Today, 21*(1), 22–37. DOI: 10.1016/j.mattod.2017.07.001

Torregrossa, D., Leopold, U., Hernández-Sancho, F., & Hansen, J. (2018). Machine learning for energy cost modelling in wastewater treatment plants. *Journal of Environmental Management, 223*, 1061–1067. DOI: 10.1016/j.jenvman.2018.06.092 PMID: 30096746

Turan, A. K., & Polat, H. (2024). Yarı denetimli makine öğrenmesi yöntemini kullanarak müzik türlerinin tespiti. Gazi University Journal of Science Part C: Design and Technology, 1-1.

Turan, S. R., Ülkir, O., Kuncan, M., & Buldu, A. (2022). STEREOLİTHOGRAFİ EKLEMELİ İMALAT YÖNTEMİYLE FARKLI DOLULUK ORANLARINDA ÜRETİLEN NUMUNELERİN MEKANİK ÖZELLİKLERİNİN İNCELENMESİ. International Journal of 3D Printing Technologies and Digital Industry, 6(3), 399-407.

U.S. Fire Administration. (2022, Nov). *Fire-related firefighter injuries (2018–2020).* https://www.usfa.fema.gov/statistics/reports/firefighters-departments/firefighter-injuries-v22i4.html

Ullah, W., Ullah, A., Hussain, T., Khan, Z. A., & Baik, S. W. (2021). An efficient anomaly recognition framework using an attention residual LSTM in surveillance videos. *Sensors (Basel), 21*(8), 2811. DOI: 10.3390/s21082811 PMID: 33923712

Uppink, L. (2022). Travel & Tourism Development Index 2021. World Economic Forum, Geneva. [https://www.weforum.org/publications/travel-and-tourism-development-index-2021]

US Forest Service. (1999). *Mann Gulch Fire Disaster.* USDA Forest Service. https://www.fs.usda.gov/manngulchfire

Van der Valk, H., Haße, H., Möller, F., & Otto, B. (2022). Archetypes of digital twins. *Business & Information Systems Engineering*, 1–17. DOI: 10.1007/s12599-022-00763-2

Van Houdt, G., Mosquera, C., & Nápoles, G. (2020). A review on the long short-term memory model. *Artificial Intelligence Review*, *53*(8), 5929–5955. DOI: 10.1007/s10462-020-09838-1

Van Schalkwyk, P. (2022). *Digital Twin Capabilities Periodic Table: A Digital Twin Consortium User Guide. Digital Twin Consortium.* https://www.digitaltwinconsortium.org/wp-content/uploads/sites/3/2022/06/Digital-Twin-Capabilities-Periodic-Table-User-Guide.pdf

Vapnik, V. (1999). *The nature of statistical learning theory*. Springer science & business media.

Velliangiri, S., Alagumuthukrishnan, S., & Thankumar Joseph, S. I. (2019). A Review of Dimensionality Reduction Techniques for Efficient Computation. *Procedia Computer Science*, *165*, 104–111. DOI: 10.1016/j.procs.2020.01.079

Veres, M., & Moussa, M. (2020). Deep Learning for Intelligent Transportation Systems: A Survey of Emerging Trends. In *IEEE Transactions on Intelligent Transportation Systems* (Vol. 21, Issue 8, pp. 3152–3168). Institute of Electrical and Electronics Engineers Inc. DOI: 10.1109/TITS.2019.2929020

Vijayakumar, A., & Vairavasundaram, S. (2024). YOLO-based object detection models: A review and its applications. *Multimedia Tools and Applications*, *83*(17), 83535–83574. DOI: 10.1007/s11042-024-18872-y

Viseras, M., Meissner, M., & Marchal, J. (2016). Wildfire front monitoring with multiple UAVs using deep Q-learning. *IEEE Access : Practical Innovations, Open Solutions*, *1*, 1–14. DOI: 10.1109/ACCESS.2021.3055651

Vymazal, J. (2007). Removal of nutrients in various types of constructed wetlands. *The Science of the Total Environment*, *380*(1-3), 48–65. DOI: 10.1016/j.scitotenv.2006.09.014 PMID: 17078997

Vymazal, J. (2010). Constructed Wetlands for Wastewater Treatment. *Water (Basel)*, *2*(3), 530–549. DOI: 10.3390/w2030530 PMID: 20795704

Walz, W. (2023). Machine Learning for Brain Disorders Series Editor. *Humana Press*.

Wang, K., Gou, C., Duan, Y., Lin, Y., Zheng, X., & Wang, F. (2017). Generative Adversarial Networks : Introduction and Outlook. *IEEE/CAA Journal of Automatica Sinica, 4*(4), 588–598.

Wang, Y., Zhang, D., Liu, Y., Dai, B., & Lee, L. H. (2019). Enhancing transportation systems via deep learning: A survey. In *Transportation Research Part C: Emerging Technologies* (Vol. 99, pp. 144–163). Elsevier Ltd. DOI: 10.1016/j.trc.2018.12.004

Wang, B., Wang, Q., Cheng, J. C., & Yin, C. (2022). Object verification based on deep learning point feature comparison for scan-to-BIM. *Automation in Construction, 142*, 104515. DOI: 10.1016/j.autcon.2022.104515

Wang, H., Meng, X., & Zhu, X. (2022). Improving knowledge capture and retrieval in the BIM environment: Combining case-based reasoning and natural language processing. *Automation in Construction, 139*, 104317. DOI: 10.1016/j.autcon.2022.104317

Wang, N., Issa, R. R., & Anumba, C. J. (2022). NLP-based query-answering system for information extraction from building information models. *Journal of Computing in Civil Engineering, 36*(3), 04022004. DOI: 10.1061/(ASCE)CP.1943-5487.0001019

Wang, P. Y., Liu, W. C., Liu, N., & You, Y. P. (2020). Digital twin-driven system for roller conveyor line: Design and control. *Journal of Ambient Intelligence and Humanized Computing, 11*(11), 5419–5431. DOI: 10.1007/s12652-020-01898-z

Wang, S., Zeng, Y., Liu, Q., Zhu, C., Zhu, E., & Yin, J. (2018). Detecting abnormality without knowing normality: A two-stage approach for unsupervised video abnormal event detection. In *Proceedings of the 26th ACM International Conference on Multimedia* (pp. 636–644). ACM. DOI: 10.1145/3240508.3240615

Wang, Y., Addepalli, S., & Zhao, Y. (2020). Recurrent neural networks and its variants in remaining useful life prediction. *IFAC-PapersOnLine, 53*(3), 137–142. DOI: 10.1016/j.ifacol.2020.11.022

Wei, W., Lu, Y., Zhong, T., Li, P., & Liu, B. (2022). Integrated vision-based automated progress monitoring of indoor construction using mask region-based convolutional neural networks and BIM. *Automation in Construction, 140*, 104327. DOI: 10.1016/j.autcon.2022.104327

Wen, Y., Li, Y., Ye, Y., Dai, Z., & Wang, H. (2022). Transformers in time series: A survey. *Journal of Systems Science and Complexity*. https://arxiv.org/abs/2202.07125

White, G. (2004). *Industrial belts and drives: Selection and applications*. Newnes.

Wilson, P. W. (2018). Dimension reduction in nonparametric models of production. *European Journal of Operational Research, 267*(1), 349–367. DOI: 10.1016/j.ejor.2017.11.020

Wu, C. M., Adetona, A., Song, C., Naeher, L., & Adetona, O. (2019). Measuring acute pulmonary responses to occupational wildland fire smoke exposure using exhaled breath condensate. *Archives of Environmental & Occupational Health, 75*(1), 65–69. DOI: 10.1080/19338244.2018.1562413 PMID: 30668286

Wu, C., Zhou, Y., Pessôa, M. V. P., Peng, Q., & Tan, R. (2021). Conceptual digital twin modeling based on an integrated five-dimensional framework and TRIZ function model. *Journal of Manufacturing Systems*, *58*, 79–93. DOI: 10.1016/j.jmsy.2020.07.006

Wu, H., Zhang, J., Ngo, H. H., Guo, W., Hu, Z., Liang, S., Fan, J., & Liu, H. (2015). A review on the sustainability of constructed wetlands for wastewater treatment: Design and operation. *Bioresource Technology*, *175*, 594–601. DOI: 10.1016/j.biortech.2014.10.068 PMID: 25453440

Wu, M., Goh, K. W., Chaw, K. H., Koh, Y. S., Dares, M., Yeong, C. F., Su, E. L. M., William, H., & Zhang, Y. (2024). An intelligent predictive maintenance system based on random forest for addressing industrial conveyor belt challenges. *Frontiers of Mechanical Engineering*, *10*, 1383202. DOI: 10.3389/fmech.2024.1383202

Wu, M., Yeong, C. F., Su, E. L. M., Holderbaum, W., & Yang, C. (2023). A review on energy efficiency in autonomous mobile robots. Robotic Intell. *Automation*, *43*(6), 648–668. DOI: 10.1108/RIA-05-2023-0060

Wu, P., Pan, C., Yan, Y., Pang, G., Wang, P., & Zhang, Y. (2024). Deep learning for video anomaly detection: A review. arXiv. https://arxiv.org/abs/2409.05383

Wu, X., Cao, J., Tian, J., Wang, Z., & Zheng, P. (2020). A critical review on the performance improvement of biological nitrogen removal processes in wastewater treatment: Influencing factors, optimization strategies, and future directions. *Journal of Environmental Chemical Engineering*, *8*(6), 104553.

Wyatt, M., Marino, J., & Kingma, D.. (2022). *AnoDDPM: Anomaly detection with denoising diffusion probabilistic models*. NeurIPS 2022 Workshop on Anomaly Detection. https://arxiv.org/abs/2206.08545

Xie, C., Yao, R., Liu, Z., & Zhu, L. X. C. (2022). Process performance prediction based on spatial and temporal feature extraction through bidirectional LSTM. *Computer Aided Chemical Engineering, 49*, 1615–1620. https://doi.org/https://doi.org/10.1016/B978-0-323-85159-6.50269-4

Xin, Y., Li, J., Wang, X., & Hampson, K. (2022). Load-carrying capacity assessment of an existing highway bridge based on hybrid finite-element model updating. *Journal of Performance of Constructed Facilities*, *36*(3), 04022028. DOI: 10.1061/(ASCE)CF.1943-5509.0001729

Xu, Q., Wang, J., Gao, W., Ren, S., & Li, Z. (2025). Digital twin: State-of-the-art and future perspectives. In *2024 5th International Conference on Computer Science and Management Technology (ICCSMT '24)* (pp. 731–740). Association for Computing Machinery. DOI: 10.1145/3708036.3708159

Xu, D., Ricci, E., Yan, Y., Song, J., & Sebe, N. (2015). Learning deep representations of appearance and motion for anomalous event detection. arXiv. https://arxiv.org/abs/1510.01553 DOI: 10.5244/C.29.8

Xu, F., & Liu, Q. (2023). Building energy consumption optimization method based on convolutional neural network and BIM. *Alexandria Engineering Journal, 77*, 407–417. DOI: 10.1016/j.aej.2023.06.084

Xu, J., Zhou, W., Fu, Z., Zhou, H., & Li, L. (2021). A survey on green deep learning. ArXiv:2111.05193.

Xu, L., Yu, H., Qin, H., Chai, Y., Yan, N., Li, D., & Chen, Y. (2024). Digital twin for aquaponics factory: Analysis, opportunities, and research challenges. *IEEE Transactions on Industrial Informatics, 20*(4), 5060–5073. DOI: 10.1109/TII.2023.3331135

Xu, S., & Wang, W. (2013). Numerical investigation on weld residual stresses in tube to tube sheet joint of a heat exchanger. *International Journal of Pressure Vessels and Piping, 101*, 37–44. DOI: 10.1016/j.ijpvp.2012.10.004

Yalçın, B., & Ergene, B. (2017). Endüstride yeni eğilim olan 3-d eklemeli imalat yöntemi ve metalurjisi. *Uluslararası Teknolojik Bilimler Dergisi, 9*(3), 65–88.

Yamashita, R., Nishio, M., Do, R. K. G., & Togashi, K. (2018). Convolutional neural networks: an overview and application in radiology. In *Insights into Imaging* (Vol. 9, Issue 4, pp. 611–629). Springer Verlag. DOI: 10.1007/s13244-018-0639-9

Yang, Y., Lee, K., Dariush, B., Cao, Y., & Lo, S.-Y. (2025). Follow the rules: Reasoning for video anomaly detection with large language models. In *European Conference on Computer Vision* (pp. 304–322). Springer. DOI: 10.1007/978-3-031-73004-7_18

Yavuz, İ., Erçek, E., & Yuran, A. F. (2021). Ulaştırma sektöründe imdat çekici tasarımı ve 3B yazıcı ile üretimi. *International Journal of Multidisciplinary Studies and Innovative Technologies, 5*(1), 46–49.

Yeşildağ, İ., & Günay, M. (2019). Eklemeli imalatin metal malzemelere uygulanması üzerine bir değerlendirme. 4th International Congress on 3D Printing (Additive Manufacturing) Technologies and Digital Industry (ss. 881-892)

Yilmaz, A., & Yolcu, U. (2023). A robust training of dendritic neuron model neural network for time series prediction. *Neural Computing & Applications*, *35*(14), 10387–10406. DOI: 10.1007/s00521-023-08240-6

Ying, H., Sacks, R., & Degani, A. (2023). Synthetic image data generation using BIM and computer graphics for building scene understanding. *Automation in Construction*, *154*, 105016. DOI: 10.1016/j.autcon.2023.105016

Yin, H., Chen, Y., Zhou, J., Xie, Y., Wei, Q., & Xu, Z. (2024, Dec 3). A probabilistic deep learning approach to enhance the prediction of wastewater treatment plant effluent quality under shocking load events. *Water Research X*, *26*, 100291. DOI: 10.1016/j.wroa.2024.100291 PMID: 39720317

Yizhou, L., Zhang, T., Ding, Y., Wadhwani, R., & Huang, X. (2024). Review and perspectives of digital twin systems for wildland fire management. *Journal of Forestry Research*, *36*(1), 1–24. DOI: 10.1007/s11676-024-01810-x

Yong, L. I., Wanni, G. A. N., Wenbin, Z. H. O. U., & Dongsheng, L. I. (2023). Review on residual stress and its effects on manufacturing of aluminium alloy structural panels with typical multi-processes. *Chinese Journal of Aeronautics*, *36*(5), 96–124. DOI: 10.1016/j.cja.2022.07.020

Yu, D., Wang, H., Chen, P., & Wei, Z. (2014). Mixed pooling for convolutional neural networks. *The 9th International Conference on Rough Sets and Knowledge Technology (RSKT'14), Springer*, 364–375. DOI: 10.1007/978-3-319-11740-9_34

Yu, X. (2019). Machine learning based predictive maintenance and regression based control of a servo motor for industrial conveyor belts. Berlin: Technische Universität. Master Thesis.

Yuan, Z., Olsson, G., Cardell-Oliver, R., van Schagen, K., Marchi, A., Deletic, A., Urich, C., Rauch, W., Liu, Y., & Jiang, G. (2019). Sweating the assets---The role of instrumentation, control and automation in urban water systems. *Water Research*, *155*, 381–402. DOI: 10.1016/j.watres.2019.02.034 PMID: 30861379

Zabin, A., González, V. A., Zou, Y., & Amor, R. (2022). Applications of machine learning to BIM: A systematic literature review. *Advanced Engineering Informatics*, *51*, 101474. DOI: 10.1016/j.aei.2021.101474

Zafar, A., Aamir, M., Mohd Nawi, N., Arshad, A., Riaz, S., Alruban, A., Dutta, A. K., & Almotairi, S. (2022). A Comparison of Pooling Methods for Convolutional Neural Networks. *Applied Sciences (Basel, Switzerland)*, *12*(17), 8643. DOI: 10.3390/app12178643

Žarković, M., & Stojković, Z. (2019). Artificial intelligence SF6 circuit breaker health assessment. *Electric Power Systems Research, 175*, 105912. DOI: 10.1016/j.epsr.2019.105912

Zhai, R., Zou, J., He, Y., & Meng, L. (2022). BIM-driven data augmentation method for semantic segmentation in superpoint-based deep learning network. *Automation in Construction, 140*, 104373. DOI: 10.1016/j.autcon.2022.104373

Zhang, D. Q., Jinadasa, K. B. S. N., Gersberg, R. M., Liu, Y., Ng, W. J., & Tan, S. K. (2014). Application of constructed wetlands for wastewater treatment in developing countries—A review of recent developments (2000–2013). *Journal of Environmental Management, 141*, 116–131. DOI: 10.1016/j.jenvman.2014.03.015 PMID: 24784754

Zhang, H., Zhang, L., & Jiang, Y. (2019). Overfitting and Underfitting Analysis for Deep Learning Based End-to-end Communication Systems. *11th International Conference on Wireless Communications and Signal Processing (WCSP)*, 1–6. DOI: 10.1109/WCSP.2019.8927876

Zhang, J., Zhu, Y., Zhang, X., Ye, M., & Yang, J. (2018). Developing a Long Short-Term Memory (LSTM) based model for predicting water table depth in agricultural areas. *Journal of Hydrology (Amsterdam), 561*, 918–929. DOI: 10.1016/j.jhydrol.2018.04.065

Zhang, K., Song, W., Liu, L., Zhao, X., & Du, C. (2019). *Bidirectional Long Short-Term Memory for Sentiment Analysis of Chinese Product Reviews*. IEEE. DOI: 10.1109/ICEIEC.2019.8784560

Zhang, K., Wen, Q., Zhang, C., Cai, R., Jin, M., Liu, Y., Zhang, J., Liang, Y., Pang, G., Song, D., & Pan, S. (2023). Self-supervised learning for time series analysis: Taxonomy, progress, and prospects. *IEEE Transactions on Pattern Analysis and Machine Intelligence*. DOI: 10.1109/TPAMI.2023.3264798 PMID: 38598381

Zhang, S., Zhai, J., Chen, J., & He, Q. (2018). Autoencoder and its various variants. *2018 IEEE International Conference on Systems, Man, and Cybernetics (SMC)*, 415–419. DOI: 10.1109/SMC.2018.00080

Zhang, T., Wang, Z., Zeng, Y., Wu, X., Huang, X., & Xiao, F. (2022). Building Artificial-Intelligence Digital Fire (AID-Fire) system: A real-scale demonstration. *Journal of Building Engineering, 62*, 105363. DOI: 10.1016/j.jobe.2022.105363

Zhang, Y., Lu, H., Zhang, L., & Ruan, X. (2016). Combining motion and appearance cues for anomaly detection. *Pattern Recognition, 51*, 443–452. DOI: 10.1016/j.patcog.2015.09.005

Zhang, Y., Sun, S., Wang, X., & Wang, Z. (2021). A digital twin-based approach for smart manufacturing system design and evaluation. *Journal of Manufacturing Systems*, *58*, 168–179. DOI: 10.1016/j.jmsy.2020.12.003

Zhang, Z., Zohren, S., & Roberts, S. (2020). Deep reinforcement learning for trading. *The Journal of Financial Data Science*, *2*(2), 25–40. DOI: 10.3905/jfds.2020.1.030

Zhao, X., & Yang, W. (2025). Prediction of mechanical property and costs of high-strength steel using machine learning. *Canadian Metallurgical Quarterly*, 1–13. DOI: 10.1080/00084433.2025.2471616

Zhao, Y., Deng, B., Shen, C., Liu, Y., Lu, H., & Hua, X.-S. (2017). Spatio-temporal autoencoder for video anomaly detection. In *Proceedings of the 25th ACM International Conference on Multimedia* (pp. 1933–1941). ACM. DOI: 10.1145/3123266.3123451

Zhong, Y., Chen, X., Jiang, J., & Ren, F. (2022). A cascade reconstruction model with generalization ability evaluation for anomaly detection in videos. *Pattern Recognition*, *122*, 108336. DOI: 10.1016/j.patcog.2021.108336

Zhou, S., Shen, W., Zeng, D., Fang, M., Wei, Y., & Zhang, Z. (2016). Spatial–temporal convolutional neural networks for anomaly detection and localization in crowded scenes. *Signal Processing Image Communication*, *47*, 358–368. DOI: 10.1016/j.image.2016.06.007

Zhou, X., Guo, J., Wang, W., & Wang, Z. (2019). Big data mining for sewage treatment: Revealing the correlation between influent characteristics and effluent water quality. *Water Research*, *163*, 114859.

Zhu, H., Zhu, W. D., & Fan, W. (2021). Dynamic modeling, simulation and experiment of power transmission belt drives: A systematic review. *Journal of Sound and Vibration*, *491*, 115759. DOI: 10.1016/j.jsv.2020.115759

Zhu, Y., Nayak, N. M., & Roy-Chowdhury, A. K. (2013). Context-aware activity recognition and anomaly detection in video. *IEEE Journal of Selected Topics in Signal Processing*, *7*(1), 91–101. DOI: 10.1109/JSTSP.2012.2234722

Zonta, T., da Costa, C. A., da Rosa Righi, R., de Lima, M. J., da Trindade, E. S., & Li, G. P. (2020). Predictive maintenance in the Industry 4.0: A systematic literature review. *Computers & Industrial Engineering*, *150*, 106889. DOI: 10.1016/j.cie.2020.106889

Zulfiqar, M., Gamage, K. A., Kamran, M., & Rasheed, M. B. (2022). Hyperparameter optimization of bayesian neural network using bayesian optimization and intelligent feature engineering for load forecasting. *Sensors (Basel)*, *22*(12), 4446. DOI: 10.3390/s22124446 PMID: 35746227

About the Contributors

Melda Yücel is a researcher in Optimization and Machine Learning for Structural Engineering area at Istanbul Aydın University. She obtained her DPhil in Structural Engineering from Istanbul University-Cerrahpaşa with a thesis based on prediction of mechanical parameters for optimum tuned mass damper designs. She has authored different papers for journals, conferences and scientific events, besides book and book chapter studies.

***

Bekir Aksoy received his M.Sc. in machine education (2001) from Suleyman Demirel University and PhD textile engineering (2016) from Suleyman Demirel University, Turkey. Now He is Assoc. Prof. Dr. of Mechatronics Engineering at Techologies Faculty in Isparta University of Applied Sciences, Turkey. His current research interests include different aspects of the artificial intelligence and image processing techniques.

Naga Sushma Allur attained her Masters in Computer Application (MCA) from Osmania University, Hyderabad, India. Following the completion of her MCA in 2005, Naga Sushma embarked on her professional journey as an IT specialist with ICICI Bank, a prominent banking institution in India. Currently she holds the position of Senior Business Analyst with Astute Solutions LLC.

Himabindu Chetlapalli has Bachelors degree in Computer science, Information technology from Jawaharlal Nehru Technological University, passed out in 2006. Currently working as QA Architect having experience of more than 16 years in the field of Information tech as a Quality Assurance Analyst.

Durga Praveen Deevi attained his Masters in Computer Application (MCA) from Osmania University, Hyderabad, India. Following the completion of his

MCA in 2005, Mr. Durga Praveen embarked on his professional journey as a Quality Assurance Engineer with eIQnetworks now known as eIQnetworks (India) Research and Development (Pvt) Limited, a prominent Cyber Security company in India. He is working as a Senior Software Quality Assurance Engineer with O2 Technologies Inc., USA.

Koteswararao Dondapati, with over 18 years of extensive experience as a Core Testing Subject Matter Expert (SME), specializes in business-critical Retail, Energy, and HCM verticals. His 10-year expertise in test automation includes Java frameworks, T Plan robot tool, Cucumber with Ruby, and SOAP UI Automation. Noteworthy is his domain expertise in technologies like POS and Recruiting Systems modules. At Abercrombie & Fitch, as QA Lead, Mr. Koteswararao spearheads software design, development, testing, and implementation.

Saahil Jamad is a final-year Mechatronics Engineering student with a strong passion for technology, specializing in design, manufacturing, and automation solutions. Through active participation in workshops and projects, I have gained hands-on experience and deepened my technical foundation. My knowledge in these areas has been further enhanced through multiple internships alongside my academic journey. Currently pursuing my B.Tech in Mechatronics Engineering at SRM Institute of Science and Technology. Beyond academics, I enjoy playing badminton and listening to music, maintaining a balance between technical pursuits and personal interests.

Electa Alice Jayarani A. is a distinguished academician and researcher with over two decades of experience in engineering education. She holds her undergraduate and postgraduate degrees from institutions in Tamil Nadu and earned her Ph.D. from REVA University. She is also a NET-qualified faculty member and currently serves as an Associate Professor at K.S. Institute of Technology. Dr. Electa Alice Jayarani A has made significant contributions to research, with around 25 publications in reputed journals and 16 presentations at national and international conferences. She holds seven patents and serves as a recognized research supervisor under Visvesvaraya Technological University (VTU), having guided numerous student projects.

Vishnu Kumar Kaliappan is an experienced Professor with a demonstrated history of working in the higher education industry. Skilled in Python, Computer Science, C++, LabVIEW, and Eclipse. Strong education professional with a PhD focused in Control system for RUAV (Rotorcraft based unmanned aerial vehicle)) from Konkuk University.

Jan Kalina is a researcher working at the intersection of statistics and machine learning, with a particular interest in methodologies for analyzing complex data contaminated by noise and outliers. He studied mathematical statistics at MFF UK in Prague and earned his Ph.D. from the University of Duisburg-Essen in 2007. He has been working at the Institute of Computer Science of the Czech Academy of Sciences since 2010. From 2013 to 2017, he served as the head of the Department of Medical Informatics and Biostatistics, and he is currently the deputy head of the Supervisory Board at ICS. He also serves as an associate editor of Sequential Analysis and is a member of the editorial board for Applications of Mathematics. He serves as an advisor to students at Charles University and the Czech Technical University.

Helen K. Joy holds a B.E. in Electronics and Communication from Anna University, Chennai. She earned an M.E. in Applied Electronics, graduating with a gold medal from Satyabhama University. She completed her PhD in Video Compression with Deep Learning at Reva University, Bengaluru. Currently, Dr. Helen is an Assistant Professor in the Computer Science Department at Christ University. Her research interests include deep learning, video compression, CNNs, and digital image processing, along with Embedded Machine Learning. Dr. Helen has authored over 11 journal articles, 41 conference papers, and 7 book chapters. She has organized more than 10 national and international conferences and participated in over 50 workshops. She is a recipient of the Dr. Madhavan Nair Award for excellence in Applied Electronics and has received multiple awards for her academic contributions.

Sharadha Kodadi is a seasoned IT professional with over 13 years of industry experience. She holds a Master's degree in Microbiology from Kakatiya University. Currently, she serves as a Lead Consultant at Infosys, where she excels in software development, cloud computing, and data analysis. Sharadha has a proven track record of working with advanced technologies such as AWS, Spring Boot, and Java, driving impactful solutions across various projects.

Pirom Konglerd has been working at the School of Science and Technology, Sukhothai Thammathirat Open University in Thailand, with a research focus on computer vision and deep learning algorithms.

Abhishek Kumar is a qualified educator with a Ph.D. (Electronics and Electrical Engineering) completed from Lovely Professional University, M.Tech from VJTI, the University of Mumbai on the qualifying GATE examination, and graduated from Institution of Engineers(India). He is currently associated as an Associate Professor at the VLSI Domain, School of Electrical and Electronics Engineering,

Lovely Professional University, India has more than 12 years of teaching and research experience. He has authored more than 25 peer reviewed research articles indexed in SCI, Scopus indexed journals and 12 proceeding in index conferences proceedings. There are 4 book with IEEE/Wiley-SP and authored 8-book chapter. There are 4 copyrights and published 11 patents. He is an Active Reviewer and an Editorial Member of numerous prominent SCI and Scopus indexed journals.

M. Sravanan is an Associate Professor at KPR Institute of Engineering and Technology, India.

N.S.C. Chaitanya is an accomplished academician and mechanical engineering expert serving as an Associate Professor in the Department of Mechanical Engineering at Raghu Engineering College. His expertise spans across robotics, alternative fuels, energy systems, and non-conventional energy sources, making him a versatile contributor to the field of mechanical engineering. His research work has been particularly impactful in the field of alternative fuels and engine performance. He has published several significant papers in Scopus-indexed journals, with groundbreaking research on binary diesel/JME blends and their emissions characteristics. His innovative work on nano-metallic oxide additives and heterogeneous base catalysts has advanced the understanding of alternative fuel technology. As an Institution Innovation Ambassador, he has played a crucial role in fostering an environment of innovation and entrepreneurship. Dr. Chaitanya continues to contribute to the academic community through his research publications, conference presentations, and organization of technical events, making him a respected figure in mechanical engineering education.

Oluseye Olugboyega is the Research Lead of the Construction Engineering and Computational Modelling Research Group and serves as the Head of Research at the Digital Construction Analytics Laboratory at Obafemi Awolowo University, Nigeria. His work focuses on advancing the frontiers of digital transformation in the built environment through innovative research in construction engineering, computational methods, and intelligent systems. He has published extensively on the integration, adoption, and application of emerging digital technologies—particularly Building Information Modelling (BIM) and Artificial Intelligence (AI)—within construction processes, project delivery systems, and infrastructure management. His research contributes to improving productivity, decision-making, and sustainability in the construction industry through data-driven approaches and smart technologies.

Koray Özsoy received his M.Sc. in machine education (2008) from Suleyman Demirel University and PhD machine engineering (2016) from Suleyman Demirel University, Turkey. Now He is Assoc. Prof. Dr. of Machine and Metal Technologies,

Isparta OSB Vocational School in Isparta University of Applied Sciences, Turkey. His current research interests include different aspects of additive manufacturing technologies and optimization techniques with applications in manufacturing and digital industry.

Thinagaran Perumal is the recipient of 2014 Early Career Award from IEEE Consumer Electronics Society for his pioneering contribution in the field of consumer electronics. He completed his Phd at Universiti Putra Malaysia, in the area of smart technology and robotics. He is currently an Associate professor attached with Department of Computer Science, Faculty of Computer Science and Information Technology, Universiti Putra Malaysia. He is also currently appointed as Chair of IEEE Consumer Electronics Society Malaysia Chapter.

R. Sridevi is an Associate Professor in the Department of Computer Science at CHRIST (Deemed to be University, Bangalore, Karnataka). She holds PhD degree from Bharathiar University and guided 7 MPhil Scholars till date. She has taught Object Oriented Programming, Relational Database Management System, Operating System, Data Visualization and Information Security at the undergraduate and post graduate levels for over 20 years. She has published several research papers in Biometric Security, Cryptography and Information Security. She has also presented around 30 research articles in National and International Conferences. She is a member of Professional Bodies like CSTA, ISTE & IAENG. She has also published 2 Indian Patents. She holds the privilege of acting as a Resource Person for Guest lectures of various colleges.

Walisa Romsaiyud is affiliated with the School of Science and Technology, Sukhothai Thammathirat Open University in Thailand. Her research focuses on data analytics and deep learning algorithms.

Anitha Kumari Sivathanu holds a PhD in control and Instrumentation from Anna University, Chennai, India. Her research has been funded by Department of Science & Technology, Government of India. She is currently working as an Assistant Professor in Mechatronics Engineering Department, SRM Institute of Science and Technology, Kattankulathur, India. Her current research interests include the field of control systems, instrumentation, computer vision and robotics.

S. Jansirani is an Assistant Professor in the Department of Artificial Intelligence and Data Science at Jansons Institute of Technology, India.

Harpreet Singh Bedi, PhD, is an Associate Professor at Lovely Professional University (LPU), Phagwāra, India. He has 15 years of experience in both teaching and industry, specializing in areas such as Communication & Signal Processing, Information and Communication Technology, Cellular Communication, Satellite Communication, Photovoltaic Systems, and Algorithms. His contributions to research and innovation include over 30 published patents, 2 granted patents, and 2 registered copyrights under his name. He has also authored a book and published more than 50 research papers in renowned scientific and technical journals, showcasing his dedication to advancing knowledge in his field.

Sabsathit Takaew is a master's student and researcher at the School of Science and Technology, Sukhothai Thammathirat Open University in Thailand. His research focuses on natural language processing and large language models.

Pirom Konglerd has been working at the School of Science and Technology, Sukhothai Thammathirat Open University, Thailand.

Sabsathit Takaew is a master's student at the School of Science and Technology, Sukhothai Thammathirat Open University.

Index

A

Additive Manufacturing 255, 256, 258, 259, 260, 261, 262, 263, 268, 269, 270, 274, 275, 277, 278, 279, 280, 281, 282, 283

AI 1, 2, 39, 40, 44, 63, 65, 70, 71, 72, 73, 75, 76, 81, 82, 92, 97, 101, 105, 175, 176, 177, 178, 179, 180, 181, 184, 187, 188, 189, 191, 192, 195, 196, 197, 198, 199, 200, 215, 217, 221, 222, 223, 224, 231, 234, 235, 236, 237, 238, 239, 240, 241, 242, 243, 244, 245, 246, 247, 248, 249, 253, 254, 255, 257, 258, 263, 264, 266, 268, 269, 270, 271, 272, 273, 274, 275, 277, 280, 281, 285, 286, 287, 288, 289, 290, 296, 300, 305, 306, 307, 309, 311, 314, 316, 317, 318, 319

AI Technology 235, 248

ANN 3, 15, 264, 269, 286

Artificial Intelligence 1, 2, 29, 31, 33, 34, 36, 39, 40, 63, 65, 67, 68, 70, 71, 72, 86, 101, 105, 106, 135, 165, 175, 176, 179, 181, 184, 195, 198, 200, 201, 202, 219, 221, 222, 223, 229, 231, 234, 243, 244, 245, 246, 250, 253, 255, 256, 258, 263, 269, 270, 273, 274, 276, 277, 278, 317, 318

Autoencoders 5, 20, 21, 23, 29, 30, 31, 137, 140, 146, 164

B

Belt-Drive System 103, 107, 123, 130, 132

BIM 175, 176, 177, 178, 179, 180, 181, 182, 183, 184, 185, 186, 187, 188, 189, 190, 191, 192, 193, 194, 195, 196, 197, 198, 199, 200, 201, 202, 203, 204

C

Complexity Reduction 40, 41

Condition Monitoring 25, 30, 104, 123, 131, 191

CUHK Avenue 138, 140, 142, 150, 152, 159, 160, 163, 164

D

Deep Learning 1, 2, 3, 4, 5, 6, 22, 24, 25, 27, 28, 29, 30, 31, 32, 33, 34, 35, 36, 37, 61, 62, 64, 66, 70, 74, 101, 130, 139, 140, 141, 166, 170, 171, 172, 183, 188, 189, 190, 191, 198, 201, 202, 203, 204, 207, 213, 224, 233, 234, 242, 246, 251, 253, 264, 267, 271, 272, 274, 276, 277, 279, 280, 281, 285, 287, 288, 289, 290, 317, 318

Digital Twin 105, 131, 132, 133, 134, 135, 248, 253, 285, 287, 288, 289, 290, 296, 297, 305, 306, 307, 308, 309, 314, 316, 318, 319, 320

E

Edge Devices 15, 24, 26, 71, 72, 73, 75, 86, 93, 95, 101

Embedded Systems 71, 72, 86, 99

Energy Complexity 60, 61

Energy Efficiency 39, 61, 62, 63, 71, 74, 75, 84, 85, 94, 135, 183, 223, 230, 238, 239, 246, 268, 280

F

Fault Detection 26, 27, 28, 31, 78, 105, 106, 120, 122, 123, 125, 129, 130, 132, 133, 273, 279

Feature Selection 5, 42, 47, 116, 118, 119, 120, 207, 212, 233, 267

Federated Learning 26, 28, 30, 34, 71, 84, 85, 99, 100

Firefighter Safety 285, 286, 288, 290, 304, 305, 309, 316

Functional Data 66, 137, 138, 141, 142, 143, 144, 146, 147, 148, 153, 155, 156, 162, 163, 164, 165, 166, 170, 176

I

Industry 4.0 2, 78, 104, 106, 131, 132, 133, 135, 255, 256, 271, 275
Interpretability 2, 20, 39, 42, 44, 47, 54, 57, 62, 114, 118, 120, 162, 183, 198, 208, 217
IoT 28, 62, 68, 71, 75, 78, 86, 101, 105, 132, 177, 180, 237, 245, 257, 271, 272, 274, 286, 287, 290, 317

M

Machine Learning 2, 4, 26, 29, 30, 31, 32, 33, 34, 35, 36, 39, 40, 41, 42, 44, 51, 53, 55, 61, 62, 63, 64, 65, 66, 67, 68, 71, 72, 73, 74, 75, 76, 77, 78, 79, 86, 87, 91, 93, 94, 95, 96, 98, 99, 101, 103, 105, 106, 114, 116, 117, 118, 119, 123, 129, 130, 131, 133, 134, 135, 171, 175, 177, 179, 180, 181, 182, 183, 184, 190, 192, 193, 194, 195, 196, 198, 200, 201, 202, 203, 204, 205, 206, 207, 208, 210, 212, 213, 215, 216, 217, 218, 219, 220, 223, 224, 232, 233, 235, 236, 238, 239, 243, 246, 251, 253, 255, 257, 258, 263, 264, 265, 266, 268, 270, 273, 274, 276, 277, 279, 280, 281, 282, 283, 286, 289
MS-Plot 138, 142, 143, 144, 146, 147, 148, 149, 151, 153, 154, 155, 156, 157, 159, 160, 161, 162, 163, 164

N

Natural Language Processing 24, 35, 175, 177, 180, 184, 198, 201, 202, 203, 264, 266, 267
Nature-Based Solutions 221, 223, 224, 226, 227, 228, 229, 234, 235, 236, 241, 242, 243, 244, 245, 246, 247, 248, 249, 253

O

Optimization 2, 8, 14, 48, 49, 54, 57, 62, 63, 71, 81, 86, 87, 90, 95, 105, 115, 133, 177, 181, 182, 190, 198, 200, 201, 203, 205, 207, 208, 213, 215, 216, 217, 218, 220, 222, 223, 224, 225, 231, 232, 233, 234, 235, 236, 238, 239, 243, 244, 245, 246, 247, 248, 251, 253, 255, 256, 257, 258, 262, 263, 264, 265, 266, 268, 269, 270, 271, 273, 274, 276, 277, 278, 288, 290, 295, 302, 303, 304, 313, 314, 315, 316

P

Predictive Maintenance 26, 27, 28, 33, 34, 77, 78, 81, 82, 103, 104, 105, 106, 113, 115, 118, 120, 123, 129, 130, 131, 132, 133, 134, 135, 177, 223, 234, 243, 247, 257, 271, 274, 277, 289
Predictive Modeling 212, 246, 270, 273
Process Optimization 222, 223, 225, 231, 232, 234, 236, 243, 255, 263, 264, 268, 269, 273

R

Real-Time Analytics 180, 277
Reconstruction-Based Methods 140
Regression Models 43, 46, 50, 207, 208, 267
Regularization Techniques 43, 59, 62
Robustness 3, 21, 27, 40, 42, 43, 44, 45, 46, 47, 49, 50, 53, 55, 60, 64, 66, 67, 114, 115, 140, 159, 187, 189, 234, 275, 316

S

Sparsity 20, 45, 60, 61, 62, 171, 191
Steel Tube Structures 205, 206
Structural Engineering 207, 217, 219
Supervised Learning 2, 27, 29, 32, 37, 48, 67, 181, 264, 266, 267, 272
SVM 103, 104, 114, 115, 119, 123, 127, 128, 130, 138, 139, 171, 185, 187,

232, 264, 265

T

Tiny Machine Learning 71, 72, 75, 76, 77, 79, 101

U

UCSD Ped2 138, 140, 141, 142, 150, 151, 152, 153, 154, 155, 156, 157, 159, 161, 162, 164
Unsupervised Learning 2, 23, 29, 181, 185, 201, 264, 266, 267

V

Video Anomaly Detection 137, 139, 140, 141, 142, 143, 145, 146, 149, 150, 151, 152, 153, 157, 162, 164, 165, 166, 167, 168, 169, 170, 171, 172, 173

W

Wastewater Treatment 221, 222, 223, 224, 225, 226, 227, 228, 229, 230, 231, 232, 233, 234, 236, 237, 238, 239, 240, 241, 242, 243, 244, 245, 246, 247, 248, 250, 251, 252, 253, 254
Wildfire Operations 286

www.ingramcontent.com/pod-product-compliance
Lightning Source LLC
LaVergne TN
LVHW080906300926
849162LV00003B/570
9798337307466